Medicinal Plants and Mushrooms of Yunnan Province of China

Natural Products Chemistry of Global Plants

Editor:
Raymond Cooper

This unique book series focuses on the natural products chemistry of botanical medicines from different countries such as Sri Lanka, Cambodia, Brazil, China, Africa, Borneo, Thailand, and Silk Road countries. These fascinating volumes are written by experts from their respective countries. The series will focus on the pharmacognosy, covering recognized areas rich in folklore as well as botanical medicinal uses as a platform to present the natural products and organic chemistry. Where possible, the authors will link these molecules to pharmacological modes of action. The series intends to trace a route through history from ancient civilizations to the modern day showing the importance to man of natural products in medicines, foods, and a variety of other ways.

Recent Titles in This Series:

Medicinal Plants and Mushrooms of Yunnan Province of China
Clara Bik-San Lau and Chun-Lin Long

Medicinal Plants of Borneo
Simon Gibbons and Stephen P. Teo

Natural Products and Botanical Medicines of Iran
Reza Eddin Owfi

Natural Products of Silk Road Plants
Raymond Cooper and Jeffrey John Deakin

Brazilian Medicinal Plants
Luzia Modolo and Mary Ann Foglio

Medicinal Plants of Bangladesh and West Bengal
Botany, Natural Products, and Ethnopharmacology
Christophe Wiart

Traditional Herbal Remedies of Sri Lanka
Viduranga Y. Waisundara

Medicinal Plants and Mushrooms of Yunnan Province of China

Edited by
Clara Bik-San Lau
Chun-Lin Long

CRC Press
Taylor & Francis Group
Boca Raton London New York

CRC Press is an imprint of the
Taylor & Francis Group, an **informa** business

First edition published 2021
by CRC Press
6000 Broken Sound Parkway NW, Suite 300, Boca Raton, FL 33487-2742

and by CRC Press
2 Park Square, Milton Park, Abingdon, Oxon, OX14 4RN

ISBN: 978-0-367-89801-4 (hbk)
ISBN: 978-1-032-02338-0 (pbk)
ISBN: 978-1-003-02277-0 (ebk)

Typeset in Times
by Deanta Global Publishing Services, Chennai, India

Contents

Section 4 Medicinal Plants among Ethnic Groups

Foreword

Yunnan Province, located in the southwest part of China, is a low-latitude plateau. Since the terrain is high in the north with a natural mountainous environment and low in the south, various types of climate arising from tropical to subtropical to temperate, and to plateau climate zones. The complex terrain and diverse climatic conditions have created a wealth of plant resources. Yunnan is therefore rich in Chinese materia medica resources and is well known as the "Kingdom of Plants." Among the 12,800+ flora of Chinese medicinal resources, over 11,100 of them are medicinal plants, and Yunnan has nearly 6500 species (including more than 310 families and 1800 genera) of medicinal plants, accounting for about 51% of the total in China. In particular, Yunnan is still the authentic growing area of *Gastrodia elata*, *Dendrobium* spp., *Panax notoginseng*, *Rhodiola* spp., and many other important and precious medicinal herbs.

In addition, Yunnan is a multi-ethnic settlement area with over 25 ethnic minorities, who inherit and retain much knowledge of traditional medicines. There are more than 1000 kinds of ethnic medicines with characteristics represented by Yi, Dai, Miao, Wa, and Tibetan medicines, which have become an important part of China's medicinal resources. Yunnan is in fact the important pharmaceutical industrial base for Dai and Yi medicines in China, such as the well-known Yunnan Baiyao (a proprietary traditional Chinese medicine currently marketed and used as an alternative hemostatic product), which originated from a Yi medical formula. Furthermore, Yunnan is also one of the richest sources of edible fungi/mushrooms in China, which accounts for about 30% of all the edible fungi/mushrooms species in the world. Among these fungi/mushrooms, many of them exhibit important medicinal properties, such as *Ganoderma* and *Cordyceps* species.

Although Yunnan's medicinal plant resources are abundant, most of them are also scattered. The large-scale production of pharmaceutical products depend upon large-scale planting. Unfortunately, only a few species such as *Gastrodia elata*, *Dendrobium officinale*, *Panax notoginseng*, *Angelica sinensis*, *Erigeron breviscapus*, and *Ginkgo biloba* have achieved large-scale planting in Yunnan. The investment in the artificial cultivation of wild medicinal plants is also insufficient, and more than 60% of the raw herbal materials still rely on collection from the wild. In the absence of a comprehensive understanding of the reserves of medicinal plant resources in Yunnan, the establishment of a medicinal plant gene bank and seedbank (such as those in Kunming Institute of Botany), together with the enhancement of research and development on cultivation technologies will possibly be the most effective and direct protection methods at present. Certainly, conservation and sustainable utilization of these medicinal plants and fungi/mushrooms in this area are of the utmost importance. Perhaps systematic research, habitats protection, and artificial cultivation are the keys to the sustainable use of traditional medicinal plants and fungi/mushrooms in Yunnan.

Since natural medicinal resources are rich in Yunnan Province, it undoubtedly provides a great source and golden opportunity for new drug development. Hence, developing Yunnan natural medicines by fully utilizing the natural medicinal resources will be the ultimate goal, and this will depend on the R&D of Yunnan natural medicines in which those of characteristic and efficacious are selected for systematic, multidisciplinary, and cooperative research. The further development of such existing traditional Chinese medicines will require the understanding and use of natural products chemistry, together with collaborative efforts among the researchers of pharmacy, pharmacology, and other disciplines. The research of the unique medicinal plant resources in China, especially on the new application and new mechanism of action of those natural products and their derivatives, may lead to the modernization and internationalization of traditional Chinese medicines and the development of innovative drugs.

To this end, I would like to congratulate Clara Bik-San Lau and Chun-Lin Long for the success of publishing this book which has included a very solid representative collection of medicinal plants and fungi/mushrooms in Yunnan. I certainly believe this book will serve as a valuable reference book on the chemistry and biological aspects of medicinal plants and mushrooms in Yunnan for interested readers.

Han-Dong Sun
Professor of Phytochemistry and Academician of Chinese Academy of Sciences (CAS)
Kunming Institute of Botany, CAS

Preface

Yunnan Province covers most of the south-western region of China, with a total area of around 394,000 square kilometers and a population of 48.3 million as of 2018. Kunming is the capital of Yunnan Province (see maps below).

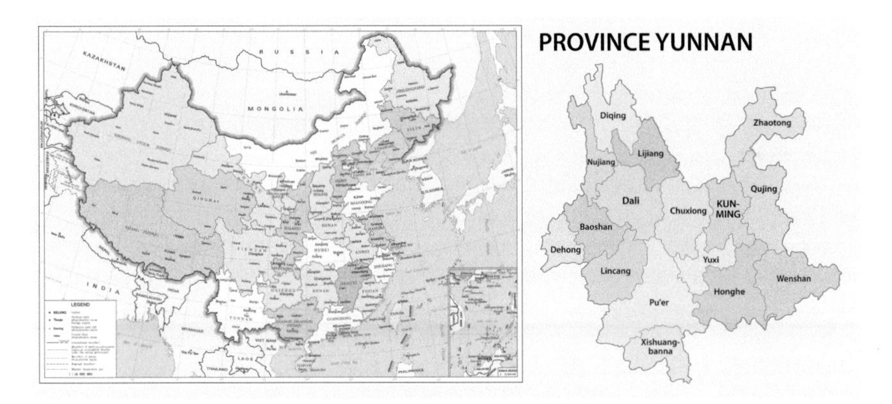

Image credits: Left: http://bzdt.ch.mnr.gov.cn/index.html with reference number GS(2019)1696; Right: Rainer Lesniewski/Shutterstock.com.

Yunnan Province is situated in a mountainous area, with low elevations in the southeast and high elevations in the northwest, with major rivers flowing through the deep valleys between the mountains. With the snow-capped mountains and true tropical environments, the province is one of the major forest zones in China.

Taken together, it is not surprising that Yunnan is rich in natural resources and has the largest diversity of plant life in China. In fact, the province has been regarded as a natural botanical garden. Among the approximately 35,000 species of higher plants in China, Yunnan, also called the "Kingdom of Plants," has more than 17,000, of which an estimated 2,500 are endemic species. Besides, Yunnan is also one of the regions in the world with the most abundant resources of wild edible mushrooms. In China, there are 938 species of edible mushrooms, and over 800 of them can be found in Yunnan.

Yunnan is in fact China's most diverse province, both biologically as well as culturally. Therefore, the focus of this book is to draw on the rich culture and environment of medicinal plants and fungi/mushrooms of Yunnan Province, especially for its resources and diversity of plants. In particular, the Kunming Institute of Botany plays an important role and is a highly regarded research institute, directly affiliated with the Chinese Academy of Sciences and dedicated to research in the fields of botany and phytochemistry. This book will cover those medicinal plants specific to Yunnan Province, and is written by world experts, mainly from the Kunming Institute of Botany.

This book is structured into four distinct sections. The first section covers medicinal plants for drug development, beginning with the *Isodon* plants (Lamiaceae), which are found in abundance in Yunnan Province. Puno et al. (Chapter 1) introduce the research progress of diterpenoids, the main secondary metabolites from *Isodon* species, including eriocalyxin B, and maoecrystal V, as well as research on endophytes of *Isodon* species. Next, *Rubia yunnanensis* (Rubiaceae) is one of the important medicinal plants and local herbs in Yunnan. Tan et al. in Chapter 2 provide a review on the botany, ethnopharmacology, phytochemistry (mainly cyclopeptides, quinones, and triterpenes), and various biological activities, particularly potential anti-tumor activity, which can be considered for future drug development. Furthermore, *Dengzhanhuasu*, a famous Chinese patent medicine authorized by the China Food and Drug Administration for the treatment of acute cerebral infarction, contains the effective component of a flavonoid extracted from *Erigeron breviscapus* (Asteraceae), a monomer containing scutellarin. In Chapter 3, Zhao and Sun present the chemical constituents and biological activities of *E. breviscapus*. Furthermore, *Paris polyphylla* var. *yunnanensis* (Melanthiaceae) is one of the main herbs included in the "Yunnan Baiyao" which is a very popular proprietary traditional Chinese medicine marketed as an alternative hemostatic product for use in both human and animals for the treatment of bleeding. Liu et al. (Chapter 4) present a systematic and comprehensive review of the ethnopharmacological usages, phytochemical constituents, and pharmacological activities of *Paris polyphylla* var. *yunnanensis*.

In Section 2, there is a strong focus on medicinal mushrooms and fungi. The natural mountainous environment of Yunnan creates ideal growing conditions for a wide variety of edible fungi/mushrooms. In particular, Chuxiong, also known as "City of Fungi," is rich in fungal resources. The edible fungi/mushrooms are in fact one of the characteristics of Yunnan, and the production is large and widespread. In Chapter 5, Wu and Yang (Part 1) introduce the speculative causes of the high biodiversity of fungi/mushrooms in Yunnan and the history of utilization of medicinal mushrooms in this province. They also emphasize the sustainable utilization of those medicinal species in Yunnan, particularly conservation measurement and resource management. This is followed in Part 2 by Liu, who discusses the various phytochemicals found in these medicinal fungi/mushrooms, their biological activities, and potential toxicities.

Section 3 reviews those important classical plants in Yunnan as health food or supplements, including *Panax notoginseng* (Araliaceae), another major herb included in the "Yunnan Baiyao" formulation. Zhang et al. (Chapter 6) give a brief introduction on the history and origin, and chemical and biological studies, as well as the clinical applications of *P. notoginseng*. There are over 17,000 higher plants species in Yunnan: more than 500 of them are orchids. Hu et al.

(Chapter 7) present the two main types of medicinal orchids, *Dendrobium* species and *Gastrodia elata* (both in the Orchidaceae family), with details of their phytochemistry and biological activities. Furthermore, among the 19 species of *Garcinia* (Clusiaceae) reported in China, 14 of them can be found in Yunnan. *Garcinia* plants are well-known for their ethnomedicinal uses apart from their edible fruits such as mangosteen. Long et al. (Chapter 8) provide a comprehensive review of the ethnomedicinal uses, phytochemistry, and bioactivity of *Garcinia* species, with special emphasis on their medicinal potential. In addition, Yunnan has several different tea-growing regions, with one of Yunnan's best-known products being Pu'er tea, named after the old tea trading town of Pu'er in the province. Zhang et al. (Chapter 9) discuss the chemical constituents and pharmacological activities of Pu-er tea, and related tea plants (*Camellia* spp., Theaceae), as well as the manufacturing process and the microbes involved during the post-fermentation process of Pu-er tea. On the other hand, Yunnan also produces most of the coffee beans grown in China, though there are much smaller plantations in Fujian and Hainan. Large-scale coffee cultivation was in fact started in Yunnan in 1988, known as Yunnan Arabica coffee (*Coffea arabica*, family Rubiaceae). Qiu et al. (Chapter 10) present the medicinal functions of Yunnan Arabica coffee and the major bioactive ingredients (e.g. caffeine, chlorogenic acids, trigonelline), and other trace natural compounds such as diterpenes.

Finally, Section 4 presents medicinal plants used among ethnic groups. Apart from the rich resources of plants and fungi/mushrooms mentioned above, Yunnan Province also has the largest population of minorities, a total of 25 ethnic minority groups of more than 5000 members. Among them, 16 are indigenous (including Bai, Dai, Naxi, Hani, Lisu, Lahu, Wa, Jingpo, Bulang, Pumi, Achang, Nu, Jinuo, De'ang, Dulong, etc.), while the biggest population is Yi, and the smallest one is Dulong. Due to these different groups, their traditional customs, cultures, and languages are all distinct and their colorful clothing often attracts attention. Since these ethnic groups have developed traditional knowledge of medicinal plants, Long et al. (Chapter 11) discuss useful ethnobotanical information of medicinal plants used by these ethnic people in Yunnan, with eight examples of the commonly used ethnomedicines with scientific data.

In conclusion, this book is an attempt to provide reliable and update information on various commonly used medicinal plants and fungi/mushrooms of Yunnan province, with special emphasis on plant parts for medicinal uses, phytochemistry, and biological activities. We sincerely hope that this book will be found useful to scientists/pharmacognosists/pharmacists/chemists/graduates/undergraduates/researchers in the fields of natural products, herbal medicines, ethnobotany, pharmacology, chemistry, and biology, and those who would like to learn more about the medicinal plants and fungi/mushrooms specific to the Yunnan Province of China.

To note, this book was written and prepared during the challenging times of the global COVID-19 pandemic, with periods of lockdown in different parts of the world. Hopefully when this book is released, the pandemic will have slowed down or even ended.

Clara Bik-San Lau, BPharm, PhD, MRPharmS
Associate Director
Institute of Chinese Medicine & State Key Laboratory of Research on Bioactivities and Clinical Applications of Medicinal Plants
The Chinese University of Hong Kong
Hong Kong SAR, China

Chun-Lin Long, PhD
Professor
College of Life and Environmental Sciences
Minzu University of China and Key Laboratory of Ethnomedicine of Ministry of Education of China
Beijing 100081, China

Acknowledgments

The editors would like to thank Raymond Cooper for inviting us to prepare this book *Medicinal Plants and Mushrooms of Yunnan Province of China* as part of the Natural Products Chemistry of Global Plants Series. His encouragement and advice, particularly his great help with the editing work, are greatly appreciated.

Our wholehearted gratitude to Han-Dong Sun who has provided invaluable suggestions on the appropriate topics and authors at the early planning stage of the book content. The editors would also like to express their sincere gratitude to all the authors who had contributed to this book, particularly during the difficult time of the outbreak of global COVID-19 pandemic. We especially appreciate the high quality of work and the authors' efficiency in preparing their book chapters which allowed this book to be completed ahead of schedule.

Also special thanks to Grace Yue and Timothy Tam for their technical support in the preparation of this book.

Finally, grateful thanks to the staff of the publisher CRC Press, Taylor & Francis Group, in particular Hilary Lafoe, Jessica Poile, and Robert Sims for their great assistance in bringing this book to fruition. Also special thanks to Keith Arnold of Deanta Global Publishing for his kind assistance in the copyediting and layout of this book.

Clara Bik-San Lau
Chun-Lin Long

About the Editors

Clara Bik-San Lau is currently the Associate Director of the Institute of Chinese Medicine and the State Key Laboratory of Research on Bioactivities and Clinical Applications of Medicinal Plants at the Chinese University of Hong Kong. With a BPharm and PhD in Pharmacy (Pharmacognosy) from King's College London, University of London, UK, she has always been interested in both western herbals and traditional Chinese medicines, and has over 27 years' experience in natural products research. Her main research areas include (a) the study of herbs and natural products with potential anti-cancer activities, including the discovery and development of anti-tumor, anti-angiogenic, and anti-metastatic natural products, and research on adjuvant potentials of Chinese herbal extracts in cancer management; (b) the study of anti-diabetic and diabetic foot ulcer-healing activities of medicinal herbs; (c) investigation of beneficial herb–drug interactions. She has published over 250 SCI journal articles and 9 book chapters. She currently serves as member of ChP-USP Advisory Group on Monographs for Traditional Chinese Medicine Ingredients and Products; Acting Secretary General and Executive Council Member of the Consortium for Globalization of Chinese Medicine (CGCM); Secretary General, President-Elect and Executive Committee Member of the Board of Directors for Good Practice in Traditional Chinese Medicine Research Association (GP-TCM RA); Council Member and Associate Chief Executive of Modernized Chinese Medicine International Association (MCMIA); Advisor for Chinese Medicine Council of Hong Kong on Registration of Proprietary Chinese Medicines; Associate Editor of the *Journal of Traditional and Complementary Medicine* (eJTCM), Consulting Editor of *Pharmacological Research*, and Editorial Board member of various journals including *Phytomedicine*, *Scientific Reports*, *Journal of Natural Products*, *Journal of Ethnopharmacology*, *Journal of Pharmacy and Pharmacology* and *Integrative Medicine Research*.

Chun-Lin Long is a full professor at Minzu University of China, and the Associate Director of Key Laboratory of Ethnomedicine of Ministry of Education, in Beijing. Prof. Long graduated from Hunan Normal University, China, and holds a master's degree from Kunming Institute of Botany, Chinese Academy of Sciences, China. He obtained his PhD in 2002 from Gifu University, Japan. He has made significant contributions in the fields of agrobiodiversity, ethnobotany and ethnoecology, ethnomedicine and ethnopharmacology, phytochemistry and natural products, and plant taxonomy and phylogeny. In particular,

Prof. Long has studied medicinal plants traditionally used by different linguistic groups in southwest China through ethnobotanical, ethnopharmacological, and phytochemical approaches. He has published 412 papers in national and international journals. He is a prolific writer and has authored 22 books and 17 patents of innovations. He has established international collaborations with 12 countries. He has been awarded 19 prizes including First-Grade prizes by the Yunnan Provincial Government, Chinese Academy of Sciences, and Society of Ethnobotanists. He is currently the Chair of the Ethnobotanical Committee of China Wild Plant Conservation Association, Vice-President of the Chinese Society of Ethnobotany, and Vice-Chair of the Committee of Medicinal Plants and Phytomedicine, Botanical Society of China. He also serves as the Associate Editor of *Journal of Ethnobiology and Ethnomedicine*, and as an Editorial Board member of various journals including *Genetic Resources and Crop Evolution*, *Archives of Pharmacal Research*, *Biodiversity Science*, and other journals.

Editors and Corresponding Authors Information

Jiang-Miao Hu
State Key Laboratory of Phytochemistry and Plant Resources in West China
Kunming Institute of Botany, Chinese Academy of Sciences
Kunming, Yunnan, People's Republic of China

Clara Bik-San Lau
Institute of Chinese Medicine and State Key Laboratory of Research on Bioactivities and Clinical Applications of Medicinal Plants
The Chinese University of Hong Kong
Shatin, New Territories, Hong Kong SAR, China

Hai-Yang Liu
State Key Laboratory of Phytochemistry and Plant Resources in West China
Kunming Institute of Botany, Chinese Academy of Sciences
Kunming, Yunnan, People's Republic of China

Ji-Kai Liu
School of Pharmaceutical Sciences
South-Central University for Nationalities
Wuhan, People's Republic of China

Chun-Lin Long
College of Life and Environmental Sciences
Minzu University of China and Key Laboratory of Ethnomedicine of Ministry of Education of China
Beijing, People's Republic of China

Pema-Tenzin Puno
State Key Laboratory of Phytochemistry and Plant Resources in West China
Kunming Institute of Botany, Chinese Academy of Sciences
Kunming, Yunnan, People's Republic of China

Ming-Hua Qiu
State Key Laboratory of Phytochemistry and Plant Resources in West China
Kunming Institute of Botany, Chinese Academy of Sciences
Kunming, Yunnan, People's Republic of China

Ninghua Tan
Department of TCMs Pharmaceuticals, School of Traditional Chinese Pharmacy
China Pharmaceutical University
Nanjing, People's Republic of China

Gang Wu
CAS Key Laboratory for Plant Diversity and Biogeography of East Asia,
Kunming Institute of Botany, Chinese Academy of Sciences
Kunming, Yunnan, People's Republic of China

Ying-Jun Zhang
State Key Laboratory of Phytochemistry and Plant Resources in West China
Kunming Institute of Botany, Chinese Academy of Sciences
Kunming, Yunnan, People's Republic of China

Qin-Shi Zhao
State Key Laboratory of Phytochemistry and Plant Resources in West China
Kunming Institute of Botany, Chinese Academy of Sciences
Kunming, Yunnan, People's Republic of China

Section 1

Medicinal Plants for Drug Development

1

Diterpenoids of *Isodon* Species in Yunnan Province

Kun Hu,* Jia-Meng Dai,* Han-Dong Sun, and Pema-Tenzin Puno†
State Key Laboratory of Phytochemistry and Plant Resources in West China, Kunming Institute of Botany, Chinese Academy of Sciences, Kunming, Yunnan, People's Republic of China

CONTENTS

* These authors contributed equally to this work.
† Corresponding author.

1.1 Introduction

Isodon (Labiatae; synonyms: *Plectranthus*, *Rabdosia*, etc.) is a plant genus containing approximately 150 species worldwide, which are mainly distributed in tropical and subtropical Asia (Figure 1.1) (Li and Hedge, 1994). Many *Isodon* species are used as important folk medicines, for example, in Henan province, China, *I. rubescens* (*dong ling cao*) is often used to treat esophageal cancer; while in Japan, the leaves of *I. japonica* and *I. trichocarpa* (*enmei-so*) are often used to treat gastrointestinal dysfunction (Fujita and Node, 1984). Modern phytochemical research on *Isodon* species was initiated in the 1910s in Japan (Yagi, 1910), and a milestone came when enmein was isolated from *enmei-so* in 1958 and was finally identified as a 6,7-*seco*-*ent*-kaurane diterpenoid in 1964 (Litake and Natsume, 1964). In China, phytochemical research on *Isodon* species began in the 1970s with the purpose of finding the constituents responsible for the anti-cancer bioactivity of *I. rubescens*. Over the past few decades, extensive research on *Isodon* species has demonstrated many of these species to be a "gold-mine" of structurally diverse and biologically fascinating diterpenoids, with *ent*-kaurane and its derivatives being the dominant members. So far, over 1200 diterpenoids have been isolated and identified from this genus and over 900 have been reported by our research group. The structures, classifications, and plausible biogenetic pathways of diterpenoids from *Isodon* species have been comprehensively presented in two reviews by our group (Sun et al., 2006, Liu et al., 2017), and this chapter builds and expands on this ongoing research. The intriguing structures of these diterpenoids have aroused great interest among synthetic organic chemists,

FIGURE 1.1 Some *Isodon* species distributed in China (photos by Pema-Tenzin Puno).

and the total syntheses of nearly 20 of them have been completed (Yan et al., 2018, Li et al., 2019a). For example, total synthesis of oridonin has been achieved by Luo's group recently (Kong et al., 2019). Furthermore, the various bioactivities of these diterpenoids, especially their desirable anti-cancer activities, make them highly appealing to pharmacologists, and the mechanism of action (MOA) of some of them has been clarified (Sun et al., 2006, Liu et al., 2017, Sarwar et al., 2020). For example, parvifoline AA was recently identified as an inhibitor of peroxiredoxins and a potential immune therapeutic agent (Zhu et al., 2019).

Yunnan Province possesses abundant resources of *Isodon* plants with an estimated 49 species identified. The medicinal value of many *Isodon* species, for example, *I. eriocalyx*, *I. scoparius*, *I. adenantha*, etc., in anti-infection, anti-inflammation, anti-cancer, etc., properties has long been recognized and employed by local inhabitants. Great efforts have been made to explore the phytochemistry of *Isodon* species grown in Yunnan Province, and a series of structurally and biologically interesting molecules has been discovered (Figures 1.2, 1.4, and 1.5; Table 1.1). For example, maoecrystal V was identified as one of the most structurally complex members within *ent*-kauranes (Li et al., 2004), and its total synthesis has been completed in a few laboratories (Smith and Njardarson, 2018). Eriocalyxin B has received considerable attention due to its remarkable anti-tumor property and its extensively studied MOA;

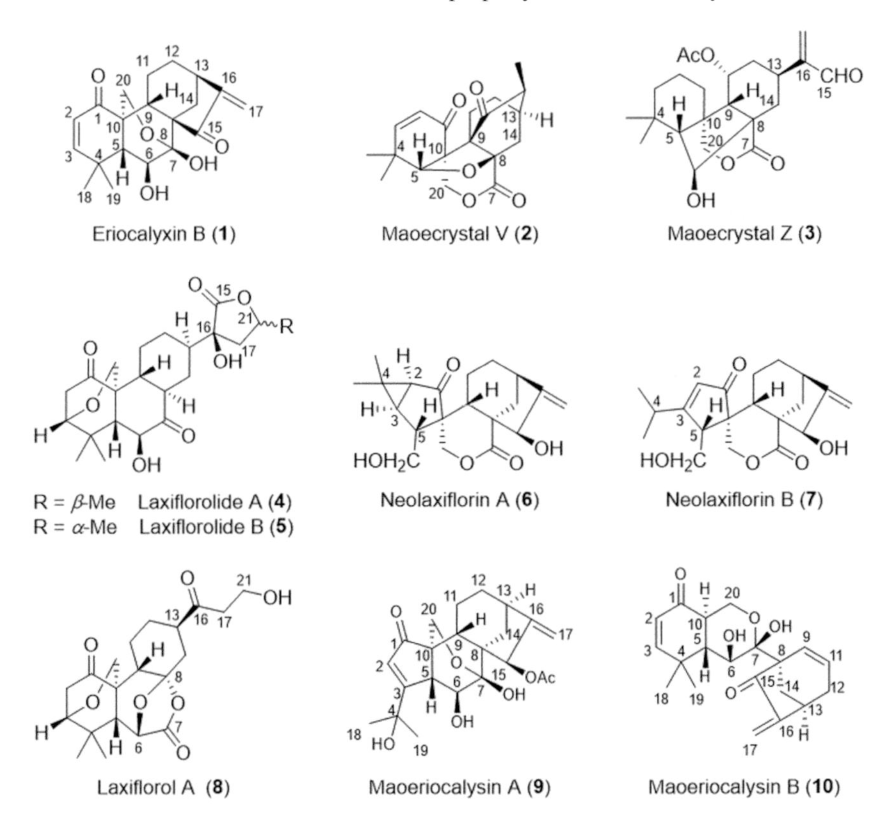

FIGURE 1.2 Representative diterpenoids from *I. eriocalyx*.

TABLE 1.1

Anti-Tumor Activities of *ent*-Kauranoids from *Isodon* Species Distributed in Yunnan

Compounds	Tumor Cell Lines	Activity (IC_{50}, μM)	Species	References
Eriocalyxin B	HL-60	0.87	*I. eriocalyx*	Shen et al., 2005
Laxiflorin J	T24	0.86	*I. eriocalyx*	Niu et al., 2002b
Laxiflorin E	K562	0.22	*I. eriocalyx* var. *laxiflora*	Niu et al., 2002a
Laxiflorin C	K562	0.57	*I. eriocalyx*	Li et al., 2004
Maoecrystal P	HL-60	1.00	*I. eriocalyx*	Wang et al., 2012b
Laxiflorin B	HL-60	0.80	*I. eriocalyx* var. *laxiflora*	Wang et al., 2014
	SMMC-7721	1.00		
	SW480	0.60		
Neolaxiflorin P	HL-60	0.95	*I. eriocalyx* var. *laxiflora*	Wang et al., 2015a
	SMMC-7721	0.69		
	MCF-7	0.83		
	SW480	0.45		
3-*epi*-Isodopharicin A	HL-60	1.00	*I. scoparius*	Jiang et al., 2017
	SW480	0.90		
Isoscoparin L	NB4	0.70	*I. scoparius*	Zhao et al., 2009
Xerophinoid B	K562	0.32	*I. xerophilus*	Weng et al., 2007
Xerophilusin I	HL-60	0.19	*I. xerophilus*	Hou et al., 2001
	MKN-28	0.07		
Xerophilusin I	K562	0.04	*I. xerophilus*	Li et al., 2007
	HepG2	0.19		
Xerophilusin II	K562	1.00	*I. xerophilus*	Li et al., 2007
Isowikstroemin A	A-549	0.88	*I. wikstroemioides*	Wu et al., 2014a
Wikstroemioidin G	SW480	0.90	*I. wikstroemioides*	Wu et al., 2014b
Wikstroemioidin P	SMMC-7721	0.80	*I. wikstroemioides*	Wu et al., 2014b
	A-549	1.00		
	SW480	1.00		
Wikstroemioidin Q	A-549	0.90	*I. wikstroemioides*	Wu et al., 2014b
	SW480	1.00		

Note: only those diterpenoids with IC_{50} values below 1 μM are listed here.

thus, it was considered as an important anti-tumor lead compound (Riaz et al., 2019). In this chapter, we will introduce the research on diterpenoids of *Isodon* species in Yunnan Province, mainly concerning their structures, bioactivities, pharmacological mechanisms, and chemical syntheses. The research on several important species, including *I. eriocalyx*, *I. scoparius*, etc., as well as a few important molecules, including maoecrystal V, eriocalyxin B, etc., will be highlighted. Furthermore, in the past few years, interesting results have been obtained in our research on endophytes of *Isodon* species, and they will be briefly introduced in this chapter.

1.2 Research on *Isodon eriocalyx*

I. eriocalyx, a perennial herb or shrub widely distributed in southwestern China, is employed as a good treatment for sore throat, influenza, hypertension, and dermatophytosis, etc., in folk medicine. It was also the active ingredient of a Chinese patent drug *mao e xiang cha cai qing re li yan pian*. Being the source of over 100 new diterpenoids, *I. eriocalyx* is one of the most extensively and phytochemically studied species, and is a rich source of highly oxygenated and skeletally rearranged *ent*-kauranes (Sun et al., 2006, Liu et al., 2017). In the following sections, research on eriocalyxin B, maoecrystal V, and a few other diterpenoids of interest is presented.

1.2.1 Research on Eriocalyxin B

Eriocalyxin B (EriB, **1**) is a 7,20-epoxy-*ent*-kaurane diterpenoid found in relatively high content in *I. eriocalyx* (Wang and Xu, 1982). EriB possess two α,β-unsaturated ketone moieties in its structure; one of them is located in ring D as many other *ent*-kauranes, and the other one is located in ring A. In a study aiming at exploring the structure activity relationship (SAR) of EriB through structural modifications, 19 derivatives were obtained and their cytotoxicities against five tumor cell lines were evaluated. The result indicated that the α,β-unsaturated ketones in rings A and D are the leading active sites. Meanwhile, the 7,20-epoxy moiety, HO-6 and HO-7, were shown to be important for cytotoxic activity (Zhao et al., 2007).

A number of studies have revealed that EriB is cytotoxic towards multiple cancer cells through acting on different molecular targets and cell signaling pathways (Riaz et al., 2019). EriB induces apoptosis and cell cycle arrest at the G2/M phase in a CAPAN-2 pancreatic cancer cell line through caspase-dependent apoptosis and the p53 pathway. Notably, its effects were comparable to that of the chemotherapeutic camptothecin, but with much lower toxicity against normal human liver cells (Li et al., 2012). EriB inhibits NF-κB activation in SMMC-7721 hepatocellular carcinoma cells by interfering with the binding of both p65 and p50 to the response element in a non-competitive manner (Kong et al., 2014). As for human SW1116 colon cancer cells, EriB blocks its progression by regulating the JAK2/STAT3 signaling pathway, resulting in the inhibition of cell proliferation, migration invasion, and angiogenesis (Lu et al., 2016). EriB induces apoptosis and autophagy of breast cancer and prostate cancer cells via targeting the Akt/mTOR pathway (Yu et al., 2019, Zhou et al., 2017). Furthermore, it also inhibits VEGF-induced angiogenesis in breast cancer (Zhou et al., 2016). It significantly inhibits lymphoma cell proliferation and induced apoptosis in association with multiple pathways including the caspase, NF-κB, AKT, and ERK pathways (Zhang et al., 2010). EriB has been reported to have the activity of inducing apoptosis of t(8;21) leukemia cells through the NF-κB and MAPK signaling pathways and triggers the degradation of AML1-ETO oncoprotein in a caspase-3-dependent manner (Wang et al., 2007). In addition, EriB was efficacious in autoimmune diseases such as experimental autoimmune encephalomyelitis (EAE), it inhibits Th1 and Th17 cell differentiation through the JAK/STAT and NF-κB signaling pathways as well as the elevation of reactive oxygen species (Lu et al., 2013b), and details of mechanisms of action studies of EriB have been recently reviewed (Riaz et al., 2019).

1.2.2 Research on Maoecrystal V

Maoecrystal V (MaoV, **2**) is a novel C_{19} diterpenoid possessing a unique 6,7-*seco*-6-*nor*-15(8→9)-*abeo*-5,8-epoxy-*ent*-kaurane scaffold. In its highly congested pentacyclic ring system, there exist six chiral centers (including three consecutive quaternary ones), spiro and bridged ring systems, etc. The structure of MaoV is so unusual that it was not fully determined until an X-ray diffraction analysis was successfully performed nearly ten years after isolating this compound. Moreover, a preliminary bioactivity evaluation showed that MaoV could inhibit HeLa cells (IC_{50}=0.02 μg/mL). All these characteristics of MaoV make it an ideal target for synthetic organic chemists (Li et al., 2004).

The first total synthesis of (±)-MaoV was completed in 2010 by Yang and coworkers, with a Wessely oxidative dearomatization, an intramolecular Diels–Alder (IMDA) reaction, and a Rh-catalyzed O–H bond insertion employed as the key synthetic strategies (Gong et al., 2010). During 2012 to 2015, several groups (Peng and Danishefsky, 2012, Lu et al., 2013a, Lu et al., 2014, Zheng et al., 2014) successively reported successful racemic or asymmetric total syntheses of MaoV (Zhang et al., 2015). This research provided the rational employment of a variety of excellent synthetic strategies: Rh-catalyzed O–H or C–H functionalization reaction, Dess–Martin periodinane, intramolecular Heck reaction, and pinacol shift, etc. In all these syntheses, the IMDA reaction was utilized to construct the [2.2.2]-bicyclooctane core. In 2016, the Baran group accomplished the total synthesis of MaoV through a quite different strategy: instead of the IMDA strategy, they chose pinacol rearrangement inspired by the proposed biosynthesis as the key strategy. The synthesis was accomplished in only 11 steps (Figure 1.3) (Cernijenko et al., 2016). In addition, it was notable that the Baran group reevaluated the biological activity of MaoV and found their result was not consistent with the original report (Cernijenko et al., 2016), and currently our original results are being reinvestigated. Readers interested in the synthetic studies of MaoV may refer to those cited papers or a recent review by Njardarson group (Smith and Njardarson, 2018).

1.2.3 Research on Other Diterpenoids from *I. eriocalyx*

1.2.3.1 Maoecrystal Z (3)

Maoecrystal Z represents the first 6,7:8,15-di-*seco*-6,8-cyclo-*ent*-kaurane diterpenoid from the genus *Isodon*. It exhibited an inhibitory effect against K562, MCF-7, and A2780 tumor cell lines (IC_{50}=2.90, 1.63, and 1.45 μg/mL, respectively) (Han et al., 2006). The first total synthesis of maoecrystal Z was achieved by the Reisman group in 2011 (Cha et al., 2011). Using (–)-γ-cyclogeraniol as the starting material, (–)-maoecrystal Z was obtained in only 12 steps. The key steps in this synthesis included a highly diastereoselective Ti^{III}-mediated reductive epoxide coupling and a Sm^{II}-mediated reductive cascade cyclization. The most recent advance in the total synthesis of maoecrystal Z came from the Liang group in 2018. Efficient cross-ring radical cyclization was employed as the key strategy for rapid construction of an all-carbon quaternary center in their synthesis (Lv et al., 2018).

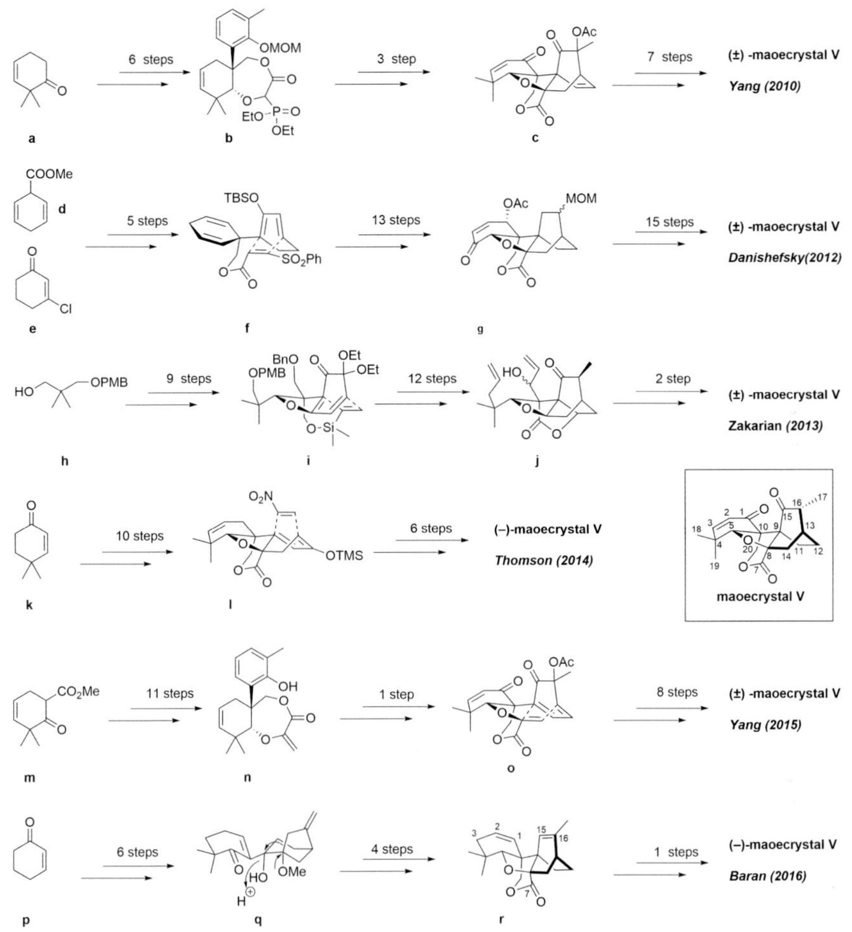

FIGURE 1.3 A brief summary of the total syntheses of maoecrystal V.

1.2.3.2 Laxiflorolides A (4) and B (5)

Laxiflorolides A and B are a pair of epimeric bishomoditerpene lactones representing the first examples of *ent*-kauranoids bearing a unique C_{22} carbon framework. Their structures were both confirmed by X-ray diffraction analysis (Wang et al., 2012b).

1.2.3.3 Neolaxiflorins A (6) and B (7)

Neolaxiflorin A is an unprecedented *ent*-kaurane diterpenoid bearing a bicyclo[3.1.0] hexane unit, and neolaxiflorin B is the 2,4-*seco*-derivative of neolaxiflorin A (Wang et al., 2012a). Seven derivatives of neolaxiflorin B have been synthesized, and one of them (compound **15** in the original paper) was found to exhibit potent cytotoxicity and relatively high selectivity (Liu et al., 2019).

1.2.3.4 Laxiflorol A (8)

Laxiflorol A is an unprecedented 7,8:15,16-di-*seco*-15-*nor*-21-homo-*ent*-kauranoid. The absolute configuration of laxiflorol A was determined by comprehensive spectroscopic analysis, as well as quantum chemical calculation of NMR chemical shifts and ECD spectra (Wang et al., 2015b).

1.2.3.5 Maoeriocalysins A–D

Maoeriocalysin A (**9**) is a novel rearranged *ent*-kaurane diterpenoid with an unprecedented 4,5-*seco*-3,5-cyclo-7,20-epoxy-*ent*-kaurane scaffold, and maoeriocalysins B–D (B: **10**) possess the rare 9,10-*seco*-7,20-epoxy-*ent*-kaurane scaffold. To cope with the difficulties in determining the complicated structures of these compounds, quantum chemical calculation in conjunction with quantitative interproton distance analysis was employed in their structural determination (Yang et al., 2019).

1.3 Research on *Isodon scoparius*

I. scoparius is a dwarf shrub endemic to Yunnan Province, and it often grows in pine-forest or on limestone at an elevation of 2300~2900 meters. *I. scoparius* is named *zhu zong cao* and is often used as an antipyretic, analgesic, and anti-inflammatory agent by the Naxi minority people in the Lijiang Naxi Autonomous County in Yunnan Province. Previous phytochemical studies, together with some of our unpublished

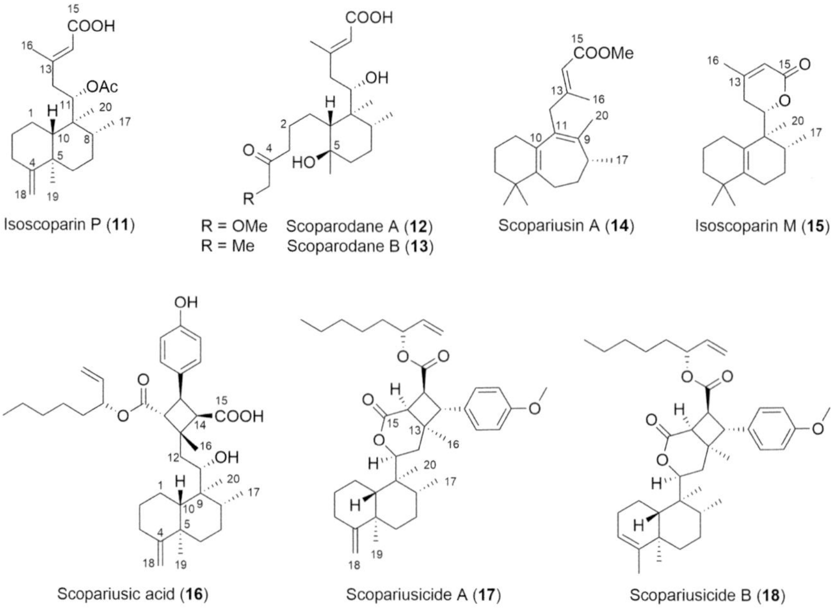

Isoscoparin P (**11**)

R = OMe Scoparodane A (**12**)
R = Me Scoparodane B (**13**)

Scopariusin A (**14**)

Isoscoparin M (**15**)

Scopariusic acid (**16**)

Scopariusicide A (**17**)

Scopariusicide B (**18**)

FIGURE 1.4 Representative diterpenoids from *I. scoparius*.

results, have revealed that, unlike other common *Isodon* species, the main constituents of *I. scoparius* consist of two types of bicyclic diterpenoids: *ent*-clerodanes and *ent*-labdanes.

1.3.1 Research on *ent*-Clerodanoids from *I. scoparius*

Isoscoparins A–C were the first *ent*-clerodanes ever reported from *Isodon* species (Xiang et al., 2004). Isoscoparins O, P (**11**), R, and S were another four *ent*-clerodanes isolated from the aerial parts of *I. scoparius*. Among them, the configurations of isoscoparins O, P, and R were validated by X-ray diffraction analysis, and isoscoparin S showed weak autophagic inhibitory activity (Li et al., 2019b). Furthermore, scoparodanes A (**12**) and B (**13**), two rare 4,5-*seco*-18-*nor*-*ent*-clerodanes, were recently reported by our group (Jiang et al., 2020). Their concise synthesis was completed starting from isoscoparin P, present in high amounts in *I. scoparius*. The synthesis featured selective hydrolysis and esterification of the non-conjugated carboxylic acid. Furthermore, scoparodane B showed significant resistance-reversal activity against fluconazole-resistant *Candida albicans* (Jiang et al., 2020).

1.3.2 Research on *ent*-Halimanoids from *I. scoparius*

Scopariusins A–C, M–N (A: **14**; M: **15**) represent five new *ent*-halimanoids isolated from *I. scoparius*, among them, scopariusins A–C are three 10(9→11)*abeo*-*ent*-halimanes featuring a bicyclo [5.4.0] undecane ring system. However, the failure to date to obtain crystals of these rearranged *ent*-halimanoids makes it difficult to determine their absolute configurations. Thus, a biomimetic transformation with isoscoparin O as the starting material was undertaken to obtain scopariusins A and N, which not only confirmed their absolute configurations, but also validated the biogenetic hypothesis (Zhou et al., 2013a).

1.3.3 Research on Meroditerpenoids from *I. scoparius*

Meroditerpenoids are not uncommon in *Isodon* species, however, scopariusic acid (**16**) and scopariusicides A (**17**) and B (**18**) were the first *ent*-clerodane-based meroditerpenoids ever reported from *Isodon* species. All these three compounds contain a unique cyclobutane ring, hypothesized to be formed via an intermolecular [2+2] cycloaddition between an *ent*-clerodane diterpenoid nucleus (part A) and an unusual ester of *trans*-4-hydroxycinnamic acid (part B). Among them, both scopariusic acid and scopariusicide A exhibited moderate immunosuppressive activity. The relative configuration of scopariusic acid was confirmed by X-ray diffraction analysis, and its absolute configuration was determined through alkali hydrolysis and ORD comparison with an authentic sample. Besides, a concise stereocontrolled semi-synthesis of scopariusicide A was efficiently achieved with isoscoparin P as the starting material. The synthesis features the combinatorial employment of intermolecular [2+2] photocycloaddition and Pd-catalyzed sp^3 C–H bond β-arylation, and this strategy was further used to generate analogues with enhanced biological activities. The synthetic strategy may be applicable to the synthesis of molecules with similar unsymmetrically substituted cyclobutanes in their structures (Zhou et al., 2013b).

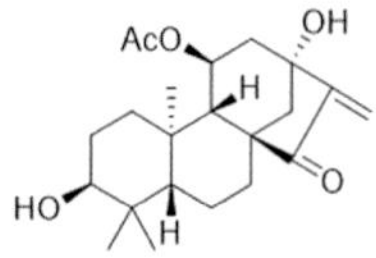

3-*epi*-Isodopharicin A (**19**) 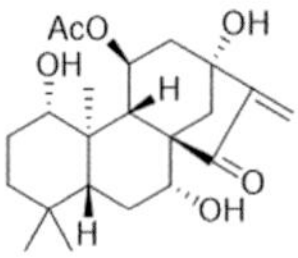 Scopariusol C (**20**) 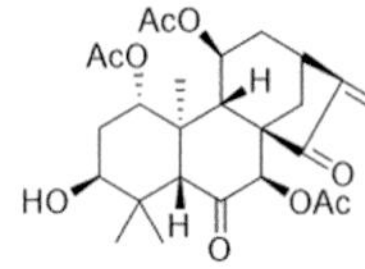 Adenanthin (**21**) 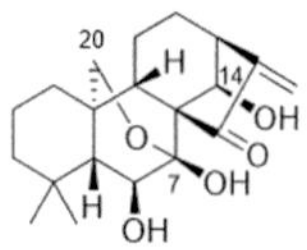 Longikaurin A (**22**)

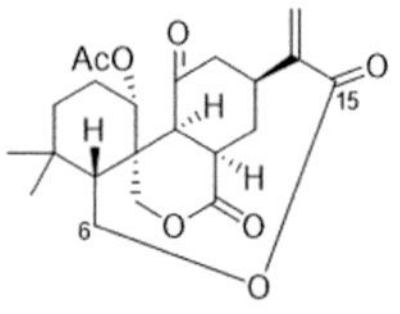

Ternifolide A (**23**) 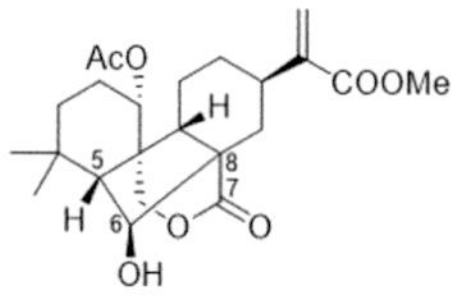 Ternifoliusin A (**24**) 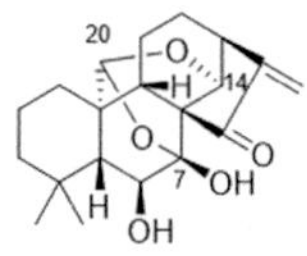 Xerophilusin B (**25**) 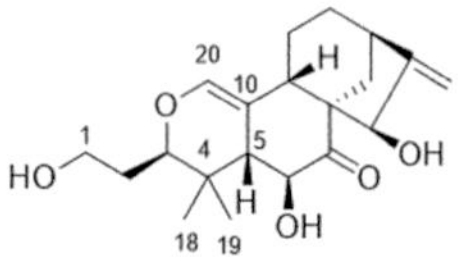 Phyllostachysin M (**26**)

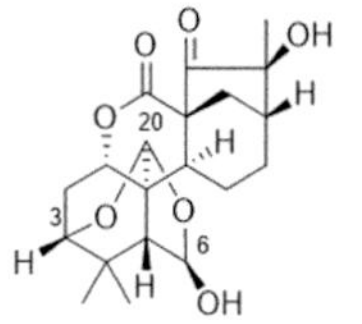

Phyllostacin J (**27**) 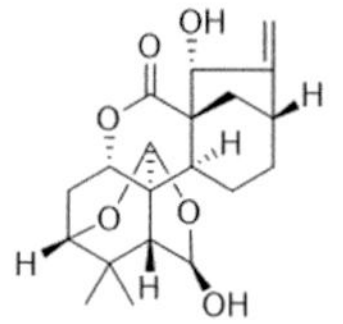 Phyllostacin K (**28**) 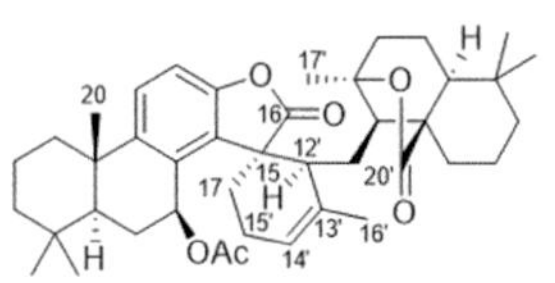 Hispidanin A (**29**)

FIGURE 1.5 Representative diterpenoids from other *Isodon* species.

1.3.4 Research on *ent*-Kauranoids from *I. scoparius*

Together with the above-mentioned interesting bicyclic diterpenoids and meroditerpenoids, *I. scoparius* also produces structurally diverse and biologically interesting *ent*-kauranoids. For example, 3-*epi*-isodopharicin A (**19**) and scopariusol C (**20**) isolated from *I. scoparius* growing in the Yulong snow mountain of Lijiang remarkably inhibited NO production in LPS-stimulated RAW264.7 cells, with IC_{50} values of 1.0 and 3.1 μM, respectively (Jiang et al., 2017).

1.4 Research on Other *Isodon* Species

1.4.1 I. Adenanthus

I. adenanthus is used as a folk medicine to treat enteritis and dysentery, and it is rich in bioactive *ent*-kauranes, among which adenanthin (Ade, **21**) is undoubtedly the most famous one (Xu et al., 1987). Ade is considered as the first natural inhibitor of peroxidases I and II (Prxs I and II) protein, and it can be used as a potential lead compound for the development of treatment of acute promyelocytic leukemia (Liu et al., 2012). Ade has been found to be able to kill hepatocellular carcinoma cells *in vitro* and xenografts as well through targeting Prxs I and II (Hou et al., 2014). In addition to Prxs I and II, Ade can also target the thioredoxin-thioredoxin reductase system and the protein disulfide isomerase. Moreover, thioredoxin-1 (Trx-1) was identified as an Ade binding protein incubated with biotinylated Ade as an affinity probe (Muchowicz et al., 2014). Research by Soethoudt et al. (2014) showed that the reaction rate of Ade with Prxs I and II was significantly lower than that with glutathione and thioredoxin reductase. In the experiment of treating erythrocytes and Jurkat T with Ade, no evidence of Prxs I and II binding to Ade was found. The author thought Ade was not a specific Prx inhibitor, and the reported anti-tumor and anti-inflammatory effects were more likely to involve more general inhibition of thioredoxin and/or glutathione redox pathways (Soethoudt et al., 2014). It is suggested that Ade can be used to treat a series of diseases whose pathogenesis is related to the abnormal activity of peroxidase, the thioredoxin-thioredoxin reductase system, and the activity of protein disulfide isomerase (Muchowicz et al., 2014). Yin and co-workers found Ade was a novel NF-κB signaling pathway and it had potent immunomodulatory activity for the treatment of multiple sclerosis and other autoimmune diseases (Yin et al., 2013, Hu et al., 2019). Another recent research study indicated Ade significantly reduces weight gain and adipose tissue quality, indicating the beneficial effects of Ade as a potential drug for preventing obesity in high fat diet-induced obesity (Hu et al., 2019).

1.4.2 I. Ternifolius

I. ternifolius is a perennial herb with relatively unique morphology: it usually has verticillate leaves and stems with six edges, and it is commonly used in folk medicine for the treatment of enteritis, icterohepatitis, and other inflammatory ailments. Besides, it is the major ingredient of a Chinese patent anti-hepatitis drug named *fu fang san ye xiang cha cai pian* (Gou et al., 2019). *I. ternifolius* is another excellent source of novel and bioactive diterpenoid. For example, longikaurin A (**22**), a previously described cytotoxic *ent*-kauranoid, was also isolated (Zou et al., 2012) and provided a few interesting results in both bioactivity and mechanism studies: longikaurin A can inhibit nasopharyngeal carcinoma cell proliferation *in vitro* and *in vivo* through the induction of cell cycle arrest and apoptosis (Zou et al., 2013). Later, longikaurin A was also found to induce cell cycle arrest and apoptosis in human HCC cell lines by dampening Skp2 expression (Liao et al., 2014). Moreover, longikaurin A can inhibit esophageal squamous cell carcinoma cells growth *in vitro* and in *vivo* by inducing ROS production and activating the JNK and p38 MAPK pathways in ESCC cells (Che et al., 2017). Notably, ternifolide A (**23**), a unique diterpenoid reported in 2012, featuring a ten-membered lactone ring formed between C-6 and C-15, was also isolated (Zou et al., 2012). In recent research, *I. ternifolius* collected from three counties in Yunnan Province led to the isolation of nine new *ent*-kauranes, ternifoliusins A–I (A: **24**), together with 26 known *ent*-kauranes: some of them exhibited good cytotoxic activity against human cancer cells (Gou et al., 2019).

1.4.3 I. Xerophilus

I. xerophilus is mainly distributed in south central Yunnan and has medicinal value in its anti-tumor, anti-inflammation, and anti-bacteria properties. Xerophilusin B (**25**) is a 7,20:14,20-diepoxy-*ent*-kaurane diterpenoid isolated from *I. xerophilus* (Hou et al., 2000a). It exhibits anti-neoplastic activity on esophageal squamous cell carcinoma (ESCC) cell lines *in vivo* in a dose-dependent manner without significant secondary adverse effects. Furthermore, xerophilusin B can induce G2/M cell cycle arrest and promote apoptosis through mitochondrial cytochrome *c*-dependent activation of the caspase-9 and caspase-3 cascade pathway in ESCC cell lines (Yao et al., 2015).

1.4.4 I. Phyllostachys

I. phyllostachys has been used as an antiphlogistic and antibiotic agent in folk medicine. Recent phytochemical investigations on *I. phyllostachys* resulted in the discovery of a number of diterpenoids with interesting structures (Hou et al., 2000b, Li et al., 2006, Yang et al., 2016). Among them, phyllostachysin M (**26**) is an unusual *ent*-kauranoid featuring a novel dihydro-2H-pyran ring motif resulting from 1,10-cleavage in ring A (Yang et al., 2017), while phyllostacins J and K (**27, 28**) are the first examples of 3,20:6,20-diepoxy enmein-type *ent*-kauranoids (Yang et al., 2016).

1.4.5 I. Hispida

I. hispida is used to treat traumatic injury. Phytochemical research on the rhizomes of *I. hispida* led to the discovery of hispidanins A–D (A: **29**), four unprecedented asymmetric dimers formed by the bonding of totarane and labdane (Huang et al., 2014). Two years later, the asymmetric total synthesis of hispidanin A was completed (Deng et al., 2017).

1.5 Study of Endophytic Fungi from *Isodon* Species

Endophytic fungi are a unique type of microorganism colonizing in plants without causing apparent disease. Extensive research on endophytic fungi during the past decades has demonstrated their tremendous capability of producing structurally and biologically interesting molecules. In 2016, we began to focus on the proprietary collection of endophytes isolated from healthy plants of genus *Isodon*, and a series of interesting fungal metabolites have been discovered in recent years (Figure 1.6).

Chemical investigation of the cultures of *Penicillium* sp. sh18 endophytic to the stems of *I. eriocalyx* var. *laxiflora* led to the isolation of penicilfuranone A (**30**), a unique furancarboxylic acid featuring a 4-(methoxycarbonyl)furan-3(2H)-one core fused with an oxygenated C_9 unit. Penicilfuranone A exhibited a significant antifibrotic effect in activated hepatic stellate cells via negative regulation of transforming growth factor-β (TGF-β)/Smad signaling (Wang et al., 2016). Further research on the same fungus resulted in the discovery of isopenicins A–C (**31–33**), three novel meroterpenoids possessing two types of unprecedented terpenoid-polyketide hybrid skeletons. Among them, isopenicin A can inhibit Wnt-dependent reporter activities and the expression of endogenous Wnt signaling target genes. Besides, isopenicin A can selectively inhibit the growth of colorectal cancer cells with constitutive Wnt signaling and cause obvious G2/M cell cycle arrest and subsequent apoptosis (Tang et al., 2019a). Furthermore, in a recent study, isopenicin A was identified as a novel natural microtubule depolymerization agent with tumor proliferation inhibitory activity by inducing G2/M cell cycle arrest and cell apoptosis (Chen et al., 2020).

A chemical study on the endophytic fungus, *Phomopsis* sp. shj2 isolated from the *I. eriocalyx* var. *laxiflora* afforded phomopchalasins A (**34**) and B (**35**), two cytochalasans featuring unprecedented 5/6/5/8-fused tetracyclic and 5/6/6/7/5-fused pentacyclic skeletons. Moreover, phomopchalasin B exhibited an anti-migratory effect against MDA-MB-231 *in vitro* with IC_{50} values of 19.1 μM (Tang et al., 2017). Research on an

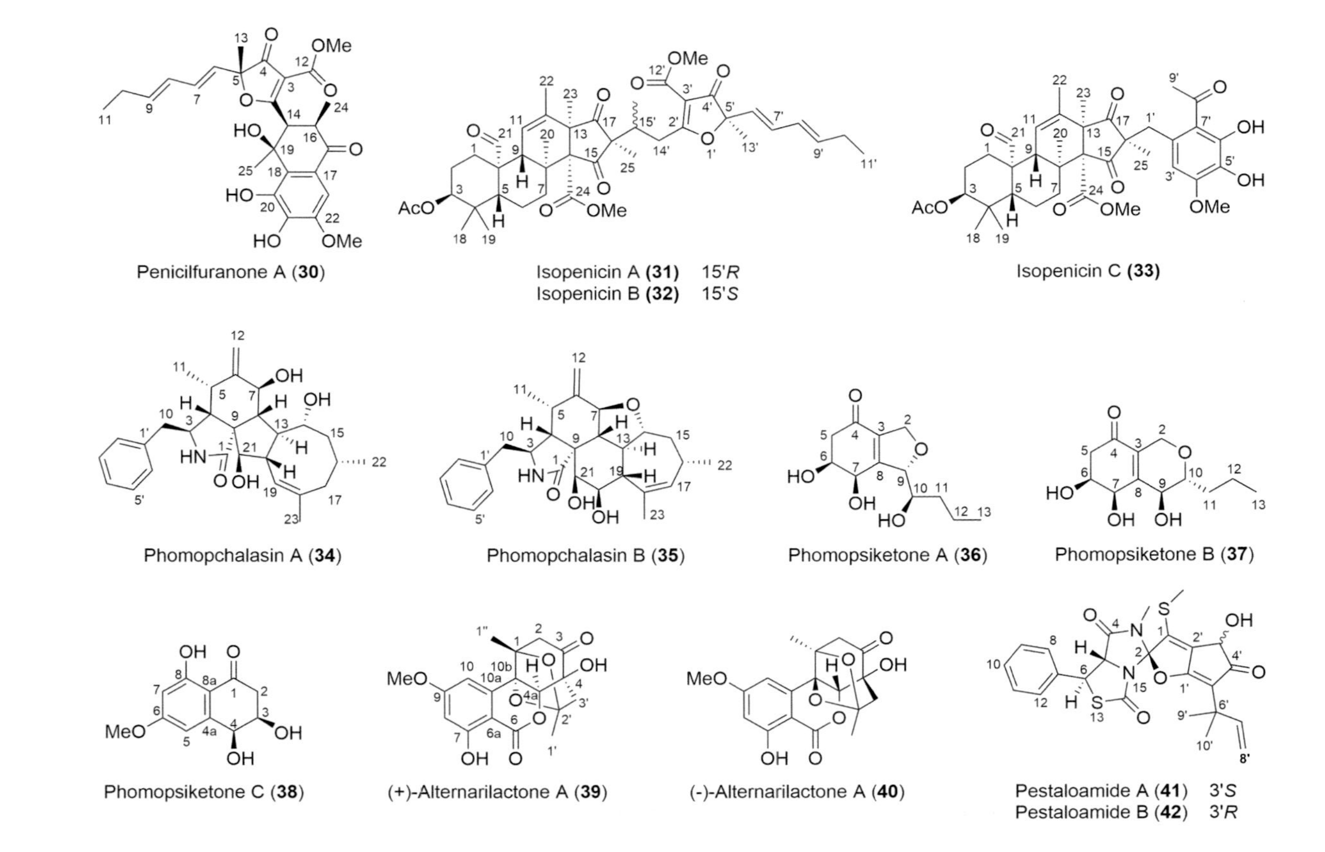

FIGURE 1.6 Representative compounds from endophytic fungus of *Isodon* species.

endophytic fungus of the same genus, *Phomopsis* sp. sh917, utilizing a one strain/many compounds (OSMAC) strategy resulted in the isolation of five new polyketides, phomopsiketones A–C (**36–38**), (10*S*)-10-O-β-D-40-methylmannopyranosyldiaporthin, and clearanol H (Chen et al., 2018).

An endophytic fungus *Alternaria* sp. hh930 colonizing in the stem of *I. sculponeatus* was also subjected to research. As a result, (+)- and (–)-alternarilactone A (**39, 40**), a pair of dibenzo-α-pyrones enantiomers bearing highly oxidized diepoxy-cage-like moieties, were obtained. Their structures and configurations were unambiguously determined using quantum chemical calculations and X-ray diffraction analysis (Tang et al., 2019b).

In another very recent research from our group, pestaloamides A and B (**41, 42**), two novel alkaloids featuring an unprecedented spiro[imidazothiazoledione-alkylidenecyclopentenone] scaffold, were obtained from *Pestalotiopsis* sp. HS30, an endophytic fungus present in the stems of *I. xerophilus*. Moreover, these two compounds both showed latent tumor immunotherapy activity through markedly promoting the cell surface engagement of NKG2D ligands involving MICA/B and ULBP1 in HCT116 cells (Su et al., 2020).

1.6 Summary and Outlook

The unlimited "creativity" of nature has made natural products quite different from various synthetic molecules and thus an invaluable source for drug discovery, which may be the major reason that natural product research is still going strongly (Newman and Cragg, 2020). Extensive research on *Isodon* species in the past decades has led to the discovery of over 1000 diterpenoids, which not only enhance our knowledge of this class of molecules, but also expand the chemical space from natural products we have access to. Moreover, numerous synthetic studies on diterpenoids from *Isodon* species promoted the development of both new synthetic methods and inspiring synthetic strategies (Yan et al., 2018, Li et al., 2019a). Most importantly, the desirable bioactivities of this class of molecules and in-depth mechanism studies on them have led to the discovery of several molecules as promising lead compounds (Sarwar et al., 2020).

Though the research on *Isodon* species has covered many decades, there are still many problems remaining to be settled and many directions remain. Firstly, several *Isodon* species have not been systematically investigated, and new species are still being identified: *I. delavayi* (Chen et al., 2014), *I. villosus* (Chen et al., 2016), *I. wui* (Xiang and Liu, 2012), *I. hsiwenii* (Chen et al., 2019), etc. Notably, to reduce the harm to valuable wild *Isodon* resources, our group has cultivated several species in our Kunming Institute of Botany using seeds collected from the wild. Secondly, the present medicinal chemical research involving diterpenoids from *Isodon* species is mostly limited to oridonin; thus, to carry out broader and more in-depth studies, more efforts should be evaluated to ensure the sample availability either by isolation from natural sources or scalable synthesis (Zhu et al., 2018). Thirdly, as for the research on endophytic fungus from *Isodon* species, it is still important to improve the strategy in hunting for "useful species." In our recent research, new techniques such as molecular networking have been employed and shown to be useful (Su et al., 2020, Tang et al.,

2019a). Finally, the ever-increasing structurally complicated molecules obtained from either *Isodon* species or their endophytes pose great challenges in their structure elucidation, which demands the introduction of more robust tools. The employment of quantum chemical calculations and advanced NMR techniques is effective and worthy of further exploration (Yang et al., 2019, Su et al., 2020).

REFERENCES

Cernijenko A, Risgaard R, Baran PS, 2016. 11-Step total synthesis of (−)-maoecrystal V. *J. Am. Chem. Soc.*, 138: 9425–9428.

Cha JY, Yeoman JTS, Reisman SE, 2011. A concise total synthesis of (−)-maoecrystal Z. *J. Am. Chem. Soc.*, 133: 14964–14967.

Che Y, Wang JN, Yuan ZY, et al., 2017. The therapeutic effects of longikaurin A, a natural *ent*-kauranoid, in esophageal squamous cell carcinoma depend on ROS accumulation and JNK/p38 MAPK activation. *Toxicol. Lett.*, 280: 106–115.

Chen L, Fan DM, Tang JW, et al., 2020. Discovery of isopenicin A, a meroterpenoid as a novel inhibitor of tubulin polymerization. *Biochem. Biophys. Res. Commun.*, 525: 303–307.

Chen R, Tang JW, Li XR, et al., 2018. Secondary metabolites from the endophytic fungus *Xylaria* sp. hg1009. *Nat. Prod. Bioprospect.*, 8: 121–129.

Chen YP, Hu GX, Xiang CL, 2014. *Isodon delavayi* (Ocimeae, Nepetoideae, Lamiaceae): a new species from Yunnan Province, Southwest China. *Phytotaxa*, 156: 291–297.

Chen YP, Wilson TC, Zhou YD, et al., 2019. *Isodon hsiwenii* (Lamiaceae: Nepetoideae), a new species from Yunnan, China. *Syst. Bot.*, 44: 913–922.

Chen YP, Xiang CL, Sunojkumar P, et al., 2016. *Isodon villosus* (Nepetoideae, Lamiaceae), a new species from Guangxi, China. *Phytotaxa*, 268: 271–278.

Deng HP, Cao W, Liu R, et al., 2017. Asymmetric total synthesis of hispidanin A. *Angew. Chem. Int. Ed.*, 56: 5849–5852.

Fujita E, Node M, 1984. Diterpenoids of *Rabdosia* species. *In*: Fujita E, Johne S, Kasai R, Node M, Tanaka O, Herz W, Grisebach H, Kirby GW & Tamm C (eds.) *Progress in the chemistry of organic natural products*. Vienna: Springer Vienna.

Gong JX, Lin G, Sun WB, et al., 2010. Total synthesis of (±)-maoecrystal V. *J. Am. Chem. Soc.*, 132: 16745–16746.

Gou LL, Hu K, Yang Q, et al., 2019. Structurally diverse diterpenoids from *Isodon ternifolius* collected from three regions. *Tetrahedron*, 75: 2797–2806.

Han QB, Cheung S, Tai J, et al., 2006. Maoecrystal Z, a cytotoxic diterpene from *Isodon eriocalyx* with a unique skeleton. *Org. Lett.*, 8: 4727–4730.

Hou AJ, Li ML, Jiang B, et al., 2000a. New 7,20:14,20-diepoxy *ent*-kauranoids from *Isodon xerophilus*. *J. Nat. Prod.*, 63: 599–601.

Hou AJ, Yang H, Jiang B, et al., 2000b. A new *ent*-kaurane diterpenoid from *Isodon phyllostachys*. *Fitoterapia*, 71: 417–419.

Hou AJ, Zhao QS, Li ML, et al., 2001. Cytotoxic 7,20-epoxy *ent*-kauranoids from *Isodon xerophilus*. *Phytochemistry*, 58: 179–183.

Hou JK, Huang Y, He W, et al., 2014. Adenanthin targets peroxiredoxin I/II to kill hepatocellular carcinoma cells. *Cell Death Dis.*, 5: e1400.

Hu J, Li X, Tian W, et al., 2019. Adenanthin, a natural *ent*-kaurane diterpenoid isolated from the herb *Isodon adenantha* inhibits adipogenesis and the development of obesity by regulation of ROS. *Molecules*, 24: 158.

Huang B, Xiao CJ, Huang ZY, et al., 2014. Hispidanins A-D: four new asymmetric dimeric diterpenoids from the rhizomes of *Isodon hispida*. *Org. Lett.*, 16: 3552–3555.

Jiang HY, Wang WG, Tang JW, et al., 2017. Structurally diverse diterpenoids from *Isodon scoparius* and their bioactivity. *J. Nat. Prod.*, 80: 2026–2036.

Jiang XC, Yan BC, Li XR, et al., 2020. 4,5-*Seco*-18-*nor*-*ent*-clerodanoids and their derivatives: structure elucidation, synthesis and resistant reversal activities against fluconazole-resistance *Candida albicans*. *Tetrahedron*, 76: 131043.

Kong L, Su F, Yu H, et al., 2019. Total synthesis of (–)-oridonin: an interrupted Nazarov approach. *J. Am. Chem. Soc.*, 141: 20048–20052.

Kong LM, Deng X, Zuo ZL, et al., 2014. Identification and validation of p50 as the cellular target of eriocalyxin B. *Oncotarget*, 5: 11354–11364.

Li L, Yue GGL, Lau CBS, et al., 2012. Eriocalyxin B induces apoptosis and cell cycle arrest in pancreatic adenocarcinoma cells through caspase- and p53-dependent pathways. *Toxicol. Appl. Pharmacol.*, 262: 80–90.

Li LM, Weng ZY, Huang SX, et al., 2007. Cytotoxic *ent*-kauranoids from the medicinal plant *Isodon xerophilus*. *J. Nat. Prod.*, 70: 1295–1301.

Li SH, Wang J, Niu XM, et al., 2004. Maoecrystal V, cytotoxic diterpenoid with a novel C_{19} skeleton from *Isodon eriocalyx* (Dunn.) Hara. *Org. Lett.*, 6: 4327–4330.

Li W, Wang JJ, Ma DW, 2019a. The total synthesis of *ent*-kaurane diterpenoids. *Prog. Chem.*, 31: 1460–1471.

Li X, Xiao WL, Pu JX, et al., 2006. Cytotoxic *ent*-kaurene diterpenoids from *Isodon phyllostachys*. *Phytochemistry*, 67: 1336–1340.

Li XR, Fu Q, Zhou M, et al., 2019b. Isoscoparins R and S, two new *ent*-clerodane diterpenoids from *Isodon scoparius*. *J. Asian Nat. Prod. Res.*, 21: 977–984.

Li XW, Hedge IC, 1994. Lamiaceae. *In*: Flora of China Editorial Committee (eds.) *Flora of China*. Beijing: Science Press.

Liao YJ, Bai HY, Li ZH, et al., 2014. Longikaurin A, a natural *ent*-kaurane, induces G2/M phase arrest via downregulation of Skp2 and apoptosis induction through ROS/JNK/c-Jun pathway in hepatocellular carcinoma cells. *Cell Death Dis.*, 5: e1137.

Litake Y, Natsume M, 1964. The X-ray study of acetyl-bromoacetyldihydroenmein. *Tetrahedron Lett.*, 5: 1257–1261.

Liu CX, Yin QQ, Zhou HC, et al., 2012. Adenanthin targets peroxiredoxin I and II to induce differentiation of leukemic cells. *Nat. Chem. Biol.*, 8: 486–493.

Liu JP, Xiao YZ, H Y, et al., 2019. Synthesis and antitumor evaluation of neolaxiflorin B inspired compounds. *Drug Des. Dev. Ther.*, 13: 3021–3028.

Liu M, Wang WG, Sun HD, et al., 2017. Diterpenoids from *Isodon* species: an update. *Nat. Prod. Rep.*, 34: 1090–1140.

Lu P, Gu ZH, Zakarian A, 2013a. Total synthesis of maoecrystal V: early-stage C-H functionalization and lactone assembly by radical cyclization. *J. Am. Chem. Soc.*, 135: 14552–14555.

Lu P, Mailyan A, Gu ZH, et al., 2014. Enantioselective synthesis of (–)-maoecrystal V by enantiodetermining C-H functionalization. *J. Am. Chem. Soc.*, 136: 17738–17749.

Lu Y, Chen B, Song JH, et al., 2013b. Eriocalyxin B ameliorates experimental autoimmune encephalomyelitis by suppressing Th1 and Th17 cells. *Proc. Natl. Acad. Sci. U. S. A.*, 110: 2258–2263.

Lu YM, Chen W, Zhu JS, et al., 2016. Eriocalyxin B blocks human SW1116 colon cancer cell proliferation, migration, invasion, cell cycle progression and angiogenesis via the JAK2/STAT3 signaling pathway. *Mol. Med. Rep.*, 13: 2235–2240.

Lv Z, Chen BL, Zhang C, et al., 2018. Total syntheses of trichorabdal A and maoecrystal. *Z. Chem.-Eur. J.*, 24: 9773–9777.

Muchowicz A, Firczuk M, Chlebowska J, et al., 2014. Adenanthin targets proteins involved in the regulation of disulfide bonds. *Biochem. Pharmacol.*, 89: 210–216.

Newman DJ, Cragg GM, 2020. Natural products as sources of new drugs over the nearly four decades from 01/1981 to 09/2019. *J. Nat. Prod.*, 83: 770–803.

Niu XM, Li SH, Li ML, et al., 2002a. Cytotoxic *ent*-kaurane diterpenoids from *Isodon eriocalyx* var. laxiflora. *Planta Med.*, 68: 528–533.

Niu XM, Li SH, Mei SX, et al., 2002b. Cytotoxic 3,20-epoxy-*ent*-kaurane diterpenoids from *Isodon eriocalyx* var. laxiflora. *J. Nat. Prod.*, 65: 1892–1896.

Peng F, Danishefsky SJ, 2012. Total synthesis of (±)-maoecrystal V. *J. Am. Chem. Soc.*, 134: 18860–18867.

Riaz A, Saleem B, Hussain G, et al., 2019. Eriocalyxin B biological activity: a review on its mechanism of action. *Nat. Prod. Commun.* (In press) https://doi.org/10.1177/1934578X19868598.

Sarwar MS, Xia YX, Liang ZM, et al., 2020. Mechanistic pathways and molecular targets of plant-derived anticancer *ent*-kaurane diterpenes. *Biomolecules*, 10: 144.

Shen YH, Wen ZY, Xu G, et al., 2005. Cytotoxic *ent*-kaurane diterpenoids from *Isodon eriocalyx*. *Chem. Biodivers.*, 2: 1665–1672.

Smith BR, Njardarson JT, 2018. Review of synthetic approaches toward maoecrystal V. *Org. Biomol. Chem.*, 16: 4210–4222.

Soethoudt M, Peskin AV, Dickerhof N, et al., 2014. Interaction of adenanthin with glutathione and thiol enzymes: selectivity for thioredoxin reductase and inhibition of peroxiredoxin recycling. *Free Radic. Biol. Med.*, 77: 331–339.

Su XZ, Zhu YY, Tang JW, et al., 2020. Pestaloamides A and B, two spiro-heterocyclic alkaloid epimers from the plant endophytic fungus Pestalotiopsis sp. HS30. *Sci. China Chem.*, 63: 1208–1213.

Sun HD, Huang SX, Han QB, 2006. Diterpenoids from *Isodon* species and their biological activities. *Nat. Prod. Rep.*, 23: 673–698.

Tang JW, Kong LM, Zu WY, et al., 2019a. Isopenicins A–C: two types of antitumor meroterpenoids from the plant endophytic fungus *Penicillium* sp. sh18. *Org. Lett.*, 21: 771–775.

Tang JW, Wang WG, Li A, et al., 2017. Polyketides from the endophytic fungus *Phomopsis* sp. sh917 by using the one strain/many compounds strategy. *Tetrahedron*, 73: 3577–3584.

Tang JW, Xu HC, Wang WG, et al., 2019b. (+)- and (−)-Alternarilactone A: enantiomers with a diepoxy-cage-like scaffold from an endophytic *Alternaria* sp. *J. Nat. Prod.*, 82: 735–740.

Wang L, Zhao WL, Yan JS, et al., 2007. Eriocalyxin B induces apoptosis of t(8;21) leukemia cells through NF-κB and MAPK signaling pathways and triggers degradation of AML1-ETO oncoprotein in a caspase-3-dependent manner. *Cell Death Differ.*, 14: 306–317.

Wang WG, Du X, Li XN, et al., 2012a. New bicyclo[3.1.0]hexane unit *ent*-kaurane diterpene and its *seco*-derivative from *Isodon eriocalyx* var. laxiflora. *Org. Lett.*, 14: 302–305.

Wang WG, Li A, Yan BC, et al., 2016. LC-MS-guided isolation of penicilfuranone A: a new antifibrotic furancarboxylic acid from the plant endophytic fungus *Penicillium* sp. sh18. *J. Nat. Prod.*, 79: 149–155.

Wang WG, Li XN, Du X, et al., 2012b. Laxiflorolides A and B, epimeric bishomoditerpene lactones from *Isodon eriocalyx*. *J. Nat. Prod.*, 75: 1102–1107.

Wang WG, Tang JW, Shi YM, et al., 2015b. Laxiflorol A, the first example of 7,8:15,16-di-seco-15-nor-21-*homo-ent*-kauranoid from *Isodon eriocalyx var. laxiflora*. *RSC Adv.*, 5: 6132–6135.

Wang WG, Yan BC, Li XN, et al., 2014. 6,7-*Seco-ent*-kaurane-type diterpenoids from *Isodon eriocalyx var. laxiflora*. *Tetrahedron*, 70: 7445–7453.

Wang WG, Yang J, Wu HY, et al., 2015a. *ent*-Kauranoids isolated from *Isodon eriocalyx var. laxiflora* and their structure activity relationship analyses. *Tetrahedron*, 71: 9161–9171.

Wang ZY, Xu YL, 1982. New diterpenoid constituents of *Rabdosia eriocalyx*. *Acta. Botanica Yunanica*, 4: 407–411.

Weng ZY, Huang SX, Li ML, et al., 2007. Isolation of two bioactive *ent*-kauranoids from the leaves of *Isodon xerophilus*. *J. Agric. Food Chem.*, 55: 6039–6043.

Wu HY, Wang WG, Jiang HY, et al., 2014a. Cytotoxic and anti-inflammatory *ent*-kaurane diterpenoids from *Isodon wikstroemioides*. *Fitoterapia*, 98: 192–198.

Wu HY, Zhan R, Wang WG, et al., 2014b. Cytotoxic *ent*-kaurane diterpenoids from *Isodon wikstroemioides*. *J. Nat. Prod.*, 77: 931–941.

Xiang CL, Liu ED, 2012. A new species of *Isodon* (Lamiaceae, Nepetoideae) from Yunnan Province, Southwest China. *Syst. Bot.*, 37: 811–817.

Xiang W, Li RT, Song QS, et al., 2004. *ent*-Clerodanoids from *Isodon scoparius*. *Helv. Chim. Acta*, 87: 2860–2865.

Xu YL, Sun HD, Wang DZ, et al., 1987. Structure of adenanthin. *Tetrahedron Lett.*, 28: 499–502.

Yagi S, 1910. On "Plectranthin", a bitter principle derived from *Plectranthus glaucocalyx* Maxim. var. *japonicus* Maxim. *J. Kyoto Med. Soc.*, 7: 30-33.

Yan BC, Hu K, Sun HD, et al., 2018. Recent advances in the synthesis of *Isodon* diterpenoids and schinortriterpenoids *Chin. J. Org. Chem.*, 38: 2259–2280.

Yang J, Wang WG, Wu HY, et al., 2016. Bioactive enmein-type *ent*-kaurane diterpenoids from *Isodon phyllostachys*. *J. Nat. Prod.*, 79: 132–140.

Yang Q, Hu K, Yan BC, et al., 2019. Maoeriocalysins A–D, four novel *ent*-kaurane diterpenoids from *Isodon eriocalyx* and their structure determination utilizing quantum chemical calculation in conjunction with quantitative interproton distance analysis. *Org. Chem. Front.*, 6: 45–53.

Yang YC, Lin PH, Wei MC, 2017. Production of oridonin-rich extracts from *Rabdosia rubescens* using hyphenated ultrasound-assisted supercritical carbon dioxide extraction. *J. Sci. Food Agric.*, 97: 3323–3332.

Yao R, Chen ZL, Zhou CC, et al., 2015. Xerophilusin B induces cell cycle arrest and apoptosis in esophageal squamous cell carcinoma cells and does not cause toxicity in nude mice. *J. Nat. Prod.*, 78: 10–16.

Yin QQ, Liu CX, Wu Y L, et al., 2013. Preventive and therapeutic effects of adenanthin on experimental autoimmune encephalomyelitis by inhibiting NF-κB signaling. *J. Immunol.*, 191: 2115.

Yu ZQ, Chen Y, Liang CZ, 2019. Eriocalyxin B induces apoptosis and autophagy involving Akt/Mammalian target of rapamycin (mTOR) pathway in prostate cancer cells. *Med. Sci. Monit.*, 25: 8534–8543.

Zhang WB, Shao WB, Li FZ, et al., 2015. Asymmetric total synthesis of (−)-maoecrystal V. *Chem. Asian J.*, 10: 1874–1880.

Zhang YW, Jiang XX, Chen QS, et al., 2010. Eriocalyxin B induces apoptosis in lymphoma cells through multiple cellular signaling pathways. *Exp. Hematol.*, 38: 191–201.

Zhao Y, Niu XM, Qian LP, et al., 2007. Synthesis and cytotoxicity of some new eriocalyxin B derivatives. *Eur. J. Med. Chem.*, 42: 494–502.

Zhao Y, Pu JX, Huang SX, et al., 2009. *ent*-Kaurane diterpenoids from *Isodon scoparius*. *J. Nat. Prod.*, 72: 125–129.

Zheng CW, Dubovyk I, Lazarski KE, et al., 2014. Enantioselective total synthesis of (−)-maoecrystal V. *J. Am. Chem. Soc.*, 136: 17750–17756.

Zhou M, Geng HC, Zhang HB, et al., 2013a. Scopariusins, A new class of *ent*-halimane diterpenoids isolated from *Isodon* scoparius, and biomimetic synthesis of scopariusin A and isoscoparin N. *Org. Lett.*, 15: 314–317.

Zhou M, Zhang HB, Wang WG, et al., 2013b. Scopariusic acid, a new meroditerpenoid with a unique cyclobutane ring isolated from *Isodon scoparius*. *Org. Lett.*, 15: 4446–4449.

Zhou XN, Yue GGL, Chan AML, et al., 2017. Eriocalyxin B, a novel autophagy inducer, exerts anti-tumor activity through the suppression of Akt/mTOR/p70S6K signaling pathway in breast cancer. *Biochem. Pharmacol.*, 142: 58–70.

Zhou XN, Yue GGL, Liu MH, et al., 2016. Eriocalyxin B, a natural diterpenoid, inhibited VEGF-induced angiogenesis and diminished angiogenesis-dependent breast tumor growth by suppressing VEGFR-2 signaling. *Oncotarget*, 7: 82820–82835.

Zhu HF, Wang B, Kong LM, et al., 2019. Parvifoline AA promotes susceptibility of hepatocarcinoma to natural killer cell-mediated cytolysis by targeting peroxiredoxin. *Cell Chem. Biol.*, 26: 1122–1132.

Zhu LZ, Ma WJ, Zhang MX, et al., 2018. Scalable synthesis enabling multilevel bio-evaluations of natural products for discovery of lead compounds. *Nat. Commun.*, 9: 1283.

Zou J, Du X, Pang G, et al., 2012. Ternifolide A, a new diterpenoid possessing a rare macrolide motif from *Isodon ternifolius*. *Org. Lett.*, 14: 3210–3213.

Zou QF, Du JK, Zhang H, et al., 2013. Anti-tumour activity of longikaurin A (LK-A), a novel natural diterpenoid, in nasopharyngeal carcinoma. *J. Transl. Med.*, 11: 200.

2

Rubia yunnanensis Diels

Ninghua Tan,* Zhe Wang, Junting Fan, and Xuejia Zhang
Department of TCMs Pharmaceuticals, School of Traditional Chinese Pharmacy, China Pharmaceutical University, Nanjing, People's Republic of China

CONTENTS

2.1 Introduction

The roots and rhizomes of *Rubia yunnanensis* (Franch.) Diels (Chinese name *Xiaohongshen*, Figure 2.1), a traditional Chinese medicine, native to Yunnan Province, China, have a long history of use for treating several diseases, such as tuberculosis, menoxenia, rheumatism, contusion, hematemesis, anemia, and lipoma (Editorial Committee of GMPA, 2003, Editorial Committee of YFDA, 2005, Wu et al., 2004). With the development of modern separation technology, the phytochemical studies of *R. yunnanensis* led to the isolation of 97 compounds from this plant, including Rubiaceae-type cyclopeptides (RAs), quinones, triterpenes, and others. Modern pharmacological studies on *R. yunnanensis* show that it possesses anti-tumor, anti-inflammatory, anti-oxidant, anti-microbial, anti-hepatitis, and anti-myocardial ischemia activities, and can increase white blood cells and modulate the immune system (Fan, 2010, Li, 2007). Of note, RA-VII, a Rubiaceae-type cyclopeptide (RA), has undergone phase I clinical trials at the NCI, and was prescribed as an anti-cancer candidate in Japan in the 1990s. RA-XII, a natural glucoside of RA, is a promising anti-tumor candidate and a preclinical study is ongoing (Song et al., 2017, Guo et al.,

* Corresponding author.

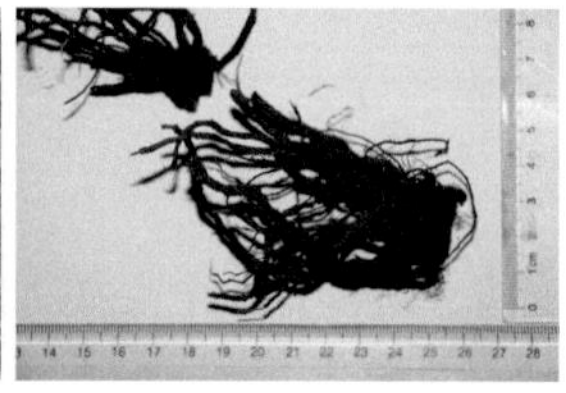

FIGURE 2.1 Photos showing the growing plant of *Rubia yunnanensis* Diels and its dried roots and rhizomes (photos by Xuejia Zhang).

2019, Wang et al., 2019). Rubidate, a naphthoquinone synthetic derivative, has been used in the clinic as a commercial remedy in China for raising levels of leukocytes. Herein, the botany, including the plants of interest, resource availability, sustainability, distribution, ethnopharmacology, phytochemistry, and biological activities of *R. yunnanensis* is described, and we hope to provide some information for the resource acquisition, quality control, and drug research of this plant.

2.2 Botany

Rubia yunnanensis Diels is a specialty traditional Chinese medicine (TCM) of Yunnan Province and is known in Chinese as *Xiaohongshen, Zishen, Dianqiancao, Yunnanqiancao, Xiaohuoxue* or *Xiaohongyao* (Wu et al., 2004). Its dried roots and rhizomes have a long history of use reported as blood activating and nourishing, blood stasis and rheumatism removing, sedative, and analgesic phytomedicine (Editorial Committee of GMPA, 2003, Editorial Committee of YFDA, 2005). *R. yunnanensis* is also an ethnobotanical medicine used by the Yi, Bai, Pumi, and Lisu nationalities in China (Li, 2007).

Rubia yunnanensis is a perennial herb, which belongs to section *Oligoneura*, genus *Rubia*, tribe Rubieae, Rubiaceae family (Chen and Friedrich, 2011). Its accepted scientific name, *Rubia yunnanensis* (Franch.) Diels, was confirmed by the Kew Science database (http://wcsp.science.kew.org/), and *Rubia ustulata* Diels became its synonym name due to the morphological variability (Chen and Friedrich, 2011). The characteristics of the original plant are described as follows: several to dozens of slightly fleshy roots with transverse wrinkles, 3–15 cm, 3–10 mm in diameter, dark red; stems up to 50 cm, clumped, suberect or prostrate, quadrangular or narrowly four-winged, smooth or rarely scabrid; leaves in whorls of four, sometimes six, subsessile; blade drying papery, shape and size variable with lanceolate, ovate, obovate, elliptic-oblong, broadly elliptic, or suborbicular, 1–4 cm in length and 0.3–2 cm in width, both surfaces hairy to scabrid, base cuneate to rounded, margins flat or often revolute, apex acuminate, shortly cuspidate, or acute; palmately three- or five-veined; inflorescences thyrsoid, paniculate, terminal and axillary cymes usually longer than subtending leaves; axes subglabrous to sparsely hirsutulous; bracteoles lanceolate, 2–5 mm; pedicels 1–3 mm, ovary 0.3–0.4 mm in diameter, glabrous; corolla yellow or pale yellow, rotate, about 3 mm in diameter, glabrous; fused basal part about 0.5 mm; lobes subovate, 1.2–2 mm, not reflexed, apex thickened, incurved, shortly rostrate; mericarp berry black at maturity, 3–5 mm in diameter. The flowering period is generally from summer to fall, and the fruiting period is usually in early winter (Chen and Friedrich, 2011).

The medicinal material of *R. yunnanensis* is generally obtained from the dried roots and rhizomes: slightly bending and slender, 3–6 mm in diameter. The surface shows a longitudinal wrinkle, reddish brown or dark red: section yellow red or dark red. The roots and rhizomes are collected in autumn and winter, followed by washing the surface and air drying (Editorial Committee of GMPA, 2003, Editorial Committee of YFDA, 2005).

As an endemic species, *R. yunnanensis* is only distributed in Yunnan and southwest Sichuan, China. It is found growing wild in thickets, grass slopes, and roadsides within the altitude range of 1700–3000 m (Chen and Friedrich, 2011). *R. yunnanensis* is mainly used in Yunnan, Sichuan, and Guizhou Provinces. Yunnan is the main origin of *R. yunnanensis*, and accounts for about 90% of resources in China.

To date, *R. yunnanensis* is rarely cultivated on a large scale, and most raw materials are derived from the wild, but relevant artificial planting investigations have been initiated. Both the roots and seeds of *R. yunnanensis* can be cultivated well in greenhouses, and its plant size and active components retain their levels compared with the wild type (Miao, 2020). Additionally, water regulation in different degrees implied that moderate drought-induced injury in *R. yunnanensis* was reversible, which indicates the root system of *R. yunnanensis* has a strong ability of drought resistance (Miao et al., 2020). Re-watering initiates beneficial changes in the photosynthesis, biomass production, and the levels of anti-tumor components, i.e. the accumulation of Rubiaceae-type cyclopeptides (RAs) in *R. yunnanensis* (Miao et al., 2020). Density and ray passing rate (RPR) are the two important factors, which could influence the yield and quality of *R. yunnanensis.* When the density is 345,000 plants/hm^2 and RPR is 75%, the yield of *R. yunnanensis* reaches up to 734.9 kg/hm^2 and its quality meets the relevant standards (Li, 2004). Additionally, the transcriptome sequencing of *R. yunnanensis* was completed and the stable internal reference genes were systematically screened and verified (Yi et al., 2020b). Thirty-two glycosyltransferases were cloned and four UGTs with high conversion efficiency were screened by enzyme function verification *in vitro*, which enriched the chemical diversity and source pathway of anthraquinones (Yi et al., 2020a). The RAs biosynthetic pathway of *R. yunnanensis* was inferred by a RiPPs (ribosomally synthesized and post-translationally modified peptides) pathway. The key enzymes most likely to be involved in this pathway were cloned and expressed, and further study is ongoing (Yi, 2019).

2.3 Ethnopharmacology

The medical records of *R. yunnanensis* have a thousand years of history. Cangqi Chen, a herbalist from the Tang dynasty, first described the medical use of *R. yunnanensis* for insect bites and scabies by soaking it in wine for drinking. During the Song dynasty, Song Su wrote the *Illustrated Classics of Materia Medica* (*Tu Jing Ben Cao* in Chinese) and recorded the use of *R. yunnanensis* for the therapy of various scabies as a folk medicine. In Shuwei Xu's *Prescriptions for Universal Relief* (*Pu Ji Ben Shi Fang* in Chinese), the plaster of *R. yunnanensis* could treat lung injury caused by blood smearing during labor, and uncontrolled motion of blood with a unique curative effect. Later, another important book, *Herbs of the Southern Yunnan* (*Dian Nan Ben Cao* in Chinese) written by Mao Lan, a native of Yunnan Province during the Ming dynasty, recorded "it tastes bitter, slightly worm in nature and the major function was dredging the channels and collaterals" (Wu et al., 2004).

Records of common folk herbs in Kunming showed that *R. yunnanensis* could relax the muscles, stimulate blood circulation to remove blood stasis, and promote tissue regeneration. The records in Chinese Herbal Medicine in Yunnan demonstrated that it could "warm channels and dredge collaterals, nourish qi and blood." Chinese Herbal Medicine in Honghe recorded its functions, such as "cooling blood and hemostasis, circulation of meridians and blood, removing and relieving pain" (Li, 2007).

Standards of *R. yunnanensis* have been established by national and local standards. It was recorded as an herbal medicine in the Chinese Pharmacopoeia (edition 1997) (Li, 2007), Standard of TCM Decoction Pieces in Yunnan Province (edition 2005) (Editorial Committee of YFDA, 2005), Standard of Chinese Medicinal Materials in Yunnan Province (edition 2005, Yi medicine) (Editorial Committee of YFDA, 2007), Quality Standard of TCM and Ethnic Medicine in Guizhou Province (edition 2003) (Editorial Committee of GMPA, 2003), and Standard of Chinese Medicinal Materials in Hunan Province (edition 2009) (Editorial Committee of HFDA, 2009). The contents evolved into pharmacognostic studies on name and source, shape, morphology, microscopic structure and physicochemical properties, processing, nature and taste, function, usage and dosage, etc.

Generally, decoctions for internal use and medicinal wine for both internal and external use are the most common preparations of *R. yunnanensis*. The modern application of *R. yunnanensis* is to treat tuberculosis, irregular menstruation, fall injury, cold and dampness, corneal cloudiness, traumatic bleeding, anemia, etc. (Li, 1990, Wu, 1991). The decoction is also used to treat anaphylactoid purpura, and a powder preparation is used to treat stubborn epistaxis with significant effect (Song, 1986, Peng, 1990). Additionally, a clinical study showed that *R. yunnanensis* was the main medicine in lotion form to treat soft tissue injury (Li, 1987). In a preliminary study, its liquor extract and syrup could treat insomnia of neurasthenia with some effect, and in injection form has an effect on coronary heart disease, angina pectoris, and psoriasis (Dai et al., 1994).

Due to its wide therapeutic effects, *R. yunnanensis* may be used together with other preparations to constitute a TCM mixture. For example, *R. yunnanensis* is one of the ingredients in a famous Chinese patent drug named *Paiduyangyan Capsule*. Prescription analysis indicated that *R. yunnanensis* acts as a detoxification agent and together with rhubarb (*Rheum officinale*) opens the whole-body detoxification pipeline (Li, 2019).

2.4 Phytochemistry and Biological Activities

2.4.1 Rubiaceae-Type Cyclopeptides (RAs)

RAs are found only in the Rubiaceae species and commonly in the *Rubia* plants. They are homodicyclohexapeptides, mainly composed of one α-D-alanine, one α-L-alanine, three *N*-methyl-α-L-tyrosines, and one other proteinogenic α-L-amino acid. The most unusual feature is a 14-membered ring formed by a phenolic oxygen linkage between 2 adjacent tyrosines with a *cis* peptide bond, and the 14-membered ring is fused to an 18-membered cyclic hexapeptide ring (Tan and Zhou, 2006, Zhao et al., 2011). Bouvardin and deoxybouvardin (RA-V, **2**) were the first two RAs isolated from *Bouvardin ternifolia*, in 1977, and they possess potential anti-tumor activities (Jolad et al., 1977). In 1993, RA-V (**2**), RY-I (RA-XII, **3**), and RY-II (**5**) were isolated from the roots of *R. yunnanensis* (He, et al., 1993, Zou and Hao, 1993). For their unique bicyclic structures and significant anti-tumor activities *in vitro* and *in vivo*, RAs attracted our interest, and by 2000, we established a special chemical detection method for plant cyclopeptides, which is a new TLC protosite reaction with ninhydrin reagent (Zhou and Tan, 2000). In 2010, eight new RAs (rubiyunnanins A–H) (**6**–**13**) and five known RAs (**1**–**5**) were isolated using this method by our group from the roots of *R. yunnanensis* (Fan et al., 2010a,b), and their structures were elucidated mainly by NMR and MS. To date, 13 RAs (**1**–**13**) have been obtained from *R. yunnanensis* and their structures are shown in Figure 2.2.

To obtain these RAs in an efficient manner, several methods have been developed using TLC (Zhou and Tan, 2000) and HPLC (Itokawa et al., 1984). Recently, two methods of ultra-performance liquid chromatography, coupled with triple quadrupole tandem mass spectrometry (UPLC-QqQ-MS/MS) with a multiple reaction monitoring (MRM) strategy based on predicted precursor ions and characteristic product ions, were developed for the qualitative and quantitative analyses of RAs (Bi et al., 2020, Zhang et al., 2018). These two methods are more sensitive, specific, and accurate, and can be used to detect unknown RAs and profile all RAs, which hopefully will lead to the isolation of new RAs.

RAs possess promising anti-tumor activity and have attracted much attention. In the last decade, our group and collaborators focused on how to clarify their anti-tumor

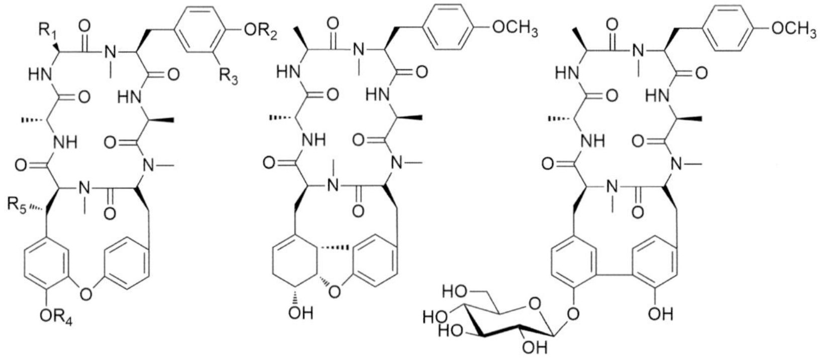

Rubiyunnanin A (**12**) Rubiyunnanin B (**13**)

No.	Name	R_1	R_2	R_3	R_4	R_5
1	RA-I	CH_2OH	CH_3	H	H	H
2	RA-V (deoxybouvardin)	CH_3	CH_3	H	H	H
3	RA-XII (RY-I)	CH_3	CH_3	H	Glc	H
4	RA-XXIV	$CH_2CH_2CONH_2$	CH_3	H	H	H
5	RY-II	CH_2OH	CH_3	H	Glc	H
6	Rubiyunnanin C	$CH_2CH_2COOCH_3$	CH_3	H	H	H
7	Rubiyunnanin D	CH_2CH_2COOH	H	H	H	H
8	Rubiyunnanin E	CH_2CH_2COOH	CH_3	OH	H	H
9	Rubiyunnanin F	$CH_2CH_2CONH_2$	CH_3	H	Glc	H
10	Rubiyunnanin G	CH_3	H	H	Glc	H
11	Rubiyunnanin H	CH_3	CH_3	H	Glc	OH

FIGURE 2.2 Rubiaceae-type cyclopeptides (RAs) isolated from *Rubia yunnanensis*.

mechanisms of action. We found that RAs are a new kind of NF-κB inhibitors, and RA-V (**2**) is the best one with an IC_{50} value of 30 nM (Fan et al., 2010b). Furthermore, RA-V (**2**) could target TAK1 and exert an anti-tumor effect by inhibiting NF-κB signaling pathway (Wang et al., 2018). RA-V (**2**) could kill human breast cancer cells by inducing mitochondria-mediated apoptosis through blocking the PDK1–AKT interaction (Fang et al., 2013). RA-V (**2**) also inhibits organ enlargement and tumorigenesis induced by YAP activation (Ji et al., 2018). Besides, RA-V (**2**) and RA-XII (**3**), as new autophagy inhibitors, could promote apoptosis and inhibit protective autophagy through AMPK/mTOR/P70S6K pathways (Yang et al., 2018, Song et al., 2017). RAs could also inhibit angiogenesis, cell adhesion, and invasion through regulating ERK1/2 phosphorylation, PI3K/AKT, and NF-κB signaling pathways (Yue et al., 2011, Leung et al., 2015a, b). Recently, we found that RA-XII (**3**) could suppress the development and growth of liver and colorectal cancers by the inhibition of lipogenesis via SCAP-dependent SREBP suppression (Guo et al., 2019, Wang et al., 2019). All these studies indicate that RAs are one kind of new anti-cancer candidates. However, their poor solubilities in physiological conditions limited their applications for cancer therapy *in vivo*. Therefore, pH-sensitive polymers were developed by us for targeted RA-V (**2**) delivery into tumor sites and for acid-triggered drug release (Qiao et al., 2015). To further enhance the anti-tumor activity of RA-V (**2**) *in vivo*, several programmed drug delivery systems were developed, such as sequential delivery of RA-V (**2**) and doxorubicin for combination therapy on resistant tumors and *in situ* monitoring of cytochrome c release (Chen et al., 2017], redox dual-responsive and O_2-evolving theranostic nanosystems for highly selective chemotherapy against hypoxic tumors (Chen et al., 2019), and programmed delivery of RA-V (**2**) and antisense oligonucleotides for combination therapy on hypoxic tumors, and for therapeutic self-monitoring (Yao et al., 2020).

2.4.2 Quinones

Quinones, including benzoquinones, anthraquinones, naphthoquinones, and phenanthraquinones, widely exist in medicinal plants and have various biological activities. In 2002, the first study of anthraquinones and naphthoquinones from the roots of *R. yunnanensis* was undertaken, leading to one new anthraquinone glycoside, rubiayannone A (**14**) (Liou et al., 2002a), and four new naphthohydroquinones, rubinaphthins A–D (**31–34**) (Liou et al., 2002b). Subsequently, one new anthraquinone, rubianthraquinone (**15**), was isolated from the roots of *R. yunnanensis* in 2003 (Tao et al., 2003). In 2011, 2 new anthraquinones (**16**, **17**) and 2 new naphthoquinones (**35**, **36**), together with 14 known anthraquinones (**15**, **18–30**) and 1 known naphthoquinone (**37**), were isolated by our group from the roots of *R. yunnanensis*, and their structures were elucidated mainly by NMR and MS (Fan et al., 2011). Recently, one new naphthoquinone dimer (**38**) and two trimeric naphthoquinones (**39**, **40**) were obtained from this plant (Suyama et al., 2017). All 27 quinone structures isolated from the roots of *R. yunnanensis* are presented in Figures 2.3 and 2.4.

Quinones are the predominant bioactive constituents of *Rubia* plants, and possess inhibitory activities on tumor cell growth, phosphatase of regenerating liver-3, platelet aggregation, and the release of hexosaminidase (Xu et al., 2014). For example, 1,3,6-trihydroxy-2-methyl-9,10-anthraquinone (**16**), one of the major components isolated from *R. yunnanensis*, exerted cytotoxicity via apoptosis induction and G_2/M cell cycle arrest in human cervical carcinoma HeLa cells (Zeng et al., 2013). Therefore,

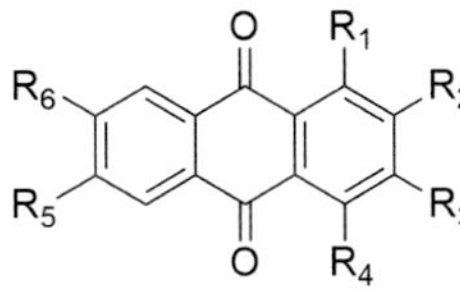

No.[a]	R_1	R_2	R_3	R_4	R_5	R_6
14	OH	OGlc(6→1)Xyl	H	OH	H	H
15	H	OH	CH_3	OCH_3	H	OH
16	OH	CH_3	OH	H	OCH_3	H
17	OH	CH_2OH	O(6'-OAc)Glc	H	OH	H
18	OH	CH_3	OH	H	OH	H
19	H	CH_2OH	OH	H	OH	H
20	OH	H	OH	H	H	H
21	OH	CH_3	H	H	OH	H
22	OH	CH_3	OH	H	H	H
23	H	CH_2OH	H	H	H	H
24	OH	CH_3	H	H	H	H
25	H	$COOCH_3$	H	H	H	H
26	OH	CH_3	O(6'-OAc)Glc	H	OH	H
27	OH	CH_3	OGlc(2→1)Rha	H	OH	H
28	OH	CH_3	O(6'-OAc)Glc(2→1)Xyl	H	OH	H
29	OH	CH_3	O(3'-OAc)Glc(2→1)Rha	H	OH	H
30	OH	CH_3	O(6'-OAc)Glc(2→1)Rha	H	OH	H

[a] **14,** Rubiayannone A; **15**, Rubianthraquinone; **16**, 1,3-Dihydroxy-6-methoxy-2-methyl-9,10-anthraquinone;
17, 1,3,6-Trihydroxy-2-hydroxymethyl-9,10-anthraquinone-3-*O*-(6'-*O*-acetyl)-β-D-glucopyranoside;
18, 1,3,6-Trihydroxy-2-methyl-9,10-anthraquinone; **19**, 2-Hydroxymethyl-3-hydroxy-9,10-anthraquinone;
20, Xanthopurpurin; **21**, 1,6-Dihydroxy-2-methyl-9,10-anthraquinone; **22**, Rubiadin; **23**, 2-Hydroxylmethyl-9,10-anthraquinone;
24, 1-Hydroxy-2-methyl-9,10-anthraquinone; **25**, 2-Carbomethoxy-9,10-anthraquinone;
26, 1,3,6-Trihydroxy-2-methyl-9,10-anthraquinone-3-*O*-(6'-*O*-acetyl)-β-D-glucopyranoside;
27, 1,3,6-Trihydroxy-2-methyl-9,10-anthraquinone-3-*O*-α-L-rhamnopyranosyl-(1→2)-β-D-glucopyranoside;
28, 1,3,6-Trihydroxy-2-methyl-9,10-anthraquinone-3-*O*-(6'-*O*-acetyl)-β-D-xylopyranosyl-(1→2)-β-D-glucopyranoside;
29, 1,3,6-Trihydroxy-2-methyl-9,10-anthraquinone-3-*O*-(3'-*O*-acetyl)-α-L-rhamnopyranosyl-(1→2)-β-D-glucopyranoside;
30, 1,3,6-Trihydroxy-2-methyl-9,10-anthraquinone-3-*O*-(6'-*O*-acetyl)-α-L-rhamnopyranosyl-(1→2)-β-D-glucopyranoside.

FIGURE 2.3 Anthraquinones isolated from *Rubia yunnanensis.*

their qualitative and quantitative analysis methods are critical to quality control, clinical, and commercial applications of these plants. Previously, several analytical methods for these quinones, including TLC and HPLC, have been developed (Boldizsár et al., 2006). However, these methods are time-consuming with low sensitivity and low resolution. Recently, a sensitive and efficient UPLC-QqQ-MS/MS method in positive and negative multiple reaction-monitor modes was developed for the quinones from *Rubia* plants (Hu et al., 2020), which now offers quick detection and isolation of bioactive quinones.

2.4.3 Triterpenoids

There are many types of triterpenoids, such as oleanane-type, arborinane-type, ursane-type, and fernane-type ones. Among them, arborinane-type triterpenoids, one

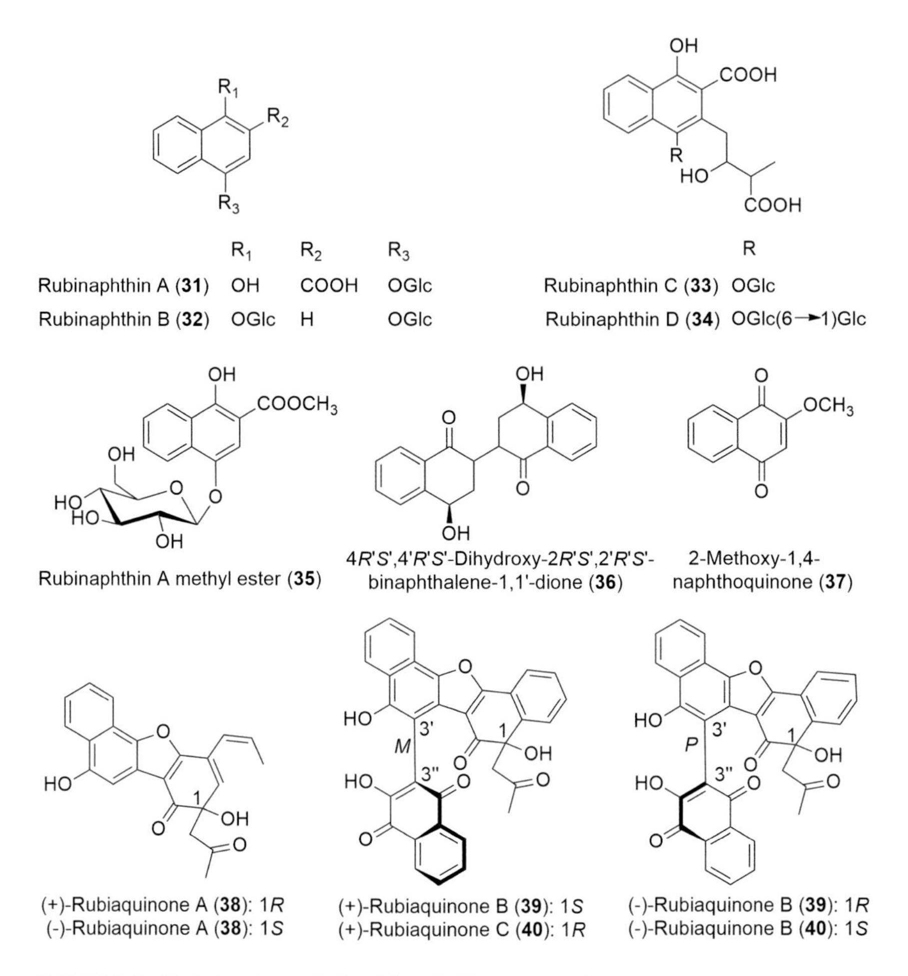

FIGURE 2.4 Naphthoquinones isolated from *Rubia yunnanensis*.

kind of rare pentacyclic triterpenes, were mainly isolated from *Rubia* plants and could be considered as their characteristic chemical constituents. To date, 36 arborinane-type compounds (**41**–**76**) and 6 other types (**77–82**) were obtained from the roots of *R. yunnanensis*. Some of them showed inhibitory activity on tumor cell growth, nitric oxide production, and antiplatelet aggregation (Fan et al., 2011, Liou and Wu, 2002, Morikawa et al., 2003, Tao et al., 2003, Xu et al., 1994, Zou et al., 1993, Zou et al., 1999). For example, rubiarbonol G (**44**), one of the cytotoxic constituents in *R. yunnanensis*, induces apoptosis and G_0/G_1 arrest of HeLa cells by NF-κB and JNK signaling pathways (Zeng et al., 2018). Rubiarbonone C (**49**), a selective inhibitor of cytochrome P450 4F enzymes, inhibits platelet-derived growth factor-induced proliferation and migration of vascular smooth muscle cells through FAK, MAPK, and STAT3 Tyr705 signaling pathways (Choi et al., 2018, Park et al., 2017). The structures of these compounds are presented in Figure 2.5.

No.	Name	R_1	R_2	R_3	R_4	R_5
41	Rubiarbonol A	H	OH	OH	OH	CH_2OH
42	Rubiarbonol B	H	OH	OH	OH	CH_3
43	Rubiarbonol F	OH	OH	OH	OH	CH_2OH
44	Rubiarbonol G	H	OH	OH	OAc	CH_2OH
45	Rubiarbonol K	H	OH	H	OH	CH_3
46	Rubiarbonol L	H	OH	H	OH	CH_2OH
47	Rubiarbonone A	H	=O	OH	OAc	CH_2OH
48	Rubiarbonone B	H	=O	OH	OH	CH_2OH
49	Rubiarbonone C	H	=O	OH	OH	CH_2OAc
50	Rubiarbonone D	H	=O	OH	OH	CH_3
51	Rubiarbonone F	H	=O	OH	OH	COOH
52	Rubianol-a	OH	=O	OH	OH	CH_2OH
53	Rubianol-b	OH	=O	OH	OH	CH_2OAc
54	Rubianol-c	H	OH	OH	OH	CH_2OAc
55	Rubianol-d	OH	OH	OH	OH	CH_2OAc
56	Rubianol-e	OAc	OH	OH	OH	CH_2OAc
57	Rubiarboside A	OAc	OGlc	OH	OH	CH_3
58	Rubiarboside F	OH	OGlc	OH	OH	CH_2OH
59	Rubiarboside G	H	OGlc(6→1)Glc	OH	OH	CH_2OH
60	Rubianoside II	H	OGlc	OH	OH	CH_3
61	Rubianoside III	H	OGlc	OH	OH	CH_2OH
62	Rubianoside IV	OH	OGlc	OH	OH	CH_2OH
63	Rubiarbonol A 7-acetate	H	OH	OAc	OH	CH_2OH
64	19,28-Didehydroxyrubiarbonol A	H	OH	OH	H	CH_3
65	Rubianol-e-3-*O*-(6'-*O*-acetyl)-β-D-gulcopyranoside	OAc	O(6'-OAc)Glc	OH	OH	CH_2OAc
66	Rubiarboside G 28-acetate	H	OGlc(1→6)Glc	OH	OH	CH_2OAc
67	2α-Acetoxy-28-acetylrubiarboside G	OAc	OGlc(1→6)Glc	OH	OH	CH_2OAc
68	Rubiarboside G 28-al	H	OGlc(1→6)Glc	OH	OH	CHO
69	Rubiarbonol A-3-*O*-β-D-gulcopyranosyl-(1→2)-β-D-gulcopyranoside	H	OGlc(1→2)Glc	OH	OH	CH_2OH

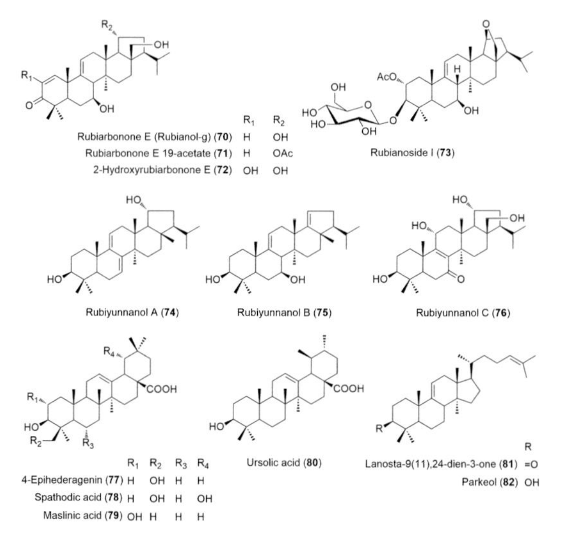

FIGURE 2.5 Triterpenoids isolated from *Rubia yunnanensis*.

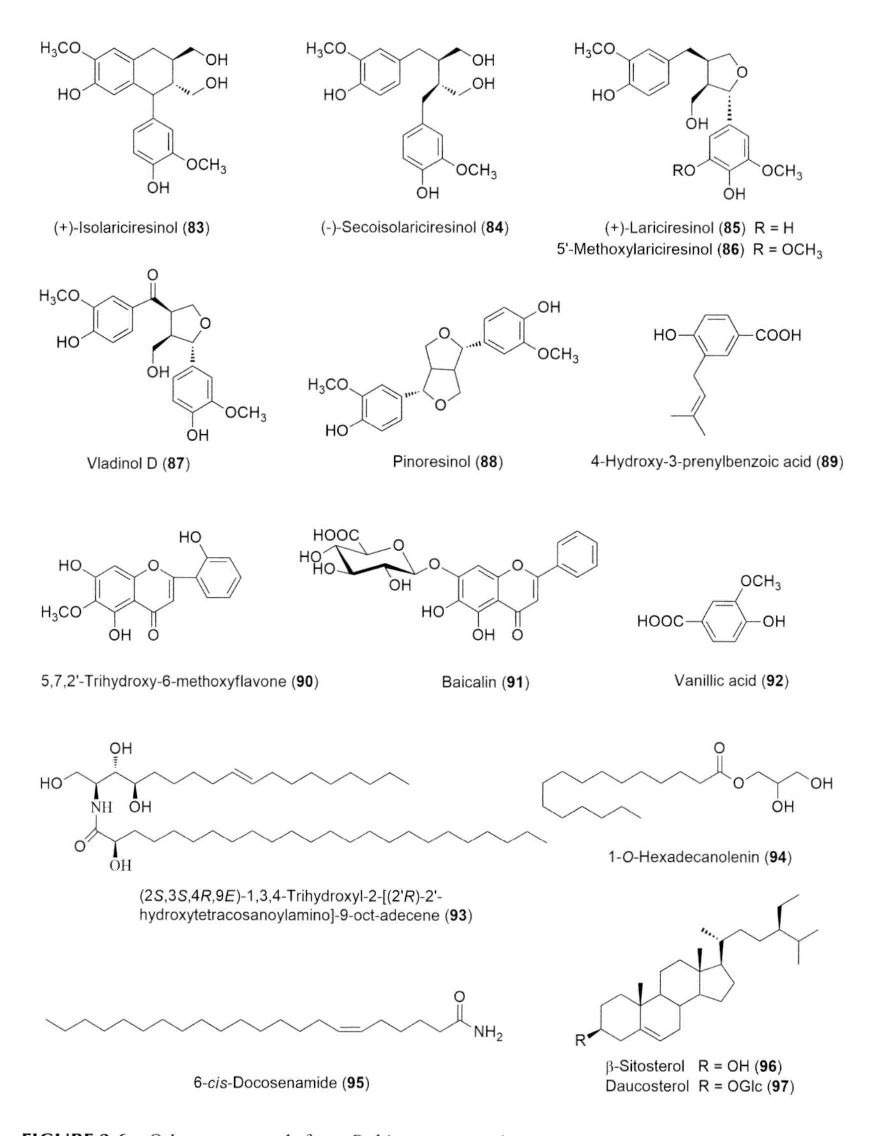

FIGURE 2.6 Other compounds from *Rubia yunnanensis*.

2.4.4 Other Compounds in *R. yunnanensis*

Fifteen other compounds, (+)-isolariciresinol (**83**), (–)-secoisolariciresinol (**84**), (+)-lariciresinol (**85**), 5'-methoxylariciresinol (**86**), vladinol D (**87**), pinoresinol (**88**), 4-hydroxy-3-prenylbenzoic acid (**89**), 5,7,2'-trihydroxy-6-methoxyflavone (**90**), baicalin (**91**), vanillic acid (**92**), (2*S*,3*S*,4*R*,9*E*)-1,3,4-trihydroxyl-2-[(2'*R*)-2'-hydroxytetracosanoylamino]-9-oct-adecene (**93**), 1-*O*-hexadecanolenin (**94**), 6-*cis*-docosenamide (**95**), β-sitosterol (**96**), and daucosterol (**97**), were isolated from the roots of *R. yunnanensis* (Fan et al., 2011, Liou et al., 2002a, b), and their structures are shown in Figure 2.6.

2.5 Conclusion

In this chapter, we focus on *Rubia yunnanensis* (Franch.) Diels (Chinese name *Xiaohongshen*) and review its botany, ethnopharmacology, phytochemistry, and biological activity. We believe that this can contribute to the resource acquisition, quality control, and drug research of *R. yunnanensis.*

REFERENCES

Bi QR, Liu Q, Han S, et al., 2020. An integrated multiple reaction monitoring strategy based on predicted precursor ions and characteristic product ions for global profiling Rubiaceae-type cyclopeptides in three *Rubia* species. *J Chromatogr A*, 1618: 460902.

Boldizsár I, Szűcs Z, Füzfai Z, et al., 2006. Identification and quantification of the constituents of madder root by gas chromatography and high-performance liquid chromatography. *J Chromatogr A*, 1133: 259–274.

Chen HC, Li F, Yao YY, et al., 2019. Redox dual-responsive and O_2-evolving theranostic nanosystem for highly selective chemotherapy against hypoxic tumors. *Theranostics*, 9: 90–103.

Chen HC, Wang YR, Yao YY, et al., 2017. Sequential delivery of cyclopeptide RA-V and doxorubicin for combination therapy on resistant tumor and in situ monitoring of cytochrome c release. *Theranostics*, 7: 3781–3793.

Chen T, Friedrich E, 2011. *Rubia*, In: Wu ZY, Raven PH, Hong DY (eds.) *Flora of China*, Vol. 19, Science Press, Beijing & Missouri Botanical Garden Press, St. Louis, pg. 305–319.

Choi YJ, Quan KT, Park I, et al., 2018. Discovery of rubiarbonone C as a selective inhibitor of cytochrome P450 4F enzymes. *Arch Toxicol*, 92: 3325–3336.

Dai F, Wang XW, Wang CF, et al., 1994. Study on the pharmaceutical basis of *R. yunnanensis* in the treatment of psoriasis. *Kunming Med*, 15: 289–290.

Editorial Committee of GMPA, 2003. Quality standard of TCM and ethnic medicine in Guizhou Province, Guizhou Science and Technology Press, Guiyang, pg. 55.

Editorial Committee of HFDA, 2009. Standard of TCM in Hunan Province, Hunan Science and Technology Press, Changsha, pg. 11–12.

Editorial Committee of YFDA, 2005. Standard of TCM decoction pieces in Yunnan Province, Vol II, Yunnan Science and Technology Press, Kunming, pg. 24–25.

Editorial Committee of YFDA, 2007. Standard of TCM in Yunnan Province, Vol II (Yi medicine), Yunnan Science and Technology Press, Kunming, pg. 11–12.

Fan JT, 2010. Study on antitumor cyclopeptides and their mechanism of *Rubia yunnanensis*. Kunming Institute of Botany, University of Chinese Academy of Sciences, Beijing, China.

Fan JT, Chen YS, Xu WY, et al., 2010a. Rubiyunnanins A and B, two novel cyclic hexapeptides from *Rubia yunnanensis*. *Tetrahedron Lett*, 51: 6810–6813.

Fan JT, Kuang B, Zeng GZ, et al., 2011. Biologically active arborinane-type triterpenoids and anthraquinones from *Rubia yunnanensis*. *J Nat Prod*, 74: 2069–2080.

Fan JT, Su J, Peng YM, et al., 2010b. Rubiyunnanins C-H, cytotoxic cyclic hexapeptides from *Rubia yunnanensis* inhibiting nitric oxide production and NF-kappa B activation. *Bioorg Med Chem Lett*, 18: 8226–8234.

Fang XY, Chen W, Fan JT, et al., 2013. Plant cyclopeptide RA-V kills human breast cancer cells by inducing mitochondria-mediated apoptosis through blocking PDK1–AKT interaction. *Toxicol Appl Pharmacol*, 267: 95–103.

Guo D, Wang YR, Wang J, et al., 2019. RA-XII suppresses the development and growth of liver cancer by inhibition of lipogenesis via SCAP-dependent SREBP supression. *Molecules*, 24: 1892.

He M, Zou C, Hao XJ, et al., 1993. A new antitumor cyclic hexapeptide glycoside from *Rubia yunnanensis*. *Acta Bot Yunnani*, 15: 408.

Hu YY, Zhang XJ, Zhang ZH, et al., 2020. Qualitative and quantitative analyses of quinones in multi-origin *Rubia* species by ultra-performance liquid chromatography-tandem mass spectrometry combined with chemometrics. *J Pharmaceut Biomed*, 189: 113471.

Itokawa H, Takeya K, Mori N, et al., 1984. Studies on antitumour cyclic hexapeptides RA obtained from Rubiae Radix, Rubiaceae (IV): quantitative determination of RA-VII and RA-V in commercial Rubiae Radix and collected plants. *Planta Med*, 50: 313–316.

Ji XY, Song LH, Sheng L, et al., 2018. Cyclopeptide RA-V inhibits organ enlargement and tumorigenesis induced by YAP Activation. *Cancers*, 10: 449.

Jolad SD, Hoffman JJ, Torrance SJ, et al., 1977. Bouvardin and deoxybouvardin, antitumor cyclic hexapeptides from *Bouvardia ternifolia* (Rubiaceae). *J Am Chem Soc*, 99: 8040–8044.

Leung HW, Wang Z, Yue GGL, et al., 2015a. Cyclopeptide RA-V inhibits cell adhesion and invasion in both estrogen receptor positive and negative breast cancer cells via PI3K/AKT and NF-kappa B signaling pathways. *BBA - Mol Cell Res*, 8: 1827–1840.

Leung HW, Zhao SM, Yue GGL, et al., 2015b. RA-XII inhibits tumour growth and metastasis in breast tumour-bearing mice via reducing cell adhesion and invasion and promoting matrix degradation. *Sci Rep*, 5: 16985.

Li GN, 1990. Records of traditional Chinese medicine in Yunnan. Vol I. Yunnan Science and Technology Press, Kunming, pg. 111.

Li HX, 1987. Qiancao lotion in the treatment of soft tissue injury. *Shaanxi Trad Chin Med*, 8: 35.

Li J, 2019. Study on quality standard of Paiduyangyan capsules. Chengdu University of Traditional Chinese Medicine, Chengdu, Sichuan, China.

Li YC, 2004. The researches on density and ray passing rate of a wild medicinal material *Xiaohongshen* in cultivation. Gansu Agricultural University, Lanzhou, Gansu, China.

Li Z, 2007. Study on the standardization of the quality standard of the ethnomedicine *Rubia yunnanensis*. Chengdu University of Traditional Chinese Medicine, Chengdu, Sichuan, China.

Liou MJ, Teng CM, Wu TS, 2002a. Constituents from *Rubia ustulata* Diels and *Rubia yunnanensis* Diels and their antiplatelet aggregation activity. *J Chin Chem Soc*, 49: 1025–1030.

Liou MJ, Wu PL, Wu TS, 2002b. Constituents of the roots of *Rubia yunnanensis*. *Chem Pharm Bull*, 50: 276–279.

Liou MJ, Wu TS, 2002. Triterpenoids from *Rubia yunnanensis*. *J Nat Prod*, 65: 1283–1287.

Miao YY, 2020. Study on water regulation and hairy root culture of *Rubia yunnanensis* based on its secondary metabolites. China Pharmaceutical University, Nanjing, Jiangsu, China.

Miao YY, Bi QR, Qin H, et al., 2020. Moderate drought followed by re-watering initiates beneficial changes in the photosynthesis, biomass production and Rubiaceae-type cyclopeptides (RAs) accumulation of *Rubia yunnanensis*. *Ind Crops Prod*, 148: 112284.

Morikawa T, Tao J, Ando S, et al., 2003. Absolute stereostructures of new arborinane-type triterpenoids and inhibitors of nitric oxide production from *Rubia yunnanensis*. *J Nat Prod*, 66: 638–645.

Park HS, Quan KT, Han, JH, et al., 2017. Rubiarbonone C inhibits platelet-derived growth factor-induced proliferation and migration of vascular smooth muscle cells through the focal adhesion kinase, MAPK and STAT3 Tyr705 signalling pathways. *Br J Pharmacol*, 174: 4140–4154.

Peng JD, 1990. Treatment of Jiawei Qiancaogen San to 32 cases of intractable epistaxi. *Jiangxi Trad Chin Med*, 21: 26.

Qiao ZY, Zhang D, Hou CY, et al., 2015. A pH-responsive natural cyclopeptide RA-V drug formulation for improved breast cancer therapy. *J Mater Chem B*, 3: 4514–4523.

Song LH, Wang Z, Wang YR, et al., 2017. Natural cyclopeptide RA-XII, a new autophagy inhibitor, suppresses protective autophagy for enhancing apoptosis through AMPK/mTOR/P70S6K pathways in HepG2 cells. *Molecules*, 22: 1934.

Song YL, 1986. Observation on the therapeutic effect of Qiancao decoction on 60 cases of henoch schonlein purpura. *Shandong Trad Chin Med*, 5: 15–16.

Suyama Y, Higashino Y, Tanaka N, et al., 2017. Stereochemical assignments of rubiaquinones A-C, naphthoquinone derivatives from *Rubia yunnanensis*. *Tetrahedron Lett*, 58: 4568–4571.

Tan NH, Zhou J, 2006. Plant cyclopeptides. *Chem Rev*, 106: 840–895.

Tao J, Morikawa T, Ando S, et al., 2003. Bioactive constituents from Chinese natural medicines. XI. Inhibitors on NO production and degranulation in RBL-2H3 from *Rubia yunnanensis*: Structures of rubianosides II, III, and IV, rubianol-g, and rubianthraquinone. *Chem Pharm Bull*, 51: 654–662.

Wang YR, Guo D, He JQ, et al., 2019. Inhibition of fatty acid synthesis arrests colorectal neoplasm growth and metastasis: anti-cancer therapeutical effects of natural cyclopeptide RA-XII. *Biochem Bioph Res Co*, 512: 819–824.

Wang Z, Zhao SM, Song LH, et al., 2018. Natural cyclopeptide RA-V inhibits the NF-kappa B signaling pathway by targeting TAK1. *Cell Death Dis*, 9: 715.

Wu YQ, Gao XL, Zhang RP, 2004. Review on the studies of *Rubia yunnanensis*. *Lishizhen Med Mater Med Res*, 15: 44–47.

Wu ZY, 1991. Xinhua compilation of Materia Medica. Vol II, Shanghai Science and Technology Press, Shanghai, pg. 456–456.

Xu K, Wang P, Wang L, et al., 2014. Quinone derivatives from the genus *Rubia* and their bioactivities. *Chem Biodivers*, 11: 341–363.

Xu XY, Zhang JQ, Fang QC, 1994. Study on the chemical constituents of *Rubia yunnanensis* (Franch) Diels. *Acta Pharm Sini B*, 23: 237–240.

Yang JH, Yang T, Yan W, et al., 2018. TAK1 inhibition by natural cyclopeptide RA-V promotes apoptosis and inhibits protective autophagy in Kras-dependent non-small-cell lung carcinoma cells. *RSC Adv*, 8: 23451–23458.

Yao YR, Feng L, Wang Z, et al., 2020. Programmed delivery of cyclopeptide RA-V and antisense oligonucleotides for combination therapy on hypoxic tumors and for therapeutic self-monitoring. *Biomater Sci*, 8: 256–265.

Yi SY, 2019. Gene mining and function of novel glycosyltransferases in the medicinal plant *Rubia yunnanensis*. China Pharmaceutical University, Nanjing, Jiangsu, China.

Yi SY, Kuang TD, Miao YY, et al., 2020a. Discovery and characterization of four glycosyltransferases involved in anthraquinone glycoside biosynthesis in *Rubia yunnanensis*. *Org Chem Front*, 7: 2442–2448.

Yi SY, Lin QW, Zhang XJ, et al., 2020b. Selection and validation of appropriate reference genes for quantitative RT-PCR analysis in *Rubia yunnanensis* Diels based on transcriptome data. *Biomed Res Int*, 2020: 5824841.

Yue GGL, Fan JT, Lee JKM, et al., 2011. Cyclopeptide RA-V inhibits angiogenesis by down-regulating ERK1/2 phosphorylation in HUVEC and HMEC-1 endothelial cells. *Br J Pharmacol*, 164: 1883–1898.

Zeng GZ, Fan JT, Xu JJ, et al., 2013. Apoptosis induction and G2/M arrest of 2-methyl-1,3,6-trihydroxy-9,10-anthraquinone from *Rubia yunnanensis* in human cervical cancer HeLa cells. *Pharmazie*, 68: 293–299.

Zeng GZ, Wang Z, Zhao LM, et al., 2018. NF-kappa B and JNK mediated apoptosis and G_0/G_1 arrest of HeLa cells induced by rubiarbonol G, an arborinane-type triterpenoid from *Rubia yunnanensis*. *J Ethnopharmacol*, 220: 220–227.

Zhang XJ, Bi QR, Wu XD, et al., 2018. Systematic characterization and quantification of Rubiaceae-type cyclopeptides in 20 *Rubia* species by ultra performance liquid chromatography tandem mass spectrometry combined with chemometrics. *J Chromatogr A*, 1581: 43–54.

Zhao SM, Kuang B, Fan JT, et al., 2011. Antitumor cyclic hexapeptides from *Rubia* plants: history, chemistry, and mechanism (2005–2011). *Chimia*, 65: 952–956.

Zhou J, Tan NH, 2000. Application of a new TLC chemical method for detection of cyclopeptides in plants. *Chin Sci Bull*, 45: 1825–1831.

Zou C, Hao XJ, 1993. Antitumor glycocyclohexapeptide from *Rubia yunnanensis*. *Acta Bot Yunnan*, 15: 399.

Zou C, Hao XJ, Chen CC, et al., 1993. New arborane type triterpenoids from *Rubia yunnanensis* (I). *Acta Bot Yunnan*, 15: 89–91.

Zou C, Hao XJ, Chen CC, et al., 1999. Structures of rubiarbonone B and C. *Acta Bot Yunnan*, 21: 256–259.

3

Chemical Constituents and Biological Activity of *Erigeron breviscapus* (Vant.) Hand.-Mazz.

Qin-Shi Zhao* and Han-Dong Sun
State Key Laboratory of Phytochemistry and Plant Resources in West China, Kunming Institute of Botany, Chinese Academy of Sciences, Kunming, Yunnan, People's Republic of China

CONTENTS

3.1 Introduction

Erigeron breviscapus (Vant.) Hand.-Mazz. (Asteraceae) is recorded in Flora of China. This plant is mainly distributed at mid-elevation mountain levels, alpine to montane meadows, forest margins, *Pinus* forests, streamsides, grasslands, disturbed slopes, and roadsides at an altitude of 1200–3600 meters above sea level in Southwest China. It grows as a perennial erect or ascending herb and can reach a height of 1–50 cm; the rhizome is woody and it has several or solitary stems sparsely to densely hirsute, strigillose, sparsely to moderately stipitate glandular, denser distally. Its leaves are mostly basal, present at anthesis, winged petiolate, blade oblanceolate to obovate-lanceolate or broadly spatulate, and the capitula solitary at the ends of stems or branches. In addition to the typical variety, there are some other varieties that have been recognized, which may represent phenotypic plasticity at the extreme limit of the species range.

As an herbal medicine, *E. breviscapus* was firstly recorded as *dengzhanhua* for treating stroke and sequelae in *Dian Nan Materia Medica* (Dian Nan Ben Cao or

* Corresponding author.

FIGURE 3.1 (left) Book cover of *Diannan Materia Medica*; (right) photo of the plant *E. breviscapus* (photo by Qin-Shi Zhao).

Southern Yunnan Materia Medica) in 1443, written by Lan Mao (AD 1397–1476), a famous scholar and physician of Yunnan during the Ming Dynasty (Figure 3.1). *Dian Nan Materia Medica* was published 153 years earlier than *Compendium for Materia Medica* (AD 1596). It is a well-known and classical regional medicinal and historical book for the research of medicine in Yunnan. This book provided the earliest clue to the national "hot spot" of medicinal research and has been quoted in many materia medica works subsequently. Therefore, *Dian Nan Materia Medica* has a wide range of influence in Yunnan, even till today.

In fact, the whole plant of *E. breviscapus* was exploited as a folk medicine by the Yi minority people of Southwest China to treat paralysis caused by stroke and joint pain from rheumatism. There is an interesting story of the rediscovery of *E. breviscapus* for treating stroke and sequelae. In the 1970s, China launched a large campaign to identify and modernize therapeutics from traditional Chinese medicine. A number of clinically successful drugs were discovered within that period. The systematic study of *E. breviscapus* was a "hot topic" during that time, when a civilian doctor Luo published his ancestral secret remedy on *E. breviscapus* for curing paralysis caused by stroke during this campaign in Yunnan. As a result, *E. breviscapus* (Vant.) Hand.-Mazz. was recorded into the Chinese Pharmacopoeia (Chinese Pharmacopoeia Commission, 1977, 2015, and 2020). The dried whole plant is the original material of the traditional Chinese medicine *Dengzhanxixin (Dengzhanhua)*.

3.2 Chemical Constituents

The chemical components of *E. breviscapus* have been the subject of extensive studies for their medicinal value. The phytochemical studies on *E. breviscapus* have identified various classes of compounds including flavonoids, sesquiterpenoids, caffeoylquinic acids, glycosides, pyrones, coumarins, essential oils, and more.

Flavones are the characteristic constituents of *E. breviscapus*. Breviscapine, identified by Zhang Renwen of the Yunnan Institute of Materia Medica, is a crude extract of flavones from *E. breviscapus* consisting mainly of flavonoids, with scutellarin as a major component and apigenin-7-O-glucuronide as the minor component (Zhang et al., 1988). Scutellarin was first isolated from the roots of *Scutellaria baicalensis* (Shibata et al., 1923). Scutellarin and scutellarein are the characteristic constituents of *E. breviscapus* and display a broad spectrum of pharmacological activities (Figure 3.2).

As a major bioactive component of *E. breviscapus*, scutellarin has been widely studied in chemistry. The scutellarin content in wild species is about 0.1%, while the content in cultivars is up to 3.0–4.0%. Scutellarin can now be isolated from cultivated planting. The cultivation of *E. breviscapus* could provide enough plant materials; however, the study of the total chemical synthesis and biosynthesis of scutellarin is important, since synthesis can provide different approaches for the development of scutellarin derivatives, while biosynthesis and synthetic biology provide other approaches for the plant resources and sustainable development. Liu et al. (2019) reported the total synthesis of scutellarin and apigenin 7-O-β-D-glucuronide (Figure 3.3).

Liu et al. (2018) reported the biosynthesis by genomic analysis and synthetic biological approaches. They identified two key enzymes in the biosynthetic pathway (flavonoid-7-O-glucuronosyltransferase and flavone-6-hydroxylase) from the *E. breviscapus* genome and engineered yeast to produce breviscapine from glucose. After the metabolic engineering and optimization of a batch-fed fermentation, scutellarin and apigenin-7-O-glucuronide, the two major active ingredients of breviscapine, reached levels of 108 and 185 mg/L, respectively (Figure 3.4). However, this yield was still considered very low. This problem may be solved by further engineering the host cells' metabolism. One approach is to improve the intracellular concentration of malonyl-CoA which was needed to form the flavonoid backbone

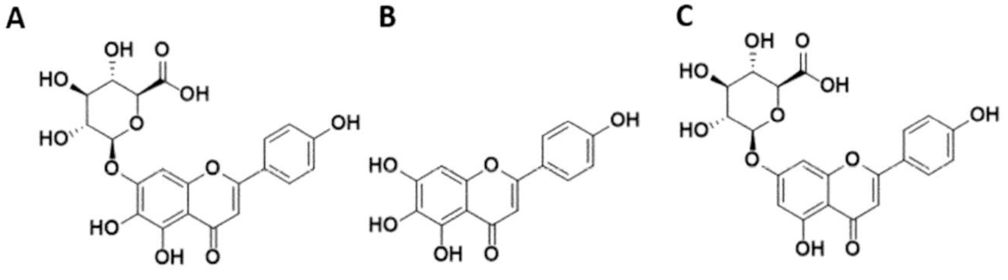

FIGURE 3.2 Structures of (A) scutellarin; (B) scutellarein; (C) apigenin-7-O-glucuronide.

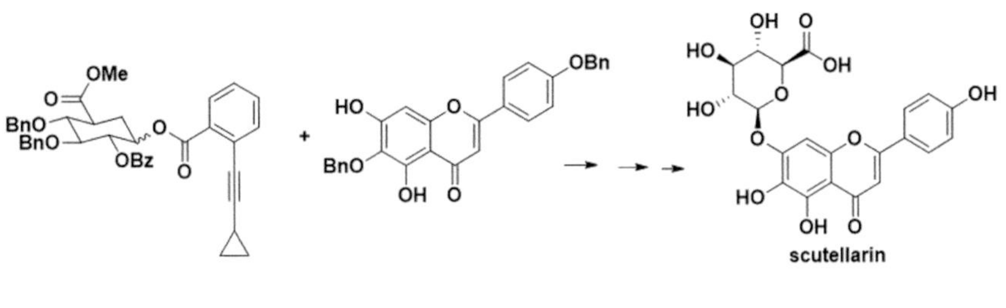

FIGURE 3.3 The total synthesis of scutellarin.

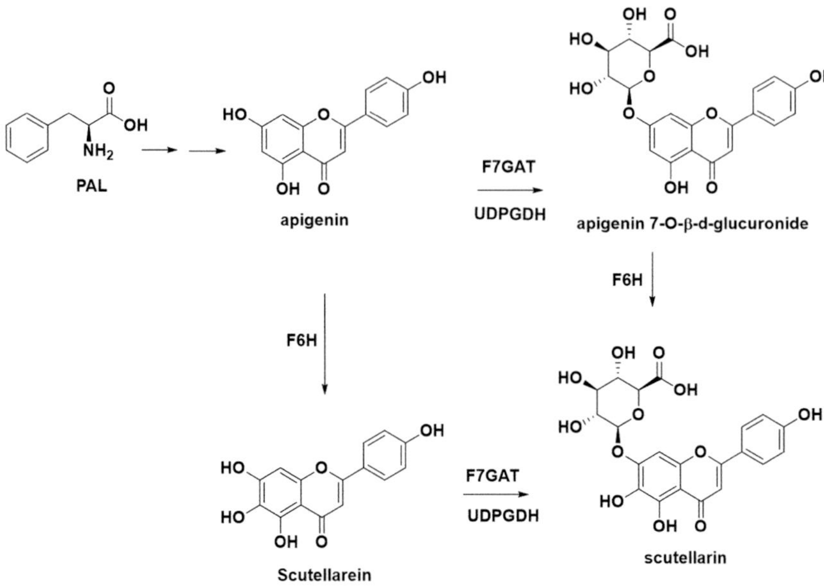

FIGURE 3.4 The synthetic biology approaches of breviscapine.

by decreasing the consumption of acetyl-CoA through deleting the cytosolic malate synthase gene (MLS1) and deleting the peroxisomal citrate synthase gene (CIT2) in the yeast strain, whereby the scutellarin yield was significantly improved. In another way, a malonyl-CoA-producing multiple-gene module was designed to increase the supply of acetyl-CoA in the metabolic network. As a result, an engineered yeast strain was obtained with improved scutellarin yield compared with the starting strain (Liu et al., 2018).

Caffeic acid esters are another class of characteristic constituents of *E. breviscapus*. Yue et al. (2000) first reported the caffeic acid esters from *E. breviscapus*. This class includes about ten members of caffeic acid esters, including 3-O-caffeoylquinic acid (chlorogenic acid), 3,5-O-dicaffeoylquinic acid, 3,4-O-dicaffeoylquinic acid, and erigoster A and B (Figure 3.5). Caffeoylquinic acid esters are widely distributed in other plants too, but erigoster A and B were only discovered in *E. breviscapus*. Erigoster A and B are a unique group derived from 2-keto-3-deoxymanno-octulosonic acid and caffeic acid. In addition, 2-keto-3-deoxymanno-octulosonic acid is an acidic monosaccharide belonging to a large family of 2-keto-3-deoxy-sugar acids which are important constituents of complex carbohydrates, playing significant roles in biological systems. Its structure is the only recurrent structural element in all bacterial lipopolysaccharides (Cipolla et al., 2010).

Pyromeconic acid and its glucosidic derivatives (Figure 3.6) are also major constituents of *E. breviscapus* (Zhang et al., 1981) and can be isolated from different species in the genus *Erigeron* (Zhang et al., 1998). Apart from those compounds already mentioned, the sesquiterpenoids and other classes have been isolated from this plant (Han et al., 2019).

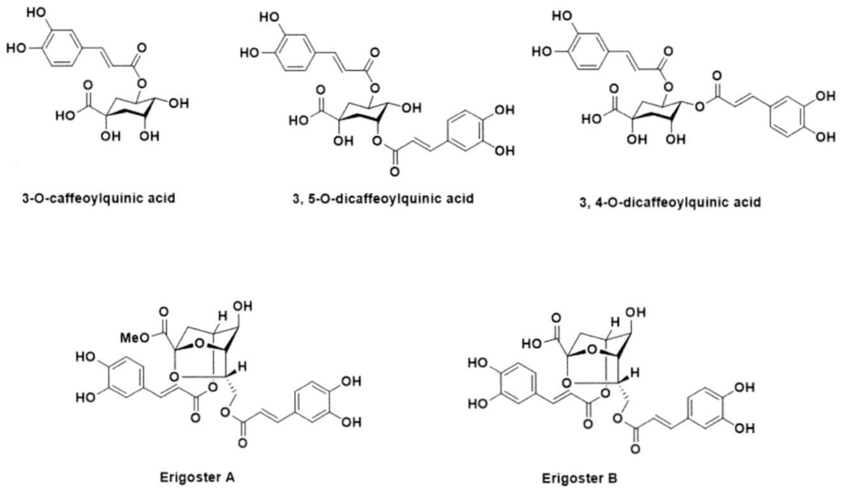

3-O-caffeoylquinic acid
3, 5-O-dicaffeoylquinic acid
3, 4-O-dicaffeoylquinic acid
Erigoster A
Erigoster B

FIGURE 3.5 Caffeic acid esters from *E. breviscapus*.

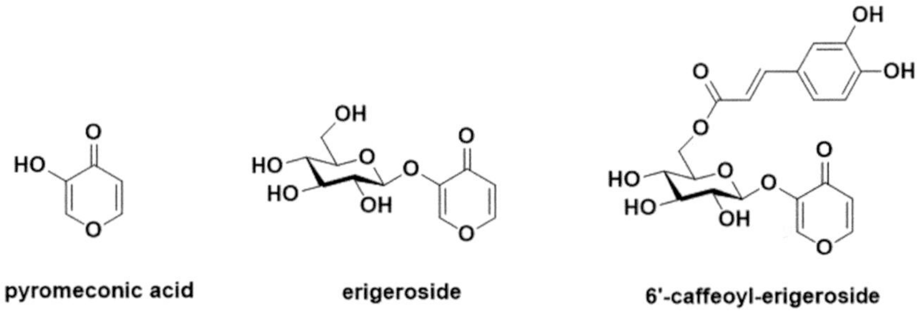

pyromeconic acid
erigeroside
6'-caffeoyl-erigeroside

FIGURE 3.6 Pyromeconic acid and its glucosidic derivatives.

3.3 Biological Activity and Drug Development

The multi-effect nature of *E. breviscapus* may derive from its mixed herbal components, which exert their effects on diseases by binding multiple targets to result in complex and synergistic therapeutic activities. It is critical to figure out the pathological development of related illnesses and identify the key bioactive components and the molecular targets of *E. breviscapus* including the dynamic interactions among multiple functions and structures.

Scutellarin is the major active component in *E. breviscapus*: several biological activities have been reported, such as anti-oxidant, anti-inflammation, vascular relaxation, anti-platelet, anti-coagulation, and myocardial protection, and scutellarin has been used clinically to treat stroke, myocardial infarction, and diabetic complications (Wang and Ma, 2018). In the clinical treatment of coronary heart disease, angina pectoris, myocardial ischemia, stroke, and cerebral thrombotic diseases, the doses for scutellarin range from 50 to 200 mg. In animal models, doses of 15 mg/kg and

50 mg/kg significantly reduced myocardial infraction, and inhibited myocardium cell apoptosis and focal brain ischemia (Chledzik et al., 2018).

Scutellarin exerts a broad range of cardiovascular pharmacological effects, as the most typical biological activity is related to cardiovascular and cerebrovascular diseases (Gao et al., 2017). In recent studies investigating the mechanism of cardioprotective effects of scutellarin, it reduced NLRP3 inflammasome activation, inhibited mTORC1 activity, and increased Akt phosphorylation (Xu et al., 2020). Scutellarin up-regulated endothelial nitric oxide synthase (eNOS) expression, down-regulated the expression of VEGF, bFGF, and iNOS, and suppressed the release of pro-inflammatory cytokines (TNFα, IL-1β, IL-6, and IL-8), and creatinine kinase. Meanwhile scutellarin inhibited the influx of extracellular calcium ions, which is related to an endothelial-independent vasorelaxant effect. Based on these mechanisms, scutellarin may improve diastolic dysfunction, ameliorate myocardial structure abnormality, inhibit myocyte apoptosis and inflammatory response, and promote autophagy. Further, published clinical studies on the therapeutic effects of scutellarin and breviscapine indicated that the addition of breviscapine to conventional treatments produced significantly better results for patients with coronary heart disease than conventional treatments alone. It can increase the cure rate, reduce mortality, improve the symptoms and ECG in angina pectoris patients, and improve recovery from myocardial infarction (Chledzik et al., 2018).

Scutellarin is also reported to have anti-tumor activity, including liver, prostate, lung, breast, cervical, colon, tongue, and renal cancer. It induces apoptosis and cell cycle arrest and inhibits the proliferation and progression of a wide spectrum of human cancer cells through multiple molecular targets and pathways (Chan et al., 2019). One study found a significant inhibition of scutellarin to the activation and nuclear translocation of the NF-κB transcription factor, for which aberrant activation is associated with various aspects of oncogenesis. Meanwhile, scutellarin has an effect on the induction of the cell cycle arrest at G0/G1 transition (Chledzik et al., 2018). Scutellarin can also be utilized in some central nervous system (CNS) diseases, such as Alzheimer's disease (AD), by exhibiting neuroprotective and neuromodulatory effects via inhibiting pro-inflammatory components and improving associative memory deficits (Pimplikar et al., 2010). These pharmacological studies suggested that scutellarin had a protective effect against oxidative neuronal damage by free radicals and increased the expression of SOD to prevent Aβ neuronal deposition and spatial memory impairment. The compound has even shown estrogenic properties without stimulating cancer cell growth (Guo et al., 2011, Guo et al., 2013).

Scutellarin inhibits inflammation and protects cartilage from degeneration in *in vitro* and *in vivo* studies (Liu et al., 2020). It down-regulates the mRNA and protein expression of MMP1, MMP13, ADAMTS-5, Wnt3a, and Frizzled7, which are related to the degradation of collagen II and proteoglycan that can maintain the structural integrity of cartilage. Meanwhile scutellarin can promote the protein expression of collagen II and aggrecan and inhibit the migration of β-catenin and phosphorylation of p38 into the nucleus, which may relate to the mediation of the Wnt/β-catenin and MAPK signaling pathway. As a result, scutellarin may effectively ameliorate cartilage degeneration in osteoarthritis.

Scutellarin also showed anti-HIV-1 effects against three strains of human immune-deficiency virus (HIV-1IIIB, HIV-174V, and HIV-1KM018) (Zhang et al., 2005), and antimicrobial activity against *Helicobacter pylori* was detected.

The safety of scutellarin has been evaluated based on acute and sub-acute investigations. Scutellarin caused dose-dependent, general behavior adverse effects, but the LD_{50} values could not be detected, and the maximum tolerated dose was more than 10 g/kg with no treatment-related mortality in the animals treated at this dose. So scutellarin can be regarded as non-toxic from acute ingestion. In the acute and sub-acute study, scutellarin had no effect on the normal growth, production, and metabolism of rats. There were no significant changes in the hematological and biochemical profile, which indicated that scutellarin was relatively non-toxic under the reported study conditions (Li et al., 2011).

The pharmacokinetics and metabolism of scutellarin in humans and animals have been intensively studied. Scutellarin may have a low absorption rate from the intestine, and it undergoes a substantial metabolic change before reaching the systemic circulation. After oral administration to humans, scutellarin (scutellarein-7-O-glucuronide, S-7-G) can hardly be detected, whereas its isomeric metabolite scutellarein-6-O-glucuronide (S-6-G) dominates in plasma (Chen et al., 2006). After oral administration to rats, S-7-G was largely hydrolyzed in the intestinal tract and was absorbed as aglycone. While passing through the intestinal wall, aglycone was extensively glucuronidated into S-7-G and S-6-G (Gao et al., 2011). S-7-G exhibited more rapid uptake in hepatocytes, was glucuronidated at a 2.7-fold higher rate in the liver, and was excreted in greater amounts through bile and urine than S-6-G (Figure 3.7).

Scutellarein can cross membrane barriers more readily and pass the BBB to reach the brain at a higher concentration than scutellarin, which if confirmed could result in a better CNS protective and therapeutic effect.

Caffeic acid esters were the second class of biologically active components other than flavones discovered by Sun et al. from *E. breviscapus* (Sun et al., 2003, Zhao et al., 2003). These components inhibited the production of NO, TNF-α, and IL-1β induced by LPS treatment in primary microglia in a dose-dependent manner, and these compounds could significantly improve neurobehavioral performance and reduce percentage infarct volume in ischemic cerebral tissues compared with the vehicle group (Wang et al., 2012). Pharmacological studies indicated that the neuroprotective effect of scutellarein and caffeic acid esters might be mediated via a stimulation of the production and release of NTFs through p-CREB and p-Akt signaling

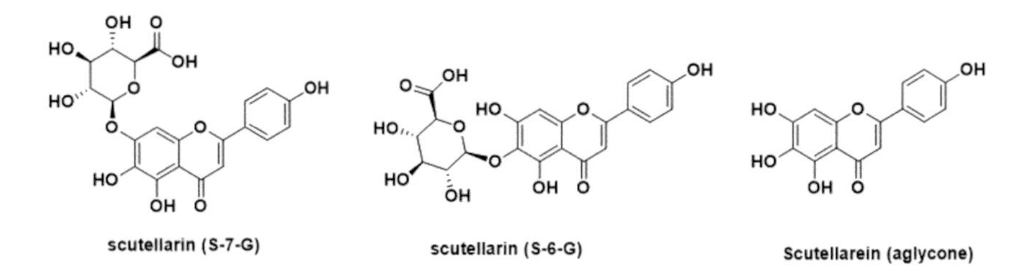

FIGURE 3.7 The metabolites of scutellarin.

and could antagonize the hypoxia-induced toxicity through astrocytes-conditioned medium, suggesting that scutellarein and caffeic acid esters might have therapeutic potential for stroke. Furthermore, quinic acid esterified with the caffeoyl group at the C-4 position showed higher antiradical activity compared to acylation at the C-3 or C-5 positions, and the dicaffeoylquinic acids showed higher antiradical activity due to the presence of an additional caffeoyl group esterified in the quinic acid than monocaffeoylquinic acids. The 3,5-O-dicaffeoylquinic acid demonstrated low selective cytotoxicity (selective index >3) against HCT-116 cancer cells compared to normal epithelial colon cells. Caffeic acid esters have beneficial potential for metabolic syndrome. The study showed that 3,5-di-o-caffeoylquinic acid inhibited amyloid β_{1-42}-induced cellular toxicity on human neuroblastoma SH-SY5Y cells and increased the mRNA expression level of glycolytic enzymes and the intracellular ATP level. The investigation of the influence of the structure–activity relationship on the accelerating activity of ATP production displayed that caffeoyl groups bound to quinic acid are important for activity and the more caffeoyl groups which are bound to quinic acid, the higher the accelerating activity on ATP production. However, the caffeic acid esters are difficult to purify due to the complexity of isolation of the isomers from caffeoylquinic acids; more biological activity data have yet to be reported.

Using a combination of scutellarin and breviscapine, an injection form of breviscapine (*Dengzhanhuasu*) has been developed and used in clinical treatment. Based on the extract, a commercial product called *Yimaikang Capsule* has been developed and used in clinical treatment. Moreover, a different product, the Dengzhan Xixin injection, has been developed and used in clinical treatment. However, based on a formula of caffeic acid esters, the new drug has not shown success in the clinic.

3.4 Chemical Modification and Derivatives through the Synthesis of Scutellarin

Scutellarin and a corresponding aglycone, scutellarein, have been isolated from *E. breviscapus*. The pharmacokinetics and metabolism of scutellarin in humans and animals indicated that scutellarin could be hydrolyzed into the corresponding aglycone scutellarein, after oral or injection administration. Although bioconversion seems promising, this application has faced some difficulties. In comparison, the acid-catalyzed cleavage of the glycosidic link is more useful and has been extensively investigated. Optimization conditions by hydrolysis resulted in up to 90% aglycone yields, minimal by-product formations, and milder hydrolysis conditions (Jiang et al., 2017) (Figure 3.8).

Scutellarin is a typical example with clinical evaluation for this class of flavonoid; it also provides a lead for structure–activity relationship studies and the development of a new generation of drug. Yan et al. (2020) developed a semi-synthetic strategy from scutellarin for the synthesis of a series of natural flavonoids and flavonoid glycosides. By taking this strategy, eight bioactive flavonoids with striking structural similarities (Figure 3.9) were synthesized efficiently and practically. Obtaining sufficient amounts of products will greatly facilitate the SAR studies and mechanistic evaluation of the *in vivo* activities.

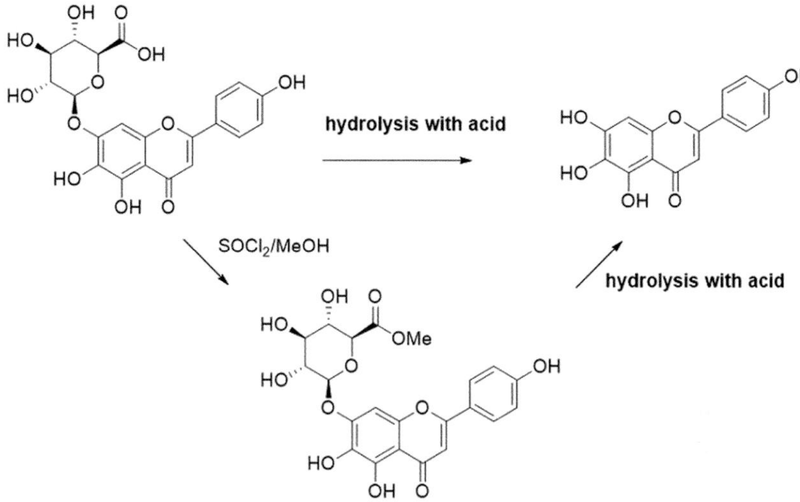

hydrolysis with acid

$SOCl_2$/MeOH

hydrolysis with acid

FIGURE 3.8 The acid-catalyzed cleavage of the glycosidic link of scutellarin.

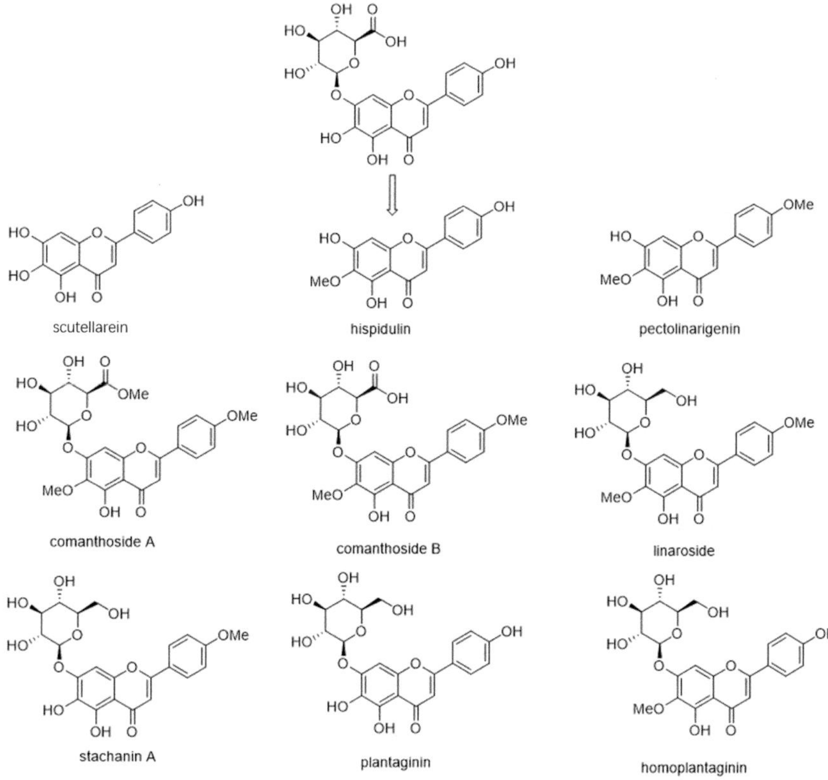

scutellarein

hispidulin

pectolinarigenin

comanthoside A

comanthoside B

linaroside

stachanin A

plantaginin

homoplantaginin

FIGURE 3.9 Nine bioactive flavonoids from scutellarin.

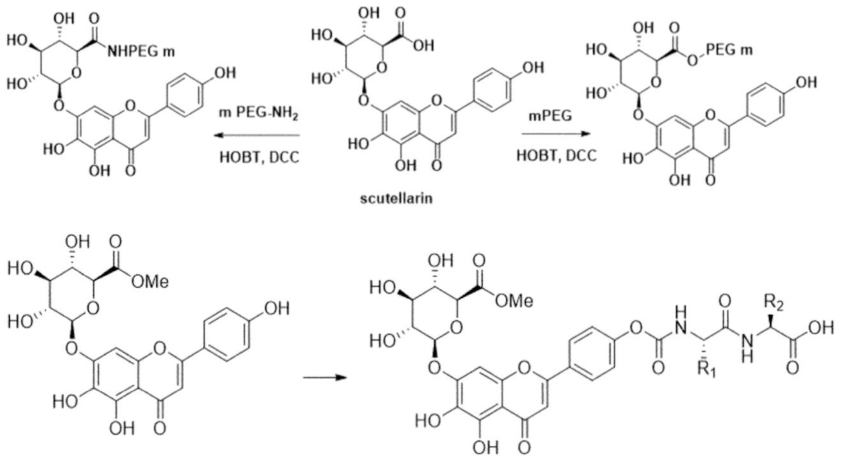

FIGURE 3.10 PEG-scutellarin prodrugs and scutellarin methyl ester-4′-dipeptide conjugates.

Although scutellarin has been clinically used for a long time, the most effective formulation is taken by intravenous injection. The oral formulation is not so effective due to the low solubility, low lipid solubility, and poor bioavailability. To improve the bioavailability of scutellarin, modification has been reported, such as the synthesis of PEG-scutellarin prodrugs (Lu et al., 2010). A series was successfully designed and synthesized by Li et al. (2020) (Figure 3.10) of novel scutellarin methyl ester-4′-dipeptide conjugates exhibiting active transport characteristics and protection against pathological damage caused by hypoxic-ischemic. The pharmacokinetics and efficacy were improved by, for example, using nanoparticle carriers with amphiphilic derivatives, and derivatives of scutellarin with morpholinyl aminomethylene substituent and NO-donating derivatives. It would be advantageous to find a more promising route of delivery of scutellarin to the body. Through the inhalation route, the absolute bioavailability of scutellarin was found to be over 77% in an *in vivo* animal study. In addition, scutellarein, the main metabolite of scutellarin, can also be synthesized to derivatives possessing a long aliphatic chain to overcome the low stability and poor lipophilicity and increase biological activity as anti-cancer agents (Wang and Ma, 2018).

3.5 Cultivation Methods and Accumulation of Active Ingredients

As the major active compounds of *E. breviscapus*, scutellarin, 1,5-dicaffeoylquinic acid, 3,5-dicaffeoylquinic acid, and erigoster B are important plant secondary metabolites that have various functions in adapting to both micro- and macro-environments. The study of Li et al. resolved the influence of genic and abiotic factors on the concentrations of these four phenolic compounds. The study showed that there is no obvious correlation between the genetic differentiation and the contents of the chemical compounds and the concentrations of 3,5-dicaffeoylquinic acid and 1,5-dicaffeoylquinic acid could be increased by decreasing the cultivation temperature and increasing the

FIGURE 3.11 GAP planting of *Erigeron breviscapus* (photos provided by Kunming Longjin Pharmaceutical Co., Ltd.).

moisture. The climate of high latitude regions can enhance the yield of 1,5-dicaffeoylquinic acid significantly. Furthermore, the positive correlation between the concentration of scutellarin and that of erigoster B as well as 3,5-dicaffeoylquinic acid maybe attributable to the same stimulus that induces the shared parts of their biosynthesis pathway (Li et al., 2013). In addition to the above factors, water stress, temperature stress, UV light, and disease resistance, CO_2 concentration, and nitrogen concentration in the soil can also make a difference to the content and the chemical composition or quantities of these compounds in plants.

Furthermore, the yield and active compounds of *E. breviscapus* from different regions were different depending on growth conditions. To date, there have been research reports on the genetic diversity and genetic relationship of *E. breviscapus*. The research team of Sheng-chao Yang analyzed the genetic diversity of *E. breviscapus*, and showed that the germplasm resources of *E. breviscapus* have a higher genetic diversity, and a line with richer germplasm genes can be bred through systematic selection as a breeding parent of *E. breviscapus*. At the same time, through systematic research, there has been research progress on *E. breviscapus* breeding technology, fertilization technology, and pest control technology, as well as the accumulation of effective ingredients (Li et al., 2013).

Based on the above studies, appropriate cultivation conditions should be applied to increase the yields of related effective compounds. One approach is to extensively collect wild germplasm resources of *E. breviscapus* and select the germplasm sources with high yield, high effective component content, and strong disease resistance as the high-quality cultivated germplasm resources. At the same time, studies on the metabolic basis and regulation of active components in *E. breviscapus* facilitate the accumulation of active components. Through the selection of suitable planting sites and the reasonable allocation of light and temperature, focusing on the quality ecological suitability based on the evaluation of the content of main active ingredients, the good agricultural practice (GAP) base of *E. breviscapus* would be established to produce high-quality medicinal materials (Figure 3.11).

3.6 Perspective

Stroke and myocardial infarction are the most common causes of mortality and disability in the world; stroke is the second foremost cause of mortality worldwide and a major cause of long-term disability. Thus, there is a compelling need to accelerate efforts to discover new drugs. *E. breviscapus* as a traditional Chinese herb has been used for more than 600 years and the chemical constituents and biological activity have been intensively studied. However, the mechanisms of biological activity of *E. breviscapus* are not fully understood. Low solubility and absorption of scutellarin, resulting in its low bioavailability, can be a limitation. The mechanisms and durability of caffeic acid esters need to be further explored, yet opportunities for new drug discovery from *E. breviscapus* still remain hopeful.

REFERENCES

Chan EWC, Lim CSS, Lim WY, et al., 2019. Role of scutellarin in human cancer - A review. *J Appl Pharm Sci*, 9: 142–146.

Chen X, Cui L, Duan X, et al., 2006. Pharmacokinetics and metabolism of the flavonoid scutellarin in humans after a single oral administration. *Drug Metab Dispos*, 34: 1345–1352.

Chinese Pharmacopoeia Commission, 1977. *Pharmacopoeia of the People's Republic of China*. Beijing: People's Medical Publishing House.

Chinese Pharmacopoeia Commission, 2015. *Pharmacopoeia of the People's Republic of China*. Beijing: China Medical Science Press.

Chinese Pharmacopoeia Commission, 2020. *Pharmacopoeia of the People's Republic of China*. Beijing: China Medical Science Press.

Chledzik S, Strawa J, Matuszek K, et al., 2018. Pharmacological effects of scutellarin, an active component of genus *Scutellaria* and *Erigeron*: A systematic review. *Am J Chin Med*, 46: 319–337.

Cipolla L, Gabrielli L, Bini D, et al., 2010. Kdo: A critical monosaccharide for bacteria viability. *Nat Prod Rep*, 27: 1618–1629.

Gao C, Chen X, Zhong D, 2011. Absorption and disposition of scutellarin in rats: A pharmacokinetic explanation for the high exposure of its isomeric metabolite. *Drug Metab Dispos*, 39: 2034–2044.

Gao J, Chen G, He H, et al., 2017. Therapeutic effects of breviscapine in cardiovascular diseases: A review. *Front Pharmacol*, 8: 289.

Guo LL, Guan ZZ, Huang Y, et al., 2013. The neurotoxicity of beta-amyloid peptide toward rat brain is associated with enhanced oxidative stress, inflammation and apoptosis, all of which can be attenuated by scutellarin. *Exp Toxicol Pathol*, 65: 579–584.

Guo LL, Guan ZZ, Wang YL, 2011. Scutellarin protects against Aβ-induced learning and memory deficits in rats: Involvement of nicotinic acetylcholine receptors and cholinesterase. *Acta Pharmacol Sin*, 32: 1446–1453.

Han S, Liu Y, Hou Z, et al., 2019. Sesquiterpenoids and γ-pyranone derivatives from the whole plant of *Erigeron breviscapus* and their neuroprotective effects. *Fitoterapia*, 138: 104288.

Jiang XY, Li XC, Liu WY, et al., 2017. An efficient, scalable approach to hydrolyze flavonoid glucuronides via activation of glycoside bond. *Tetrahedron*, 73: 1895–1903.

Li T, Wu D, Yang Y, et al., 2020. Synthesis, pharmacological evaluation and mechanistic study of scutellarin methyl ester –4'-dipeptide conjugates for the treatment of hypoxic-ischemic encephalopathy (HIE) in rat pups. *Bioorg Chem*, 101: 103980.

Li X, Peng LY, Zhang SD, et al., 2013. The relationships between chemical and genetic differentiation and environmental factors across the distribution of *Erigeron breviscapus* (Asteraceae). *PLoS One*, 8: e74490.

Li X, Wang L, Li Y, et al., 2011. Acute and subacute toxicological evaluation of scutellarin in rodents. *Regul Toxicol Pharmacol*, 60: 106–111.

Liu F, Li L, Lu W, et al., 2020. Scutellarin ameliorates cartilage degeneration in osteoarthritis by inhibiting the Wnt/β-catenin and MAPK signaling pathways. *Int Immunopharmacol*, 78: 105954.

Liu X, Cheng J, Zhang G, et al., 2018. Engineering yeast for the production of breviscapine by genomic analysis and synthetic biology approaches. *Nat Commun*, 9: 448.

Liu X, Wen GE, Liu JC, et al., 2019. Total synthesis of scutellarin and apigenin 7-*O*-β-D-glucuronide. *Carbohydr Res*, 475: 69–73.

Lu J, Cheng C, Zhao X, et al., 2010. PEG-scutellarin prodrugs: Synthesis, water solubility and protective effect on cerebral ischemia/reperfusion injury. *Eur J Med Chem*, 45: 1731–1738.

Pimplikar SW, Nixon RA, Robakis NK, et al., 2010. Amyloid-independent mechanisms in Alzheimer's disease pathogenesis. *J Neurosci*, 30: 14946–14954.

Shibata K, Iwata S, Nakamura M, 1923. Baicalin, a new flavone-glucuronic acid compound from the roots of *Scutellaria baicalensis*. *Acta Phytochim*, 1: 105–139.

Sun HD, Zhao Q, Peng L, et al., 2003. PRC patent titled "Dengzan Xixin Fen", its formulation and application in preventing and treating cardiovascular and cerebrovascular diseases, senile dementia, and hyperlipidemia. Patent publication no: CN1462620A. 24 December 2003.

Wang L, Ma Q, 2018. Clinical benefits and pharmacology of scutellarin: A comprehensive review. *Pharmacol Ther*, 190: 105–127.

Wang SX, Guo H, Hu L-M, et al., 2012. Caffeic acid ester fraction from *Erigeron breviscapus* inhibits microglial activation and provides neuroprotection. *Chin J Integr Med*, 18: 437–444.

Xu LJ, Chen RC, Ma XY, et al., 2020. Scutellarin protects against myocardial ischemia-reperfusion injury by suppressing NLRP3 inflammasome activation. *Phytomedicine*, 68: 153169.

Yan S, Xie M, Wang Y, et al., 2020. Semi-synthesis of a series natural flavonoids and flavonoid glycosides from scutellarin. *Tetrahedron*, 76: 130950.

Yue JM, Zhao Q-S, Lin Z-W, et al., 2000. Phenolic compounds from *Erigeron breviscapus* (Compositae). *Zhi wu Xue bao*, 42: 311–315.

Zhang GH, Wang Q, Chen J-J, et al., 2005. The anti-HIV-1 effect of scutellarin. *Biochem Biophys Res Commun*, 334: 812–816.

Zhang R, Zhang Y, Wang J, et al., 1988. Isolation and identification of flavonoids from shortscape fleabane (*Erigeron breviscapus*). *Zhongcaoyao*, 19: 199–201.

Zhang RW, Yang SY, Lin YY, 1981. Studies on chemical constituents of dengzhanhua (*Erigeron breviscapus* (Vant.) Hand-Mazz). I. The isolation and identification of pyromeconic acid and a new glucoside (author's translation). *Yao Xue Xue Bao*, 16: 68–69.

Zhang Y, Li L, Yang P, et al., 1998. Isolation and structure of 6′-*O*-caffeylerigeroside from *Erigeron multiradiatus*. *Yao Xue Xue Bao*, 33: 836–838.

Zhao Q, Sun H, Peng L, 2003. PRC patent titled "Erigoster B, its preparation and application in pharmaceutical." Patent publication no: CN1462750A. 24 December 2003.

4

Ethnopharmacology, Phytochemistry and Pharmacology of *Paris polyphylla* Smith var. *yunnanensis* (Franch.) Hand.-Mazz.

Hai-Yang Liu,* Ling-Ling Yu, Xu-Jie Qin, Wei Ni, and Chang-Xiang Chen
State Key Laboratory of Phytochemistry and Plant Resources in West China, Kunming Institute of Botany, Chinese Academy of Sciences, Kunming, Yunnan, People's Republic of China

CONTENTS

* Corresponding author.

4.1 Introduction

Paridis Rhizoma (*Chonglou* in Chinese) is one of the traditional Chinese medicines, especially in Southwest China, and is defined as the dried rhizomes of *Paris polyphylla* Smith var. *yunnanensis* (Franch.) Hand.-Mazz. (PPY) or *Paris polyphylla* Smith var. *chinensis* (Franch.) Hara (PPC). Because of its appearance, like worms and stairs, it has usually been called *Chonglou* (Chinese name: 虫楼 or 重楼). Based on the morphology of *Paris* species, Paridis Rhizoma is often known as *Qiyeyizhihua* in folk medicine (Chinese name: 七叶一枝花). Till now, the genus *Paris* contains over 30 species throughout the world, and most of them are found in Southwest China. This plant belongs to the Melanthiaceae family (earlier Liliaceae or Trilliaceae) in the Angiosperm Phylogeny Group IV (APG IV) classification system. PPY, as one of the more important species in the genus *Paris*, is a perennial plant with creeping and thickened underground rhizomes, a cluster of erect and unbranched stems (generally 10~100 cm), a whorl of usually 4~15 net-veined leaves. A solitary, terminal flower blooms in summer followed by green capsules containing dozens of red seeds when ripe (Figure 4.1).

Paris species are commonly used as folk medicines with diverse medicinal properties in Nepal, India, Vietnam, Myanmar, Thailand, Laos, and other countries. In Nepal, *Paris* species are locally known as *Satuwa* and the rhizomes have long been used for the treatment of snake and insect bites, internal and external wounds, fever, and intestinal parasites (Kunwar et al., 2020). *Paris polyphylla* Smith, an important medicinal plant growing in the Indian Himalayan region, has been traditionally used to treat burns, wounds, fever, diarrhea, dysentery, gastritis, skin diseases, and stomach pain (Shah et al., 2012). It is thus called "jack of all trades" in India due to the characteristic of curing diverse diseases (Shah et al., 2012). In China, *Paris polyphylla* was first recorded in *Shennong Bencao Jing* (《神农本草经》) during the Eastern Han Dynasty, in which it was known as *Zaoxiu* (蚤休) and was used for swelling and snakebites (Li, 1998). In 1436 (Ming Dynasty), named *Dian Chonglou*, PPY was first recorded as an herbal medicine possessing the efficacies

FIGURE 4.1 The morphological features of PPY (a–c: Paridis Rhizoma; d: whole plant; e: over-ground part of wild *P. polyphylla* var. *yunnanensis*; f–h: over-ground parts and fruits of cultivated *P. polyphylla* var. *yunnanensis*) (photos by Hai-Yang Liu).

of detoxification, detumescence, and pain relief in *South Yunnan Materia Medica* (*Diannan Bencao*, 《滇南本草》) by Mao Lan (Li, 1998). Later, it was called *Qiyeyizhihua* (七叶一枝花) in *Compendium of Materia Medica* (*Bencao Gangmu*, 《本草纲目》) by Shizhen Li in 1578. As recorded in the Chinese Pharmacopeia (from 1977 to 2020 editions), the dried rhizomes of both PPY and PPC are used as the TCM Paridis Rhizoma, which is commonly used for the treatment of furuncle, injuries, wounds, snake and insect bites, epilepsy, and sore throat.

PPY is one of the more important species of the genus *Paris*, mainly distributed in Southwest China. Its dried rhizomes are indispensable ingredients in over 100 patent medicines, such as *Yunnan Baiyao*, *Gongxuening Capsule*, *Chonglou Jieduding*, and so on (Huang et al., 2011). Due to over-harvesting of the plant growing in the wild, slow growth, and increasing demand, the natural populations of PPY have declined dramatically, and it became a rare and endangered medicinal plant in China (Huang et al., 2011). Since the 2000s, the Chinese government strengthened the protection of the wild species as a resource. Nowadays, the challenges in planting technology for PPY are being overcome step by step and the comprehensive utilization of PPY herbs is improving, especially the usage of its non-medicinal parts, such that the problem of resource shortage has been basically solved.

Over the past decades, the chemical components of this species have been comprehensively investigated. More than 150 compounds have been isolated and identified from PPY, and the steroidal saponins are the predominant compounds found in the plant and are regarded as the main active ingredients. Apart from steroidal saponins, phytochemicals identified from PPY comprise of sterols, flavonoids, flavonoid glycosides, triterpenoid saponins, and fatty acid esters. On this basis, the quality standards of Paridis Rhizoma have been established from qualitative and quantitative analysis. Ethnopharmacological researches reveal that the extracts and phytochemical components obtained from PPY possess various medicinal effects, such as anti-tumor, hemostatic, uterine contractile agonistic, antimicrobial, anti-viral, and anti-inflammatory activities, and protective effects on ethanol- or indomethacin-induced gastric mucosal lesions. Further pharmacokinetic and toxicological studies have indicated that *Paris* saponins possess low oral bioavailability, but excessive ingestion of Paridis Rhizoma might cause some side effects, such as hepatotoxicity and gastric stimulus.

4.2 Taxonomy and Distribution

4.2.1 Taxonomy

The botanical name, *Paris*, is derived from the Latin root; "par-" refers to whorled sepals parallel to a whorl of leaves, whereas *polyphylla* means many leaves (Shah et al., 2012). *Paris* species have been used in many regions with different local names. For example, it was named *Zaoxiu* (蚤休), *Chonglou* (重楼), *Ziheche* (紫河车), *Dujiaolian* (独脚莲), and *Chongtai* (重台) in China, but the term *Chonglou* has been widely used until now; *Paris* species are also locally known as *Satuwa* in Nepal (Kunwar et al., 2020), *Satwa* by Garhwalis of Uttarakhand (Shah et al., 2012), *Singpan* in Manipuri (Shah et al., 2012), *Tangma* in Kham (Kunwar et al., 2020), *Haimavati* in Sanskrit (Kunwar et al., 2020), *Dhumbi Mendo* in Sherpa (Kunwar et al., 2020),

Mauro, *Bajuro*, *Natardap*, or *Kalchang* in Tamang language (Kunwar et al., 2020), and *Bảy lá một hoa* or *trọng lâu* in Vietnamese (Nga et al., 2016).

Traditionally, the taxonomy and species identification of the genus *Paris* are established based on their morphological features, but wide intraspecific variation of their morphological features makes the classification complicated (Ji et al., 2006). Additionally, different opinions about *Paris* systematic taxonomy were proposed. The first *Paris* species, *P. quadrifolia* Linn., was established by Linneaus in 1753. Based on specimens collected by Delavay J. M. and David A. in Southwest China, Franchet published three new species, *P. chinensis*, *P. thibetica*, and *P. yunnanensis*, along with a new variety, *P. polyphylla* var. *stenophylla* (Franchet, 1888). But Handel-Mazzetti proposed that the morphologies of *P. yunnanensis* were largely homologous to those of *Paris polyphylla*; thus, the former was recognized as a variety of *P. polyphlla* in 1936 (Handel-Mazzetti, 1936). According to the morphological diversities for the fruits, seeds, ovaries, and rhizomes of the genus *Paris*, Takhtajan argued that the genus *Paris* was a "collective genus" that consisted of three distinct genera, *Paris* L. s. str., *Kinugasa* Tatewaki and Suto, and *Daiswa* Raf., of which *Daiswa* Raf. contains 15 species, and *D. yunnanensis* was probably one of the most archaic species in the genus *Daiswa* (Takhtajan, 1983). In terms of the morphological features of ovary and placenta, Li suggested that the genus *Paris* should include two subgenera, the subgenus *Daiswa* (Rafinesque-Schmaltz) H. Li and the subgenus *Paris*, and these two subgenera were further divided into 8 sections, which consisted of 24 species; *P. yunnanensis*, a member of the section *Euthyra* Franch of the subgenus *Daiswa* (Rafinesque-Schmaltz) H. Li, was recognized as a variety of *P. polyphlla* (Li, 1998). Recently, on the basis of ultra-barcoding analyses and morphological differences, Ji et al. (2020) confirmed that *P. yunnanensis* was a separate species, which is made up of two distinct genetic lineages, corresponding to "typical" and "high stem" phenotypes and the latter should be accepted as a previously unrecognized species (*P. liiana* sp. *nov.*).

4.2.2 Geographic Distribution

PPY is mainly distributed in Southwest China, such as Yunnan Province, Guizhou Province, Sichuan Province, Chongqing Municipality, Guangxi Zhuang Autonomous Region, and Tibet Autonomous Region (Figure 4.2). It is widely distributed in most parts of Yunnan Province and the south of Sichuan Province, which is recognized as the typical natural habitat with better quality of this species. It is also found in Mandalay, Myanmar (Burma). Ecologically, PPY is naturally found in moist and shady places within an altitude range of ca. 1400~3100 m, such as evergreen (or deciduous) broad-leaved and coniferous forests, bamboo forests, and scrub thickets (Li, 1998).

4.3 Resources and Sustainable Utilization

As a traditional Chinese medicine, Paridis Rhizoma has been in great demand over the past decade with the trade data showing harvesting of about 3,000 tons/year (Yang et al., 2012). Since the 1980s, the market price of Paridis Rhizoma has increased

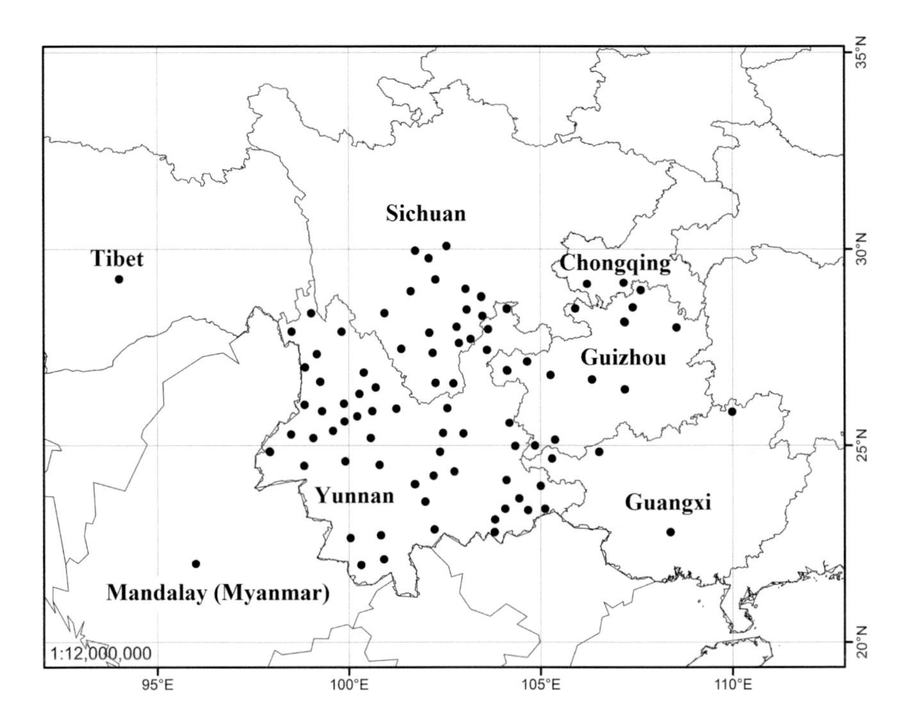

FIGURE 4.2 Geographic distribution of PPY.

400-fold (from 2.7 to 1100 CNY per kg) in China (Cunningham et al., 2018). Under the pressure of strong commercial interests, the wild *Paris* resources had been excessively excavated in China and even in neighboring countries. For example, in 2015, China imported up to 1250 tons of Paridis Rhizoma from Southeast Asia and South Asia countries, such as Vietnam, Laos, Thailand, Myanmar, India, Nepal, and Pakistan (Chinese herbal medicine network of heaven and earth, 2015). Additionally, *Paris* seeds have a long dormancy period with a low germination percentage in the wild, while the natural propagation and growth of *Paris* species are slow. The aforementioned factors not only indicate that the growth period of *Paris* rhizomes needs to be more than seven years, but also result in the increasing shortage of wild populations of *Paris* species. At present, 19 *Paris* species (*P. axialis*, *P. cronquistii* var. *cronquistii*, *P. delavayi*, *P. dunniana*, *P. polyphylla* var. *alba*, *P. polyphylla* var. *chinensis*, *P. vietnamensis*, *P. cronquistii* var. *xichouensis*, *P. daliensis*, *P. fargesii* var. *petiolate*, *P. forrestii*, *P. mairei*, *P. polyandra*, *P. rugosa*, *P. vaniotii*, *P. dulongensis*, *P. luquanensis*, *P. undulata*, *P. wenxianensis*) were considered as threatened species. Among them, *P. axialis*, *P. cronquistii* var. *cronquistii*, *P. delavayi*, *P. dunniana*, *P. polyphylla* var. *alba*, *P. polyphylla* var. *chinensis*, and *P. vietnamensis* were vulnerable (VU) species; *P. cronquistii* var. *xichouensis*, *P. daliensis*, *P. fargesii* var. *petiolate*, *P. forrestii*, *P. mairei*, *P. polyandra*, *P. rugosa*, and *P. vaniotii* were endangered (EN) species; and *P. dulongensis*, *P. luquanensis*, *P. undulata*, and *P. wenxianensis* were critically endangered (CR) species (Qin et al., 2017). PPY, one of the most important *Paris* plants, was a rare and endangered plant (Huang et al., 2011).

FIGURE 4.3 Large-scale planting of PPY in Yunnan Province (a: seedlings; b–e: planting under a shade net; f: underwood planting) (photos by Hai-Yang Liu).

Artificial cultivation is an effective approach for the conservation and sustainable utilization of medicinal plants to some extent. Several researchers specializing in agronomy have begun to resolve the large-scale plantation of *Paris* species (especially for PPY) since the 2000s. The key technologies of PPY, including seed germination, domestication, and cultivation, are gradually being solved. In recent years, the fast development of tissue culture technology has accelerated the breeding process and increased the survival rates of PPY. To date, there is an ongoing large-scale cultivation of PPY in Yunnan Province (estimated more than 154,000 *mu* (1 *mu* = 667 square meters)) (Figure 4.3), which could basically meet the demand of the pharmaceutical industry. However, good agricultural practice (GAP) in the planting and cultivation is needed as soon as possible.

In order to take full advantage of the resources, systematic phytochemical and biological studies on the aerial parts of PPY that can be renewable yearly are needed. More than 50 steroidal saponins were obtained and structurally characterized from the aerial parts of PPY in our group (Chen et al., 1990a, Chen et al., 1990b, Chen and Zhou, 1992, Chen et al., 1995a, Chen et al., 1995b, Qin et al., 2012, Qin et al., 2013, Qin et al., 2016). Among them, some of the saponins are the same or similar to the main ingredients found in the rhizomes. Furthermore, the total content of the four spirostanol saponins (paris saponins I, II, VI, and VII) in the aerial parts (especially its leaves) meet the requirements stipulated in the 2015 edition of the Chinese Pharmacopoeia. More importantly, both the main components and saponin-rich moieties from the aerial parts and rhizomes of PPY exhibited similar hemostatic, cytotoxic, and antimicrobial effects (Qin et al., 2018). Compared to the rhizomes, the above-ground parts can be used as an alternative and more sustainable source for Paridis Rhizoma. Currently, even though the medicinal use of the above-ground parts is not allowed, wide use for by-products like household healthcare products could be highly encouraged based on the above bioactive assays.

4.4 Traditional Uses (Commercial Products) and Ethnopharmacology

PPY is an important raw material for more than 100 Chinese patent medicines, which are used for the treatment of traumatic injury, bleeding, rheumatism, anti-cancer, inflammation, and skin disease (Huang et al., 2011). Some representative Chinese patent medicines are summarized in Table 4.1. *Yunnan Baiyao*, the most well-known medicine containing the rhizomes of PPY, was developed by a Chinese herbalist in Yunnan named Huanzhang Qu in 1916. It had even been used as a "miracle drug" to treat swelling, joint and bone pain, inflammation, gynecological blood stasis, and ulcers. In 1956, industrial-scale production of this medicine began at the Yunnan Baiyao Pharmaceutical Factory. In the past decades, the products and drugs of *Yunnan Baiyao* series have been developed into powder, capsule, tablet, tincture, hemorrhoid cream, aerosol, adhesive bandage, and toothpaste. Till today, the *Yunnan Baiyao* formula is kept a state secret and is obscure. This "miracle drug" has attracted great interest from medicinal chemists, pharmacologists, and therapeutic practitioners throughout the world. In 1979, *Paris* saponins I and II were obtained from *Yunnan Baiyao* by researchers at the University of Wisconsin (Ravikumar et al., 1979). *Gongxuening Capsule*, a drug composed of total steroidal saponins extracted from Paridis Rhizoma (PPY), was studied and developed by the Kunming Institute of Botany, CAS, and Yunnan Baiyao Pharmaceutical Factory from 1981 to 1985 and approved in 1985. It is used to treat menorrhagia, metrorrhagia, puerperal bleeding caused by the stagnation of vital energy, and blood stasis. This capsule has been used as the most basic medicine for gynecological bleeding and listed as one of the national protected traditional medicines in China. *Yunnan Hongyao* (including powder, tablet, and capsule preparation) are used to effectively treat traumatic injury. *Jinpin* series (including liniment, gel, and aerosol preparation) is used for the treatment of swelling, pain, and rheumatism. *Jidesheng Sheyao Tablet* has been traditionally used to treat snake and insect bites, and *Chonglou Jieduding* is widely used for the treatment of herpes zoster, mumps, and skin pruritus (Huang et al., 2011). Representative Chinese patent medicines containing Paridis Rhizoma are shown in Figure 4.4.

4.5 Phytochemicals

4.5.1 Isolation Techniques

Steroidal saponins, the main constituents of PPY, usually possess high polarity since they contain one or more sugar chains. Thus, polar solvents including methanol, ethanol, and water are chosen to extract these steroidal saponins. Apart from the steroidal saponins, lots of other minor components, such as polysaccharides, flavonoids, and fatty acid esters, can be found in the crude extract. Macroporous adsorption resins like D-101 and HPD-100 are often selected to eliminate these non-target constituents and to enrich the target steroidal saponins. As for steroidal saponins, the first traditional purification process was repetitive separation using silica gel columns with a

TABLE 4.1

Representative Chinese Patent Medicines Using Paridis Rhizoma as One of the Raw Materials[a]

Names	Main Ingredients	Usages
Biyan Qingdu Granules	Chrysanthemi Indici Flos, Xanthii Fructus, **Paridis Rhizoma**, Zanthoxyli Radix, Prunellae Spica, Gentianae Radix et Rhizoma, Codonopsis Radix	Chronic inflammation of nasopharynx
Chonglou Jiedu Tincture	**Paridis Rhizoma**, Aconiti Kusnezoffii Radix, Artemisiae Argyi Folium, Acori Tatarinowii Rhizoma, Allii Sativi Bulbus, Borneolum	Herpes zoster, mumps and skin pruritus
Fufang Chonglou Tincture	**Paridis Rhizoma**, Aconiti Kusnezoffii Radix, Artemisiae Argyi Folium, Taraxaci Herba, Angelicae Sinensis Radix, Carthami Flos, Allii Sativi Bulbus, Borneolum	Swelling and pain, mumps, mastitis
Fufang Shedan Chuanbei Power	Fel Serpentis, Fritillariae Cirrhosae Bulbus, Fructus Aristolochiae, Pinelliae Rhizoma Praeparatum Cum Zingibere et Alumine, **Paridis Rhizoma**, Glycyrrhizae Radix et Rhizoma	Cough, asthma, bronchitis
Ganfule Capsule	Codonopsis Radix, Trionycis Carapax, **Paridis Rhizoma**, Atractylodis Macrocephalae Rhizoma, Astragali Radix, Citri Reticulatae Pericarpium, Eupolyphaga Steleophaga, Persicse Semen, Scutellariae Barbatae Herba, Poria, Coicis Semen, Curcumae Radix, Sappan Lignum, Ostreae Concha, Artemisiae Scopariae Herba, Akebiae Caulis, Cyperi Rhizoma, Aquilariae Lignum Resinatum, Bupleuri Radix	Liver cancer
Hongwei Sheyao Tablet	Dioscoreae Bulbiferae Rhizoma, **Paridis Rhizoma**, Rhizoma Dysosmatis, Realgar	Insect and snake bites
Jidesheng Sheyao Tablet	**Paridis Rhizoma**, Corium Bufonis, Scolopendra, Euphorbiae Humifusae Herba	Insect and snake bites
Jinfukang Oral Liquid	Astragali Radix, Glehniae Radix, Ophiopogonis Radix, Ligustri Lucidi Fructus, Corni Fructus, Epimedii Folium, **Paridis Rhizoma**, Asparagi Radix	Non-small cell lung cancer
Kangbingdu Granules	Isatidis Radix, Lonicerae Japonicae Caulis, Sophorae Tonkinensis Radix et Rhizoma, Houttuyniae Herba, **Paridis Rhizoma**, Dryopteridis Crassirhizomatis Rhizoma, Angelicae Dahuricae Radix, Artemisiae Annuae Herba, Iridis Tectori Rhizoma	Virus flu
LouLian Capsule	Semiauqilegiae Radix, Polygoni Orientalis Fructus, **Paridis Rhizoma**, Trionycis Carapax, Curcumae Rhizoma, Lobeliae Chinensis Herba, Eupolyphaga Steleophaga, Hirudo, Ginseng Radix et Rhizoma Rubra, Polygoni Multiflori Radix Praeparata, Galli Gigerii Endothelium Corneum, Scutellariae Barbatae Herba	Primary hepatic carcinoma

(Continued)

TABLE 4.1 (CONTINUED)

Representative Chinese Patent Medicines Using Paridis Rhizoma as One of the Raw Materials[a]

Names	Main Ingredients	Usages
Qingre Zhike Granule	Scutellariae Radix, Fritillariae Thunbergii Bulbus, **Paridis Rhizoma**, Commelinae Herba, Anemarrhenae Rhizoma, Gypsum Fibrosum, Citri Reticulatae Pericarpium, Aurantii Fructus, Xanthii Fructus, Armeniacae Semen Amarum, Platycodonis Radix, Pogostemonis Herba, Perillae Folium, Glycyrrhizae Radix et Rhizoma Praeparata Cum Melle	Cough, fever, pharyngalgia, bronchitis
Qizhen Capsule	Margarita, Astragali Radix, Notoginseng Radix et Rhizoma, Isatidis Folium, **Paridis Rhizoma**	Lung cancer, breast cancer, gastric cancer
Reduqing Tablet	**Paridis Rhizoma**, Baphicacanthis Cusiae Rhizoma et Radix, Borneolum, Taraxaci Herba, Glycyrrhizae Radix et Rhizoma	Mumps, tonsillitis
Sanqi Xueshangning Powder	Notoginseng Radix et Rhizoma, **Paridis Rhizoma**, Aconiti Kusnezoffii Radix, Callicarpae Macrophyllae Folium, Dioscoreae Rhizoma, Borneolum, Cinnabaris	Gastrointestinal bleeding, tuberculosis hemoptysis, dysfunction uterine bleeding, hemorrhoidal bleeding, irregular menstruation
Tongshu Tablet	Erigerontis Herba, Notoginseng Radix et Rhizoma, Panacis Majoris Rhizoma, Gardeniae Fructus, **Paridis Rhizoma**, Glycyrrhizae Radix et Rhizoma	Traumatic injury, rheumatic arthrodynia
Tongxuekang Capsule	**Paridis Rhizoma**, Aconiti Kusnezoffii Radix, Psammosilenes Radix	Traumatic injury, inflammation, gastrointestinal ulcer
Yili Zhitong Pill	Radix Veratri Mengzeani, **Paridis Rhizoma**, Olibanum, Myrrha, Psammosilenes Radix, Moschus	Pain, dysmenorrhea
Yinbing Xiaocuo Tincture	**Paridis Rhizoma**, Ginkgo Semen, Borneolum	Acne
Yunnan Hongyao Powder	Notoginseng Radix et Rhizoma, **Paridis Rhizoma**, Radix Aconiti Vilmoriniani, Cannabis, Psammosilenes Radix, Acori Tatarinowii Rhizoma, Scutellariae Radix, Aconiti Kusnezoffii Radix Cocta	Traumatic injury, hemoptysis, hemorrhoids, metrorrhagia
Zilou Weikang Capsule	Sepiae Endoconcha, Bletillae Rhizoma, **Paridis Rhizoma**, Glycyrrhizae Radix et Rhizoma	Gastrointestinal disease

[a] Data were collected from https://www.yaozh.com/.

FIGURE 4.4 Representative Chinese patent medicines containing Paridis Rhizoma.

mixture of $CHCl_3$–MeOH–H_2O as eluent to obtain different fractions. Subsequently, medium-pressure liquid chromatography (MPLC) equipped with an RP-18 column eluted with different ratios of aqueous MeOH, EtOH, or MeCN was applied to further purify the subfractions and saponins. In addition, Sephadex LH-20 chromatography was also a good choice to remove lower molecular weight compounds, such as flavones, from saponin-rich moieties. Finally, semi-preparative or preparative HPLC was widely adopted for the effective separation of steroidal saponins and in some cases final purification by recrystallization. With the rapid development of MS technology, the molecular network is also gradually applied to avoid obtaining the same saponins from the plants of the same genus, which undeniably makes the isolation work more productive (Wang et al., 2020).

4.5.2 Chemical Constituents

More than 150 compounds, including steroidal saponins, sterols, flavonoids, flavonoid glycosides, triterpenoid saponins, polysaccharides, and fatty acid esters, have been isolated and identified from PPY. The main research progress on chemical compositions of PPY is summarized in Figure 4.5. The initial investigation of the chemical compositions of PPY can be traced back to 1962 when Huang and Zhou isolated diosgenin from the rhizomes of PPY (Huang and Zhou, 1962). Subsequently, three spirostanol and one pregnane saponins were isolated from the rhizomes of *Paris polyphylla* Sm. (Nohara et al., 1973). Further research led to the isolation of five new steroidal saponins including one new diosgenin, two new pennogenin, and two new furostanol saponins (Miyamura et al., 1982). From 1981 to 1983, Chen and Zhou successively obtained *Paris* saponins I, II, V, VI, and VII, dioscin, β-ecdysone, and sucrose from the rhizomes of PPY (Chen and Zhou, 1981, Chen et al., 1983).

Systematic phytochemical studies on the aerial parts (considered as non-medicinal parts) of this species were carried out. Ten steroidal saponins (including pennogenin,

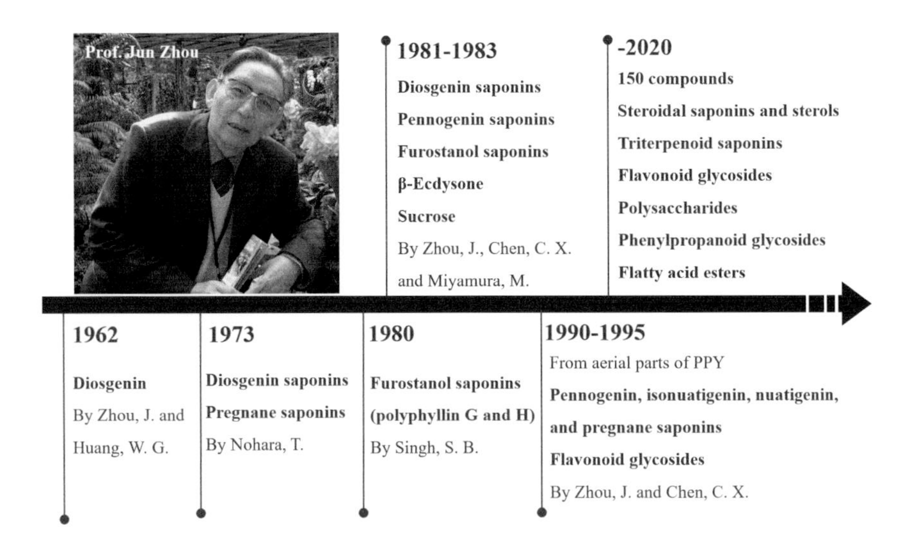

FIGURE 4.5 Research progress on phytochemicals of PPY.

isonuatigenin, nuatigenin, and pregnane saponins) were isolated from the aerial parts of PPY by Chen and Zhou et al. (Chen et al., 1990a, Chen et al., 1990b, Chen and Zhou, 1992, Chen et al., 1995a, Chen et al., 1995b). Studies by Qin et al. (2012, 2013, 2016, 2018) resulted in the isolation of 53 compounds including 20 new steroidal saponins from the stems and leaves of PPY and their bioactivity has been comprehensively evaluated. These studies revealed that the main chemical components from aerial parts of PPY were similar to those of the rhizomes and the nuatigenin-type steroidal saponins can be characteristic chemicals of the stems and leaves of PPY. After generations of unremitting efforts, the chemical composition of PPY has gradually been revealed.

4.5.2.1 Steroidal Saponins

Most steroidal saponins from PPY are composed of a 27-carbon atom aglycone skeleton with one or several glycosidic linkages through the hydroxyl group at the C-1, C-3, or C-27 position. D-glucose, L-rhamnose, L-arabinose, D-xylose, D-apiose, D-galactose, and L-fucose are the main glycosides connected to the aglycones, and these are shown in Figure 4.6. According to the cyclization state of the aglycone skeleton, steroidal saponins can be classified into four main subclasses: spirostane (including diosgenin, pennogenin, and nuatigenin types), furostane, cholestane, and pregnane saponins and representative steroidal saponins and their derivatives are shown in Figure 4.7.

Spirostane-type saponins, composed of a hexacyclic A/B/C/D/E/F-ring system, are the main steroidal saponins in PPY. Diosgenin is an important spirostane, which usually contains a β- hydroxyl at C-3 and a double bond at C-5 and C-6. Pennogenin is a 17α-hydroxylated derivate of diosgenin. To date, more than 80 spirostane-type saponins have been isolated from the aerial parts and rhizomes of PPY, including 53 diosgenin saponins, 16 pennogenin saponins, and 12 nuatigenin-type saponins. Ten

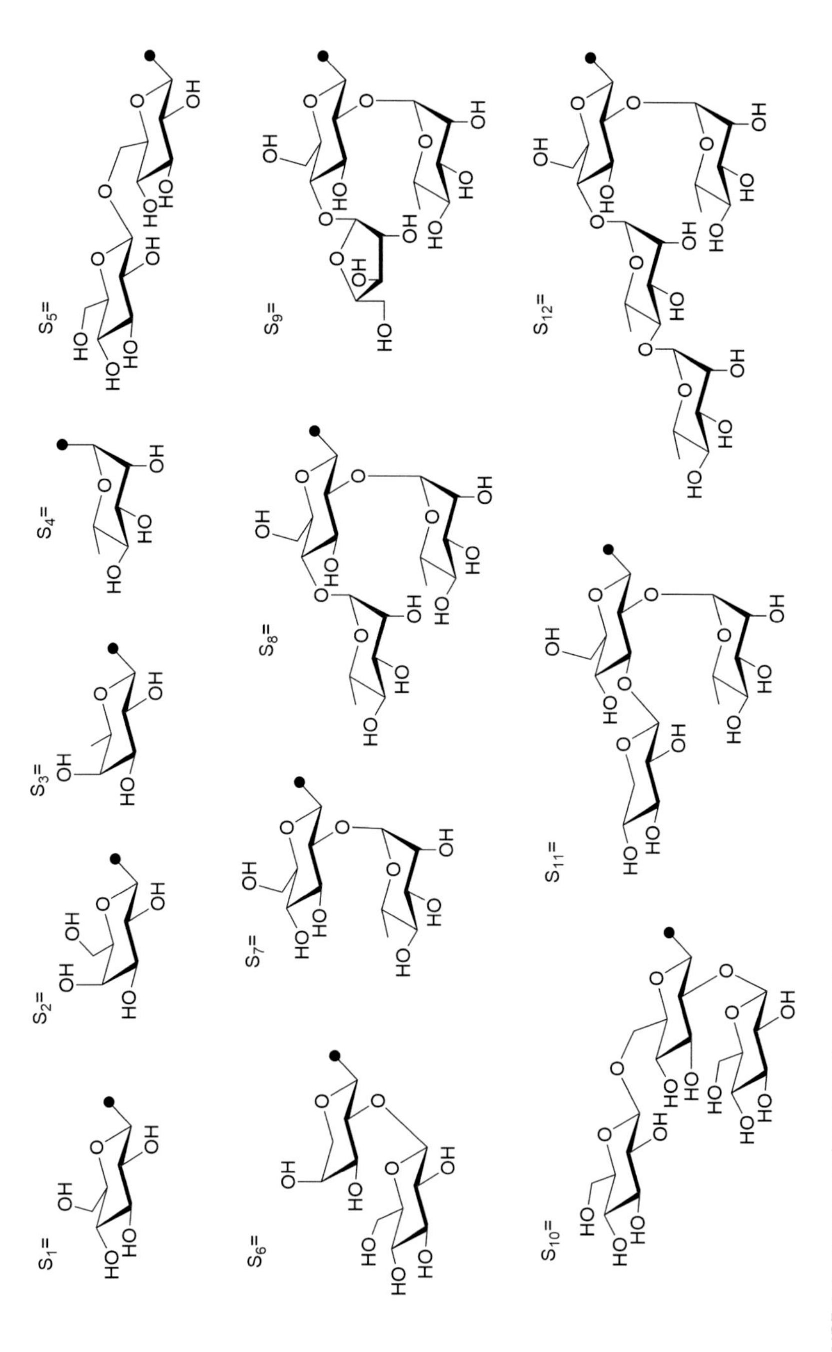

FIGURE 4.6 Representative glycosyl groups in PPY.

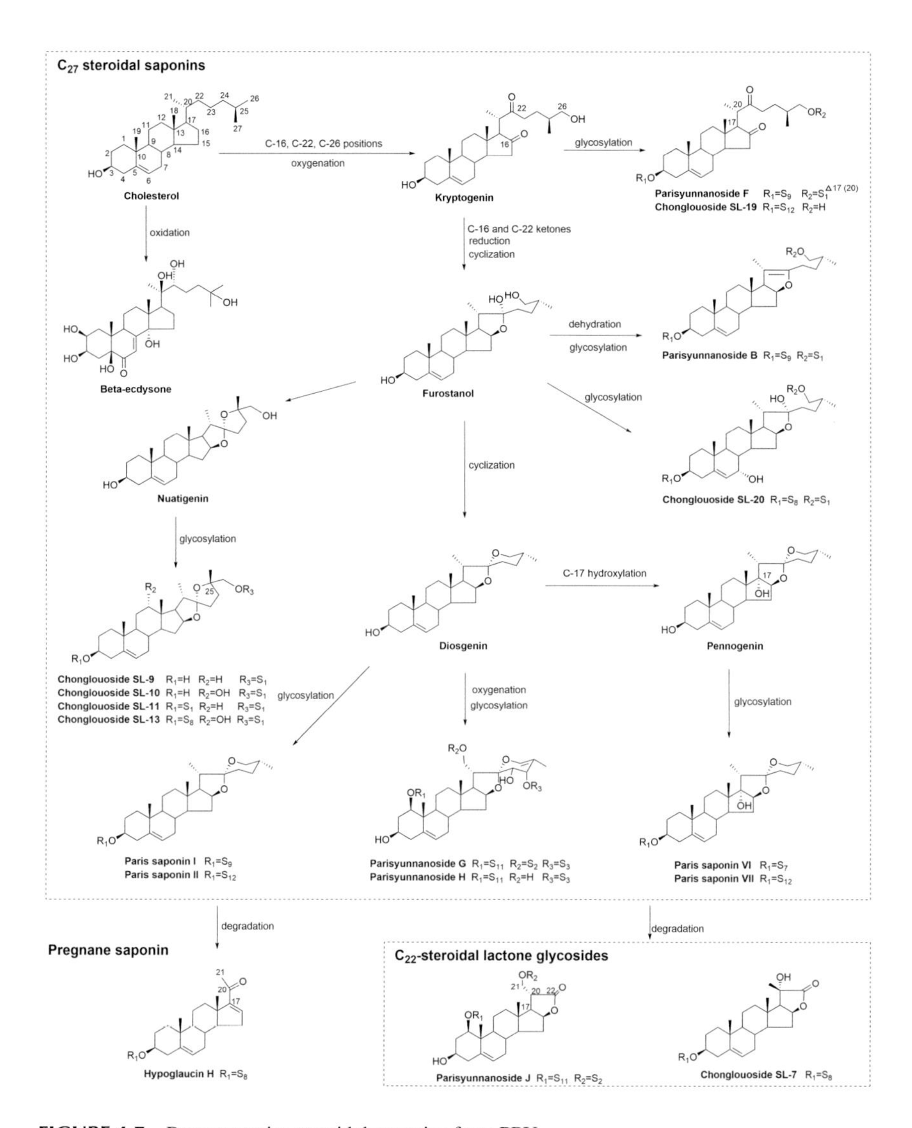

FIGURE 4.7 Representative steroidal saponins from PPY.

spirostane-type saponins (*Paris* saponins I, II, V, VI, VII, and H, dioscin, pennogenin 3-*O*-*β*-chacotrioside, gracillin, and 17-hydroxygracillin) are the main bioactive compounds present in both rhizomes and aerial parts of PPY. Among them, *Paris* saponins VI, VII, and H, pennogenin 3-*O*-*β*-chacotrioside, and 17-hydroxygracillin possess 17*α*-hydroxy, which belongs to pennogenin saponins, and the rest belong to diosgenin saponins. Polyhydroxylated steroidal saponins, linking two or more sugar units, were isolated from both rhizomes and aerial parts of PPY. Parisyunnanosides G-I, three new polyhydroxylated steroidal saponins, with two or three sugar units, were isolated from the rhizomes of PPY (Kang et al., 2012). Qin et al. obtained one new steroidal saponin (chonglouoside SL-6), bearing a sugar unit at C-1, from the stems and leaves of PPY. Nuatigenin-type steroidal saponins are rare furano-spirostanol saponins, which may be transformed from diosgenin saponins using *Streptomyces virginiae* IBL-14 transformation by C-25 tertiary hydroxylation and $[H]^+$ rearrangement reactions (Wang et al., 2007a). Chen et al. firstly isolated nuatigenin-type saponins from the aerial parts of PPY (Chen et al., 1995a). Then Qin et al. (2016) isolated seven new nuatigenin saponins along with four known nuatigenin saponins from the stems and leaves of PPY. Among them, chonglouosides SL-10 and SL-13 are rare 12*α*-hydroxynuatigenin-type saponins, while chonglouoside SL-15 is considered as an uncommon (22*R*,25*S*) nuatigenin-type saponin.

Furostanol saponin, with A/B/C/D/E rings skeleton and a *β*-glucosyl unit at the C-26 position, is considered a biosynthetic precursor of the spirostane saponin, which is easily transformed into spirostanol saponins by an enzymatic reaction with *β*-glucosidase (Inoue and Ebizuka, 1996). Fourteen furostanol saponins were isolated from the aerial parts and rhizomes of PPY. Polyphyllins G and H, the first two new furostanol saponins, were separated from *P. polyphylla* by repeated chromatography over silica gel and preliminary identified by thin-layer chromatography in 1980 (Singh and Thakur, 1980). Subsequently, their structures were further confirmed by mass spectral data and ^{1}H-NMR in 1982 (Singh et al., 1982). A new furostanol-type steroid saponin, parisaponin I, together with two known furostanol-type steroid saponins, trigofoenoside A and protogracillin, were isolated through bioassay-guided separation from the active fraction of PPY (Matsuda et al., 2003). Zhao et al. (2009) obtained two new furostanol saponins, parisyunnanosides A and B, along with five known furostanol saponins from the rhizomes of PPY. Qin et al. (2012, 2016) found one new furostane-type saponin, chonglouoside SL-20, having a hydroxy group linked to C-7, together with three known furostanol saponins from the stems and leaves of PPY.

From a biosynthetic perspective, the cholestane-type steroidal saponin can be transformed into furostane and spirostane. 16, 26-Dihydroxy-22-keto-cholesterol is considered as a common biosynthetic precursor of spirostane saponin (Bai et al., 2014). A new dehydrokryptogenin saponin, parisyunnanoside F, with sugar chains at C-3 and C-26 was isolated from the rhizomes of PPY (Zhao et al., 2007). Qin et al. obtained one new kryptogenin-type saponin only having a sugar chain at C-3 (chonglouoside SL-19) from aerial parts of PPY (Qin et al., 2016).

Pregnane saponins may be degraded components of the C_{27} steroidal saponins, which consist of a C_{21} glycoside with one or two glycosyl groups linking at C-3 or C-21 generally. Seven pregnane saponins were isolated from the aerial parts and rhizomes of PPY. In 1973, Nohara et al. (1973) obtained the first pregnane saponin (hypoglaucin H) from the rhizomes of *P. polyphylla* Sm. Chen, and Qin et al. reported hypoglaucin H and other two pregnane saponins from the PPY aerial parts, among which pregna-5,16-dinen-3*β*-ol-20-one-3*β*-*O*-*α*-L-rhamnopyranosyl-(1→2)-[*α*-L-rhamnopyranosyl-(1→4)-*α*-L-rhamnopyranosyl-(1→4)]-*β*-D-glucopyranoside was a new compound (Chen et al., 1990b, Qin et al., 2012).

C_{22}-Steroidal lactone glycosides are rare saponins, which may also be degradation products of C_{27} steroidal saponins. Parisyunnanoside J, one new C_{22}-steroidal glycoside with two glycosyl groups linked at C-1 and C-21, was isolated from the rhizomes of PPY (Kang et al., 2012). Further investigation by Qin et al. (2013) revealed three C_{22}-steroidal lactone glycosides from the stems and leaves of PPY, of which chonglouoside SL-7 and chonglouoside SL-8 were new compounds.

4.5.2.2 Phytoecdysones

Phytoecdysone usually contains multiple hydroxyl groups in its structure. β-Ecdysone as one of the main phytoecdysones was isolated from the rhizomes of PPY in 1981 (Chen and Zhou, 1981). Subsequently, Singh and Thakur (1982) reported a new phytoecdysone, paristerone, from the rhizomes of *P. polyphylla*.

4.5.2.3 Phytosterols

Phytosterols are widely found in plants and mainly include β-sitosterol, stigmasterol, daucosterol, and their derivatives. Wu et al. (2012b) obtained two new sterol glycosides, pariposides E–F, together with two known sterol glycosides from the roots of PPY. Qin et al. (2016) got two known phytosterols with 7α-hydroxyl group from the stems and leaves of this species.

4.5.2.4 Flavonoids

Flavonoids and their glycosides are commonly found in *Paris* species, especially in their aerial parts, with the skeletal structures of flavonols, biflavones, and chalcones: glycosylation sites are mostly at C-3, C-5, or C-7 in flavonols. Nine flavonoid glycosides were isolated from the aerial parts and rhizomes of PPY. Two known kaempferol glycosides were reported firstly from the aerial parts of PPY in 1995 (Chen et al., 1995a) (Figure 4.8). Wang et al. (2007b) obtained two known flavonoid glycosides from the EtOAc and *n*-BuOH extracts of the rhizomes of PPY.

4.5.2.5 Triterpenoid Saponins

Six new oleanane-type triterpenoid saponins, paritrisides A–F, together with 11 known triterpenoid saponins were isolated from the rhizomes of PPY (Wu et al., 2013, Wu et al., 2017) (Figure 4.9). This was the first report of triterpenoid saponins from the genus *Paris*.

4.5.2.6 Other Compounds

Except for the above ingredients, several other types of compounds, such as polysaccharides, phenylpropanoid glycosides, and fatty acid esters, have been reported. Zhou et al. (2003) isolated two oligosaccharides, heptasaccharide (HS) and octasaccharide

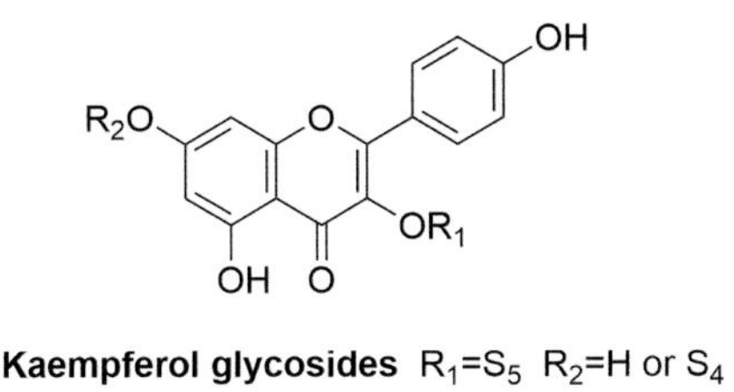

Kaempferol glycosides R_1=S_5 R_2=H or S_4

FIGURE 4.8 Kaempferol glycosides isolated from PPY.

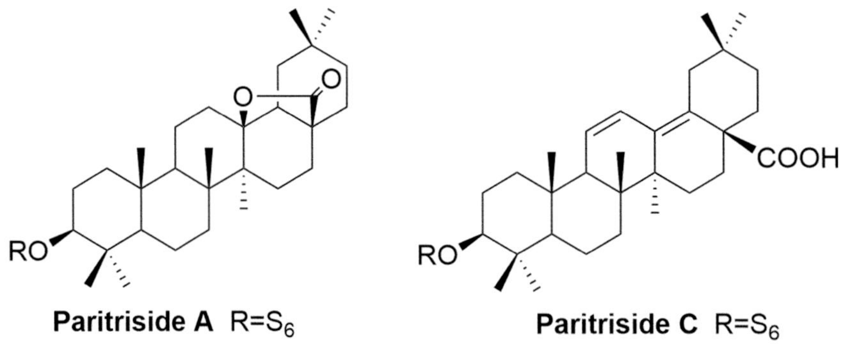

Paritriside A R=S_6 **Paritriside C** R=S_6

FIGURE 4.9 Representative triterpenoid saponins isolated from PPY.

(OS), from the water extract of the rhizomes of PPY, which were identified as linear oligomers and composed of glucose and mannose monomers. A new phenylpropanoid glycoside, named parispolyside F, and a novel derivation of phenolic glycoside, named parispolyside G, were isolated by Wang et al. (2007b) from the rhizomes of PPY. Later, the other phenylpropanoid glycoside (2-feruloyl-*O*-α-D-glucopyranoyl-(1′→2)-3,6-*O*-feruloyl-β-D-fructofuranoside) was obtained from this species (Yan et al., 2008).

4.6 Quality Control and Evaluation Studies

Paridis Rhizoma, the dry rhizomes of PPY or PPC, was documented in the Chinese Pharmacopoeia in 1977 for the first time. Until 2005, *Paris* saponins I and II were considered as the quality control (QC) components of Paridis Rhizoma and the minimum total content of these two saponins was 0.8%. Subsequently, the Chinese Pharmacopoeia (2010 and 2015 Editions) stipulated that *Paris* saponins I, II, VI, and VII were the components of the quality standard and the minimum total content of the four saponins was 0.6%. However, some researchers proposed that the *Paris* saponin Ⅵ was not suitable as a quality control marker, because it was not detected in most of the Paridis Rhizoma samples (Ju et al., 2019, Jia et al., 2020). Thus, *Paris* saponins I, II, and VII only are the official quality control standards and the total content of these three saponins should not be less than 0.6% in the latest published Chinese Pharmacopoeia 2020 Edition (Chinese Pharmacopoeia Commission, 2020). The study by Ju et al. showed that *Paris* saponin H can also be regarded as a marker due to the total amount of *Paris* saponins Ⅰ, Ⅱ, Ⅶ, and H in the analyzed samples of PPC and PPY ranging from 0.051% to 3.99% and 0.12% to 3.23%, respectively (Ju et al., 2020). Therefore, it will be necessary to continually explore the quality control standard for the rational use of Paridis Rhizoma.

The lack of supply and subsequent high price of Paridis Rhizoma have led to adulteration, which will in turn lead to quality control problems in the herbal medicine industry, such as inconsistent therapeutic effects and side effects in clinical applications (Liu and Ji, 2012) and thus, a lack of confidence in TCM preparations. Therefore, there is pressure to establish a reliable and robust quality control method for accurate identification of Paridis Rhizoma and other adulterants. The polymerase chain reaction-restriction fragment length polymorphism (PCR-RFLP), based on the internal transcribed spacer (ITS) sequence, has been proved to be a simple approach for identifying PPY from its adulterants (Liu and Ji, 2012). In 2012, Liu and Ji (2012) identified PPY and its 11 congeners through DNA sequence analysis of nuclear ITS regions. A recent study also indicated internal transcribed spacer 2 (ITS 2) molecular region coupled with high resolution melting (HRM) analysis can effectively differentiate two authentic origins of Paridis Rhizoma from their seven common adulterants (Duan et al., 2018).

4.7 Pharmacological Effects

PPY has been traditionally used to treat multiple diseases since ancient times. Both extracts and individual compounds from PPY possess broad-spectrum biological activities, such as anti-tumor, hemostatic, immune regulation, antiviral, antibacterial, and anti-oxidation activities, etc.

4.7.1 Anti-Tumor Activity

Traditional herbs may be a potential and superior approach to treat cancer rather than radiation and chemotherapy drugs because of lower toxicity and fewer side effects. More than 100 Chinese patent medicines contain Paridis Rhizoma as one of the main raw materials: 10% of them are used to treat cancer, such as *LouLian Capsules*, *Jinfukang Oral Liquid*, *Ganfule Capsules*, etc. (Huang et al., 2011). Modern pharmacological research indicates that both total steroidal saponins (TSS) and monomeric compounds, such as *Paris* saponins I, II, VI, and VII from *Paris* species, exhibit anti-tumor activity.

Several studies show that the EtOH, MeOH, and H_2O extracts of *P. polyphylla* can induce apoptosis or cell cycle arrest in multiple cancer cells, such as breast cancer (Lu et al., 2011), gastric cancer (Sun et al., 2007), prostate cancer (Zhang et al., 2018), colorectal cancer (Lin et al., 2019), ovarian cancer (Wang et al., 2016), liver cancer (Liu et al., 2016), lung cancer (Man et al., 2011), etc. Sun et al. (2007) proved that *P. polyphylla* ethanol extract had more powerful anti-tumor activity than an aqueous extract on six human digestive tumor cell lines, including human liver carcinoma cell lines (HepG-2 and SMMC-7721), human gastric cancer cell lines (BGC-823), human colon adenocarcinoma cell lines (LoVo and SW-116), and an esophagus adenocarcinoma cell line (CaEs-17) with IC_{50} values ranging from 10 μg/mL to 30 μg/mL. Zhang et al. (2018) reported that *P. polyphylla* ethanol extract induced PC-3 cell cycle arrest in G0/G1 and G2/M phases, and induced DU-145 cell cycle arrest in the G0/G1 phase, and further induced apoptosis of prostate cancer cells *in vitro* and *in vivo*. Lin et al. (2019) showed that the ethanol extract of PPY can induce DLD-1 colorectal cancer cell death by upregulation of the autophagy marker. Furthermore, the extract can combine with doxorubicin to enhance cytotoxicity and exert a potent anti-cancer effect against these tumor cells (Lin et al., 2019).

Paris saponins I, II, and V, which belong to the diosgenin skeleton with a trisaccharide chain, a tetrasaccharide chain, and a disaccharide chain, respectively, showed remarkable anti-tumor activity on multiple cancers including non-small-cell lung cancer (Wu et al., 2020), nasopharyngeal carcinoma (Hong et al., 2019), gastric carcinoma (He et al., 2019), colorectal cancer (Yu et al., 2018, Chen et al., 2019a), prostate cancer (Liu et al., 2018a, Xiang et al., 2018), lung cancer (Zhao et al., 2019, Yang et al., 2018), breast carcinoma (Qin et al., 2020), and so on. He et al. (2019) reported that *Paris* saponin I could inhibit PDK1/Akt/mTOR signal and downregulate cyclin B1 in human gastric carcinoma HGC-27 cells, inducing autophagy and cell cycle arrest with an IC_{50} of 0.34 μM *in vitro.* Chen et al. (2019a) revealed *Paris* saponin II induced apoptosis and inhibited colony formation in human colorectal adenocarcinoma HT-29 and human colorectal carcinoma HCT-116 cells by regulating mitochondrial fission and the NF-κB pathway. Recently, *Paris* saponin V was shown to affect the cell cycle distribution and induce G2/M phase arrest in human breast carcinoma MDA-MB-231 cells (Qin et al., 2020).

*Paris s*aponins VI, VII, and H, possessing a pennogenin skeleton, can also inhibit the growth of human osteosarcoma (Yuan et al., 2019), breast cancer (Wang et al., 2019), and human hepatocellular carcinoma (Chen et al., 2019b). Liu et al. (2018b) revealed that *Paris* saponin VI changed cell morphology and induced apoptosis in HepaRG cells through the mitochondrial pathway by the generation of reactive oxygen species (ROS) and the Fas death-dependent pathway. He et al. (2020) showed that *Paris* saponin VII induced apoptotic cell death in A549 human lung cancer cells by inhibiting the PI3K/Akt and NF-κB pathways. Chen et al. (2019b) demonstrated that *Paris* saponin H suppressed human hepatocellular carcinoma (HCC) by the inactivation of the Wnt/β-catenin pathway *in vitro* and *in vivo.*

4.7.2 Hemostatic Activity

Yunnan Baiyao, *Baibaodan Capsule*, and *Tongxuekang Capsule* are effective Chinese patent medicines for the treatment of bleeding. Several studies explored the hemostatic effect and mechanism of steroidal saponins from *Paris* species and proved that the hemostasis of Paridis Rhizoma is a complex process that mainly involves platelet aggregation. Liu et al. demonstrated that the 70% ethanol extract of PPY could significantly shorten the tail bleeding time and blood clotting time in mice by reducing the prothrombin time (PT) and activating partial thromboplastin time (APTT) (Liu et al., 2012b). Fu et al. (2008) revealed that the total steroidal saponin extract from PPY promoted hemostasis *in vivo* and dose-dependently induced rat or human platelet aggregation *in vitro.* Using bioassay-guided separation, four known pennogenin saponins were identified as the active ingredients, which had a synergy with other platelet agonists. A further mechanism study revealed that aggregation in response to the pennogenin saponins involved $\alpha_{IIb}\beta_3$ activation, was dependent upon extracellular calcium, and secreted ADP and thromboxane synthesis, and this process was mediated by phosphatidylinositol-3-kinase (Fu et al., 2008). Subsequent research showed that *Paris* saponin VII, as one main pennogenin saponin in PPY, induced platelet activation depending on dense granule secretion of ADP, which in turn activates the P2Y1 and P2Y12 receptor signaling pathways (Cong et al., 2012). Then Qin et al. compared the biological activity of total steroidal saponins from rhizomes and aerial parts of PPY. The results demonstrated that total steroidal saponins from both parts displayed almost equivalent maximal platelet aggregation rates of 45% and 43%

at a concentration of 1.5 mg/mL, respectively, suggesting the above-ground parts can be an alternative and more sustainable source of active ingredients compared to the rhizomes in hemostatic activity (Qin et al., 2018).

4.7.3 Uterine Contractile Agonistic Activity

Gongxuening Capsule, a drug composed of total steroidal saponins extracted from Paridis Rhizoma, has been widely used as an effective drug in the treatment of abnormal uterine bleeding (AUB). Tian et al. (1986) first used total steroidal saponins to treat 300 cases of uterine hemorrhage of various etiology with excellent results and little side effect. Guo et al. (2008) reported that total steroidal saponins of PPY dose-dependently induced phasic myometrial contractions *in vitro*, which was mediated by an increase in $[Ca^{2+}]_i$ via the influx of extracellular calcium and the release of intracellular calcium. Through bioassay-guided separation, it was found that total spirostanol saponins exhibited contractile activity in myometrium, and *Paris* saponin H was identified as the effective constituent. A further study by Yu et al. (2010) indicated that spirostanol saponins exhibited inducible or inhibitory activity in rat uterine contraction, and the synergistic actions were observed among pennogenin or diosgenin saponins as well as with the known inherent agonist PGF-2α, which indicated that they may share similar pathways with PGF-2α in stimulating myometrial contractions to some extent.

4.7.4 Antimicrobial Activity

Spirostanol saponins, *Paris* saponins V, VI, and pennogenin 3-*O*-β-chacotrioside exhibited significant anti-*Candida albicans* effects with MIC values of 21.6, 21.1, and 8.8 μM, respectively, compared with the fluconazole (MIC values of 52.3 μM) (Chen et al., 2016). Qin et al. reported that chonglouoside SL-6, a saponin with the sapogenin of a 27-hydroxyruscogenin glycoside bearing 1,27-di-*O*-sugar units, isolated from the stems and leaves of PPY, showed good anti-*Propionibacterium acnes* with an MIC value of 3.9 μg/mL (Qin et al., 2012). Further research indicated that the total steroid saponins from the rhizomes and aerial parts (stems and leaves) of PPY showed certain anti-fungal activity against *Aspergillus fumigatus*, *Candida albicans* (5314 and Y0109), and *Candida parapsilosis* (22019) (Qin et al., 2018). *Paris* saponins I, II, V, VI, and VII, dioscin, and chonglouoside SL-5, may be responsible for the anti-fungal activities. In addition, the C-27 steroidal saponins with four or five monosaccharide units exhibited significant activity against *Cryptococcus neoformans* and *Aspergillus fumigatus*, comparable to the positive control amphotericin B, which suggests that the C-27 steroidal saponins may be potential anti-fungal leads for further preclinical study (Yang et al., 2006).

4.7.5 Other Activities

Chinese patent medicines containing Paridis Rhizoma, such as *Xiaoer Tuire Oral Liquid*, *Reduqing*, and *Qingre Zhike Granule*, were used for the treatment of cold and inflammation; while *Zilou Weikang Capsule* and *Shenqixuedan Gastrointestinal Pill* were used to treat gastrointestinal diseases (Huang et al., 2011).

These data suggest that besides anti-tumor, hemostasis, and antimicrobial activities, the extract and individual compounds from PPY may possess other activities, such as anti-inflammation, immune regulation, and protective effects on gastric mucosal lesions. A pharmacological study demonstrated that *Paris* saponin VII could inhibit xylene-induced ear edema and cotton pellet-induced granuloma formation in mice and suppressed lipopolysaccharide (LPS) and $CuSO_4$-induced inflammation in zebrafish embryos (Zhang et al., 2019). Some researchers have demonstrated the anti-inflammatory and antiviral activities of *Paris* saponin I. Zhu et al. (2019) revealed that this compound significantly suppressed the secretion of inflammatory cytokines, including interleukin (IL)-6, IL-8, and tumor necrosis factor (TNF)-α, and the expression of Toll-like receptor 2 (TLR2) in *Propionibacterium acnes*-treated cells. Pu et al. (2015) showed that *Paris* saponin I had a similar inhibition effect on influenza A virus in Madin–Darby canine kidney (MDCK) cells compared with oseltamivir. In addition, the PPY plays an important role in gastrointestinal diseases. Matsuda et al. (2003) indicated the methanol extract from the rhizomes of PPY had protective effects on ethanol- or indomethacin-induced gastric mucosal lesions in rats, with *Paris* saponins I, II, VI, and H as effective ingredients.

4.8 Pharmacokinetics Studies

A pharmacokinetic study of herbal medicinal components is of great importance and plays an increasingly crucial role in understanding biological effects and safety. Therefore, studies on processes of absorption, metabolism, and excretion of Paridis Rhizoma saponins *in vivo* are important. For example, Wu et al. (2012a) separated and determined *Paris* saponins I and H in rat plasma using ginsenoside Rh2 as the internal standard (IS), and found that *Paris* saponins I and H maintained low levels in plasma and were eliminated slowly in rats. The plasma concentration maximum of *Paris* saponin I was 11.5 ng/mL occurring at 8 h post-dose, while the plasma concentration maximum of *Paris* saponin H was 15.2 ng/mL occurring at 10 h post-dose (Wu et al., 2012a). Wang et al. studied the pharmacokinetics of six main steroidal saponins, *Paris* saponins I, II, VI, and VII, dioscin, and gracillin, after intragastric administration of *P. polyphylla* extract in Sprague Dawley rats, and demonstrated that these compounds had an extremely low oral bioavailability and a long half-life (more than 10 h), suggesting that the steroidal saponins had a long residence time and a slow excretion in the body (Wang et al., 2013). Yang et al. revealed that *Paris* saponin II had low clearance, moderate $t_{1/2}$ (8.34–13.37 h), and low bioavailability (6.1–8.2%), which indicated that the absorption of this compound may primarily occur via passive diffusion in rats (Yang et al., 2020).

All the above studies suggest that *Paris* saponins were slowly eliminated with low oral bioavailability. It is speculated that the low oral bioavailability of the molecule is attributed to its high polarity and poor intestinal mucosal permeability (Dahlgren and Lennernäs, 2019). Liu et al. detected and identified seven metabolites, including six phase I and one phase II metabolites after intragastric administration of *Paris* saponin I in Sprague Dawley rats, which indicated that *Paris* saponin I underwent oxidation, deglycosylation, and glucuronidation metabolic reactions in rats (Liu et al., 2017). The molar ratios of $AUC_{metabolite}/AUC_{paris\ saponin\ I}$ were 0.9, 0.8, and 1.3 for three deglycosylation metabolites, prosapogenin A, trillin, and diosgenin, respectively, suggesting less polar saponin and sapogenin (diosgenin) were prone to being absorbed into the blood circulation (Liu et al., 2017). Zhu et al. (2015) showed that the oral bioavailability of *Paris* saponin I could increase from 0.62% to 3.52% and 3.79%, when co-administered with verapamil (VPL) and cyclosporine A (CYA), respectively. In addition, *in vitro* studies showed that *Paris* saponin I, with poor oral bioavailability, is greatly impeded by *P*-glycoprotein (*P*-gp) efflux, and the inhibition of *P*-gp can enhance its bioavailability.

4.9 Toxicological Studies

Even though steroidal saponins show many biological activities and play an important role in various diseases, some studies on the potential toxicity of PPY have been reported. Excessive ingestion of Paridis Rhizoma could cause some side effects such as hepatotoxicity, diarrhea, hypoactivity, and even heart palpitations and convulsions (Liu et al., 2012a).

4.9.1 Toxicity to Normal Cells and Hemolytic Effects

The study by Chan et al. (2011) demonstrated that *Paris* saponin I inhibited the proliferation of human microvascular endothelial cell line HMEC-1 at a concentration range of 0.1–0.4 μM without toxic effects, while at concentrations higher than 0.4 μM, cytotoxic effect was observed since the number of cells decreased from day 2 to day 3. *Paris* saponin I can also cause hemolysis and eryptosis in human red blood cells (RBCs) through increasing the intracellular Ca^{2+} concentration ($[Ca^{2+}]_i$) and caspase-3 activity in human RBCs with EC_{50} value of 0.4 μM *in vitro* (Gao et al., 2012).

4.9.2 Hepatotoxicity

A global metabolic profiling study indicated 90-day administration of Paridis Rhizoma saponins possessed certain liver toxicity in male Wistar rats, and found that Paridis Rhizoma saponins inhibited the oxidation of fatty acids, glycolysis, and the TCA cycle pathway, and affected glycine, serine, and threonine metabolism (Man et al., 2017). Zhao et al. (2020) revealed that Paridis Rhizoma could increase hepatic lipid accumulation and exert toxic effect through up-regulating lipogenesis and down-regulating lipolysis, which was caused by oxidative stress and mitochondrial dysfunction.

4.9.3 Gastrointestinal Toxicity

The gastrointestinal toxicity was carried out by intragastrically (*i.g.*) administering Paridis Rhizoma saponins (100, 250, and 500 mg/kg) to adult mice (Liu et al., 2012a). The results showed that single administration of Paridis Rhizoma saponins significantly inhibited gastric emptying but did not affect the intestinal transit. Man et al. (2016) demonstrated that combination with curcumin not only alleviated the toxicity and gastric stimulus induced by Paridis Rhizoma saponins, but also improved the quality of life of mice bearing tumor cells and enhanced their anti-cancer effect.

4.10 Conclusion

Due to a wide range of pharmacological effects, PPY has attracted great attention from researchers in the fields of phytochemistry, pharmacology, and pharmacy. This chapter reviewed the taxonomy, traditional uses, chemical constituents, quality control, pharmacology, pharmacokinetics, and toxicological studies of PPY, which provide guidance for the rational protection, utilization, and development of Paridis Rhizoma. Although comprehensive studies of PPY have laid the foundation for exploring its multifarious medicinal effects in treating various diseases, further studies on pharmacological mechanisms, clinical pharmacokinetics, and pharmacodynamics are indispensable and urgently required to better validate its safety and exact efficacy. It is still a long way to realize the high-value utilization of PPY and further develop new drugs with less toxicity and side effects.

REFERENCES

Bai HW, Zhao HX, et al., 2014. Isolation and structural elucidation of novel cholestane glycosides and spirostane saponins from *Polygonatum odoratum*. *Steroids*, 80: 7–14.

Chan JY, Koon JC, Liu X, et al., 2011. Polyphyllin D, a steroidal saponin from *Paris polyphylla*, inhibits endothelial cell functions *in vitro* and angiogenesis in zebrafish embryos *in vivo*. *J Ethnopharmacol*, 137: 64–9.

Chen CX, Lian HB, Li YC, et al., 1990a. Steroid saponins of the seed from *Paris polyphylla* var. *yunnanensis*. *Acta Bot Yunnanica*, 12: 452.

Chen CX, Zhang YT, Zhou J, 1983. Studies on the saponin components of plants in Yunnan VI. steroid glycosides of *Paris polyphylla* SM. var. *yunnanensis* (FR.) H-M. *Acta Bot Yunnanica*, 5: 91–97.

Chen CX, Zhang YT, Zhou J, 1995a. The glycosides of aerial parts of *Paris polyphylla* var. *yunnanensis*. *Acta Bot Yunnanica*, 17: 473–478.

Chen CX, Zhou J, 1981. Studies on the saponin components of plants in Yunnan V. steroid glycosides and β-ecdysone of *Paris polyphylla* SM. var. *yunnanensis* (FR.) H-M. *Acta Bot Yunnanica*, 3: 89–93.

Chen CX, Zhou J, 1992. Two new steroid sapogenins of *Paris polyphylla* var. *yunnanensis*. *Acta Bot Yunnanica*, 14: 111–113.

Chen CX, Zhou J, Hiromichi N, et al., 1995b. Two minor steroidal saponins from the aerial parts of *Paris polyphylla* var. *yunnanensis*. *Acta Bot Yunnanica*, 17: 215–220.

Chen CX, Zhou J, Zhang YT, et al., 1990b. Steroid saponins of aerial parts of *Paris polyphylla* var. *yunnanensis*. *Acta Bot Yunnanica*, 12: 323–329.

Chen MH, Ye K, Zhang BY, et al., 2019a. Paris saponin II inhibits colorectal carcinogenesis by regulating mitochondrial fission and NF-kappa B pathway. *Pharmacol Res*, 139: 273–285.

Chen TZ, Lin J, Tang DX, et al., 2019b. Paris saponin H suppresses human hepatocellular carcinoma (HCC) by inactivation of Wnt/beta-catenin pathway *in vitro* and *in vivo*. *Int J Clin Exp Pathol*, 12: 2875–2886.

Chen Y, Ni W, Yan H, et al., 2016. Spirostanol glycosides with hemostatic and antimicrobial activities from *Trillium kamtschaticum*. *Phytochemistry*, 131: 165–173.

Chinese herbal medicine network of heaven and earth, 2015. Domestic Chinese herbal medicine industry blue book in 2015 [Online]. Chinese herbal medicine network of heaven and earth. Available: http://www.zyctd.com/zixun/205/223267.htm [Accessed 30 Sept 2015].

Chinese Pharmacopoeia Commission, 2020. *Pharmacopoeia of the People's Republic of China*, China Medical Science Press, Beijing.

Cong Y, Liu XL, Kang LP, et al., 2012. Pennogenin tetraglycoside stimulates secretion-dependent activation of rat platelets: Evidence for critical roles of adenosine diphosphate receptor signal pathways. *Thromb Res*, 129: e209–e216.

Cunningham AB, Brinckmann JA, Bi YF, et al., 2018. *Paris* in the spring: A review of the trade, conservation and opportunities in the shift from wild harvest to cultivation of *Paris polyphylla* (Trilliaceae). *J Ethnopharmacol*, 222: 208–216.

Dahlgren D, Lennernäs H, 2019. Intestinal permeability and drug absorption: Predictive experimental, computational and *in vivo* approaches. *Pharmaceutics*, 11: 411.

Duan BZ, Wang YP, Fang HL, et al., 2018. Authenticity analyses of Rhizoma Paridis using barcoding coupled with high resolution melting (Bar-HRM) analysis to control its quality for medicinal plant product. *Chin Med*, 13: 8.

Franchet A, 1888. *Monographie du Genere Paris*, Memoire de la Societe Philomathique Centaire, Paris.

Fu YL, Yu ZY, Tang XM, et al., 2008. Pennogenin glycosides with a spirostanol structure are strong platelet agonists: Structural requirement for activity and mode of platelet agonist synergism. *J Thromb Haemost*, 6: 524–533.

Gao MH, Cheung KL, Lau IP, et al., 2012. Polyphyllin D induces apoptosis in human erythrocytes through Ca^{2+} rise and membrane permeabilization. *Arch Toxicol*, 86: 741–752.

Guo L, Su J, Deng BW, et al., 2008. Active pharmaceutical ingredients and mechanisms underlying phasic myometrial contractions stimulated with the saponin extract from *Paris polyphylla* Sm. var. *yunnanensis* used for abnormal uterine bleeding. *Hum Reprod*, 23: 964–971.

Handel-Mazzetti HRE, 1936. *Symbolae Sinicae: Botanische Ergebnisse der Expedition der Akademie der Wissenschaften in Wein nach Südwest-China*, Springer, Wien.

He H, Xu C, Zheng L, et al., 2020. Polyphyllin VII induces apoptotic cell death via inhibition of the PI3K/Akt and NF-kappa B pathways in A549 human lung cancer cells. *Mol Med Rep*, 21: 597–606.

He JL, Yu S, Guo CJ, et al., 2019. Polyphyllin I induces autophagy and cell cycle arrest via inhibiting PDK1/Akt/mTOR signal and downregulating cyclin B1 in human gastric carcinoma HGC-27 cells. *Biomed Pharmacother*, 117: 109189.

Hong F, Gu W, Jiang J, et al., 2019. Anticancer activity of polyphyllin I in nasopharyngeal carcinoma by modulation of lncRNA ROR and P53 signalling. *J Drug Target*, 27: 806–811.

Huang LQ, Xiao PG, Wang YY 2011. *Investigation on Resources of Rare and Endangered Medicinal Plants in China*, Shanghai Press of Science and Techhnology, Shanghai.

Huang WG, Zhou J, 1962. Studies on the aglycone of steroidal saponins from *Paris polyphylla* var. *yunnanensis*. *Med Pharm Yunnan*, 1: 64–65.

Inoue K, Ebizuka Y, 1996. Purification and characterization of furostanol glycoside 26-*O*-beta-glucosidase from *Costus speciosus* rhizomes. *Febs Lett*, 378: 157–160.

Ji Y, Fritsch PW, Li H, et al., 2006. Phylogeny and classification of *Paris* (Melanthiaceae) inferred from DNA sequence data. *Ann Bot*, 98: 245–256.

Ji YH, Liu CK, Yang J, et al., 2020. Ultra-barcoding discovers a cryptic species in *Paris yunnanensis* (Melanthiaceae), a medicinally important plant. *Front Plant Sci*, 11: 411.

Jia TY, Zhang XN, Su T, et al., 2020. A study on quality of Paridis Rhizoma and consideration on standards of Paridis Rhizoma in Chinese Pharmacopoeia. *China J Chin Mater Med*, 45: 2425–2430.

Ju BY, Li YM, Zhu HD, et al., 2019. Improving quality standard of processed slices of Paridis Rhizoma of Chinese Pharmacopoeia. *Chin J Exp Tradit Med Formulae*, 25: 93–101.

Ju BY, Zhu HD, Li YM, et al., 2020. Determination of five steroidal saponins in Paridis Rhizoma and its adulterants as well as consideration on its quantitative method described in Chinese Pharmacopoeia (2015 edition). *China J Chin Mater Med*, 45: 1745–1755.

Kang LP, Liu YX, Eichhorn T, et al., 2012. Polyhydroxylated steroidal glycosides from *Paris polyphylla*. *J Nat Prod*, 75: 1201–1205.

Kunwar RM, Adhikari YP, Sharma HP, et al., 2020. Distribution, use, trade and conservation of *Paris polyphylla* Sm. in Nepal. *Global Ecol Conserv*, 23: e01081.

Li H, 1998. *The Genus Paris (Trilliaceae)*, Science Press, Beijing.

Lin LT, Uen WC, Choong CY, et al., 2019. *Paris Polyphylla* inhibits colorectal cancer cells via inducing autophagy and enhancing the efficacy of chemotherapeutic drug doxorubicin. *Molecules*, 24: 2102.

Liu J, Man SL, Li J, et al., 2016. Inhibition of diethylnitrosamine-induced liver cancer in rats by Rhizoma paridis saponin. *Environ Toxicol Pharmacol*, 46: 103–109.

Liu T, Ji YH, 2012. Molecular authentication of the medicinal plant *Paris polyphylla* Smith var. *yunnanensis* (Melanthiaceae) and its related species by polymerase chain reaction–restriction fragment length polymorphism (PCR-RFLP). *J Med Plants Res*, 6: 1181–1186.

Liu XW, Sun ZT, Deng JK, et al., 2018a. Polyphyllin I inhibits invasion and epithelial-mesenchymal transition via CIP2A/PP2A/ERK signaling in prostate cancer. *Int J Oncol*, 53: 1279–1288.

Liu Y, Dong XX, Wang WP, et al., 2018b. Molecular mechanisms of apoptosis in HepaRG cell line induced by polyphyllin VI via the Fas death pathway and mitochondrial-dependent pathway. *Toxins*, 10: 201.

Liu YC, Zhu H, Shakya S, et al., 2017. Metabolic profile and pharmacokinetics of polyphyllin I, an anticancer candidate, in rats by UPLC-QTOF-MS/MS and LC-TQ-MS/MS. *Biomed Chromatogr*, 31: e3817.

Liu Z, Gao W, Man S, et al., 2012a. Pharmacological evaluation of sedative-hypnotic activity and gastro-intestinal toxicity of Rhizoma Paridis saponins. *J Ethnopharmacol*, 144: 67–72.

Liu Z, Li N, Gao WY, et al., 2012b. Comparative study on hemostatic, cytotoxic and hemolytic activities of different species of *Paris* L. *J Ethnopharmacol*, 142: 789–794.

Lu C, Li CJ, Wu DM, et al., 2011. Induction of apoptosis by Rhizoma Paridis saponins in MCF-7 human breast cancer cells. *Afr J Pharm Pharmacol*, 5: 1086–1091.

Man SL, Qiu PY, Li J, et al., 2017. Global metabolic profiling for the study of Rhizoma Paridis saponins-induced hepatotoxicity in rats. *Environ Toxicol*, 32: 99–108.

Man SL, Gao WY, Zhang YJ, et al., 2011. Paridis saponins inhibiting carcinoma growth and metastasis *in vitro* and *in vivo*. *Arch Pharm Res*, 34: 43–50.

Man SL, Li J, Liu J, et al., 2016. Curcumin alleviated the toxic reaction of Rhizoma Paridis saponins in a 45-Day subchronic toxicological assessment of rats. *Environ Toxicol*, 31: 1935–1943.

Matsuda H, Pongpiriyadacha Y, Morikawa T, et al., 2003. Protective effects of steroid saponins from *Paris polyphylla* var. *yunnanensis* on ethanol- or indomethacin-induced gastric mucosal lesions in rats: Structural requirement for activity and mode of action. *Bioorg Med Chem Lett*, 13: 1101–1106.

Miyamura M, Nakano K, Nohara T, et al., 1982. Steroid saponins from *Paris-Polyphylla* Sm-Supplement. *Chem Pharm Bull*, 30: 712–718.

Nga NQ, Huyen PT, Truong PV, et al., 2016. Taxonomy of the genus *Paris* L. (Melanthiaceae) in Vietnam. *Tạp Chí Sinh Học*, 38: 333–339.

Nohara T, Yabuta H, Suenobu M, et al., 1973. Steroid glycosides in *Paris Polyphylla* Sm. *Chem Pharm Bull*, 21: 1240–1247.

Pu XY, Ren J, Ma XL, et al., 2015. Polyphylla saponin I has antiviral activity against influenza A virus. *Int J Clin Exp Med*, 8: 18963–18971.

Qin HN, Yang Y, Dong SY, et al., 2017. Threatened species list of China's higher plants. *Biodivers Sci*, 25: 696–744.

Qin XJ, Chen CX, Ni W, et al., 2013. C-22-steroidal lactone glycosides from stems and leaves of *Paris polyphylla* var. *yunnanensis*. *Fitoterapia*, 84: 248–251.

Qin XJ, Ni W, Chen CX, et al., 2018. Seeing the light: Shifting from wild rhizomes to extraction of active ingredients from above-ground parts of *Paris polyphylla* var. *yunnanensis*. *J Ethnopharmacol*, 224: 134–139.

Qin XJ, Sun DJ, Ni W, et al., 2012. Steroidal saponins with antimicrobial activity from stems and leaves of *Paris polyphylla* var. *yunnanensis*. *Steroids*, 77: 1242–1248.

Qin XJ, Yu MY, Ni W, et al., 2016. Steroidal saponins from stems and leaves of *Paris polyphylla* var. *yunnanensis*. *Phytochemistry*, 121: 20–29.

Qin XJ, Zhang LJ, Zhang Y, et al., 2020. Polyphyllosides A-F, six new spirostanol saponins from the stems and leaves of *Paris polyphylla* var. *chinensis*. *Bioorg Chem*, 99: 103788.

Ravikumar PR, Hammesfahr P, Sih CJ, 1979. Cytotoxic saponins from the Chinese herbal drug Yunnan-Bai-Yao. *J Pharm Sci*, 68: 900–903.

Shah SA, Mazumder P, Duttachoudhury M, 2012. Medicinal properties of *Paris Polyphylla* Smith: A review. *J Herb Toxicol*, 6: 27–33.

Singh SB, Thakur RS, 1980. New furostanol and spirostanol saponins from tubers of *Paris polyphylla*. *Planta Med*, 40: 301–303.

Singh SB, Thakur RS, 1982. Structure and stereochemistry of paristerone, a novel phytoecdysone from the tubers of *Paris-Polyphylla*. *Tetrahedron*, 38: 2189–2194.

Singh SB, Thakur RS, Schulten HR, 1982. Furostanol saponins from *Paris polyphylla*: Structures of polyphyllin G and H. *Phytochemistry*, 21: 2079–2082.

Sun J, Liu BR, Hu WJ, et al., 2007. *In vitro* anticancer activity of aqueous extracts and ethanol extracts of fifteen traditional chinese medicines on human digestive tumor cell lines. *Phytother Res*, 21: 1102–1104.

Takhtajan A, 1983. A revision of *Daiswa* (Trilliaceae). *Brittonia*, 35: 255–270.

Tian YH, Zheng LH, Xu ZY, et al., 1986. Clinical and pharmacological study of the hemostatic action of rhizoma paridis by contraction of uterus. *J Tradit Chin Med*, 6: 178–182.

Wang BW, Ji SG, Zhang H, et al., 2013. Liquid chromatography tandem mass spectrometry in study of the pharmacokinetics of six steroidal saponins in rats. *Steroids*, 78: 1164–1170.

Wang CW, Tai CJ, Choong CY, et al., 2016. Aqueous extract of *Paris polyphylla* (AEPP) inhibits ovarian cancer via suppression of peroxisome proliferator-activated receptor-gamma coactivator (PGC)-1 alpha. *Molecules*, 21: 727.

Wang FQ, Li B, Wang W, et al., 2007a. Biotransformation of diosgenin to nuatigenin-type steroid by a newly isolated strain, *Streptomyces virginiae* IBL-14. *Appl Microbiol Biotechnol*, 77: 771–777.

Wang J, Li D, Ni W, et al., 2020. Molecular networking uncovers steroidal saponins of *Paris tengchongensis*. *Fitoterapia*, 145: 104629.

Wang PW, Yang QB, Du XY, et al., 2019. Targeted regulation of Rell2 by microRNA-18a is implicated in the anti-metastatic effect of polyphyllin VI in breast cancer cells. *Eur J Pharmacol*, 851: 161–173.

Wang Y, Gao WY, Zhang TJ, et al., 2007b. A novel phenylpropanoid glycosides and a new derivation of phenolic glycoside from *Paris polyphylla* var. *yunnanensis*. *Chin Chem Lett*, 18: 548–550.

Wu S, Gao W, Qiu F, et al., 2012a. Simultaneous quantification of Polyphyllin D and Paris H, two potential antitumor active components in *Paris polyphylla* by liquid chromatography-tandem mass spectrometry and the application to pharmacokinetics in rats. *J Chromatogr B*, 905: 54–60.

Wu X, Chen NH, Zhang YB, et al., 2017. A new steroid saponin from the rhizomes of *Paris polyphylla* var. *yunnanensis*. *Chem Nat Compd*, 53: 93–98.

Wu X, Wang L, Wang GC, et al., 2013. Triterpenoid saponins from rhizomes of *Paris polyphylla* var. *yunnanensis*. *Carbohydr Res*, 368: 1–7.

Wu X, Wang L, Wang GC, et al., 2012b. New steroidal saponins and sterol glycosides from Paris *Polyphylla* var. *yunnanensis*. *Planta Med*, 78: 1667–1675.

Wu Y, Si Y, Xiang Y, et al., 2020. Polyphyllin I activates AMPK to suppress the growth of non-small-cell lung cancer via induction of autophagy. *Arch Biochem Biophys*, 687: 108285.

Xiang ST, Zou PL, Wu JJ, et al., 2018. Crosstalk of NF-kappa B/P65 and LncRNA HOTAIR-mediated repression of MUC1 expression contribute to synergistic inhibition of castration-resistant prostate cancer by Polyphyllin 1-enzalutamide combination treatment. *Cell Physiol Biochem*, 47: 759–773.

Yan L, Gao W, Zhang Y, et al., 2008. A new phenylpropanoid glycosides from *Paris polyphylla* var. *yunnanensis*. *Fitoterapia*, 79: 306–307.

Yang CR, Zhang Y, Jacob MR, et al., 2006. Antifungal activity of C-27 steroidal saponins. *Antimicrob Agents Chemother*, 50: 1710–1714.

Yang LY, Yang B, Wang X, et al., 2012. The advances on breeding of *Paris polyphylla* Smith var. *Yunnanensis* (Franch.) Hand.-Mazz. *J Agric*, 2: 22–24.

Yang Q, Chen WY, Xu YF, et al., 2018. Polyphyllin I modulates MALAT1/STAT3 signaling to induce apoptosis in gefitinib-resistant non-small cell lung cancer. *Toxicol Appl Pharmacol*, 356: 1–7.

Yang Q, Li H, Gui M, et al., 2020. Development and validation of a rapid and sensitive LC-MS/MS method for the determination of polyphyllin II in rat plasma and its application in a pharmacokinetic study. *Biomed Chromatogr*, 34: e4861.

Yu S, Wang LJ, Cao ZX, et al., 2018. Anticancer effect of polyphyllin in colorectal cancer cells through ROS-dependent autophagy and G2/M arrest mechanisms. *Nat Prod Res*, 32: 1489–1492.

Yu ZY, Guo L, Wang B, et al., 2010. Structural requirement of spirostanol glycosides for rat uterine contractility and mode of their synergism. *J Pharm Pharmacol*, 62: 521–529.

Yuan YL, Jiang N, Li ZY, et al., 2019. Polyphyllin VI induces apoptosis and autophagy in human osteosarcoma cells by modulation of ROS/JNK activation. *Drug Des Dev Ther*, 13: 3091–3103.

Zhang C, Li CY, Jia XJ, et al., 2019. *In vitro* and *in vivo* anti-inflammatory effects of Polyphyllin VII through downregulating MAPK and NF-kappa B pathways. *Molecules*, 24: 875.

Zhang D, Li K, Sun C, et al., 2018. Anti-cancer effects of *Paris polyphylla* ethanol extract by inducing cancer cell apoptosis and cycle arrest in prostate cancer cells. *Curr Urol*, 11: 144–150.

Zhao CJ, Wang MS, Jia Z, et al., 2020. Similar hepatotoxicity response induced by Rhizoma Paridis in zebrafish larvae, cell and rat. *J Ethnopharmacol*, 250**:** 112440.

Zhao Y, Kang LP, Liu YX, et al., 2009. Steroidal saponins from the rhizome of *Paris polyphylla* and their cytotoxic activities. *Planta Med*, 75: 356–363.

Zhao Y, Kang LP, Liu YX, et al., 2007. Three new steroidal saponins from the rhizome of *Paris polyphylla*. *Magn Reson Chem*, 45: 739–744.

Zhao YY, Tang XJ, Huang YH, et al., 2019. Interaction of c-Jun and HOTAIR- increased expression of p21 converge in Polyphyllin I-inhibited growth of human lung cancer cells. *Oncotargets Ther*, 12: 10115–10127.

Zhou LA, Yang CZ, Li JQ, et al., 2003. Heptasaccharide and octasaccharide isolated from *Paris polyphylla* var. *yunnanensis* and their plant growth-regulatory activity. *Plant Sci*, 165: 571–575.

Zhu H, Zhu SC, Shakya S, et al., 2015. Study on the pharmacokinetics profiles of Polyphyllin I and its bioavailability enhancement through co-administration with P-glycoprotein inhibitors by LC-MS/MS method. *J Pharmaceut Biomed*, 107: 119–124.

Zhu TT, Wu WJ, Yang SY, et al., 2019. Polyphyllin I inhibits propionibacterium acnes-induced inflammation *in vitro*. *Inflammation*, 42: 35–44.

Section 2

Medicinal Mushrooms and Fungi

5.1

Medicinal Mushrooms and Fungi from Yunnan Province, Part 1: Resources and Diversity

Gang Wu* **and Zhu-Liang Yang**
CAS Key Laboratory for Plant Diversity and Biogeography of East Asia, Kunming Institute of Botany, Chinese Academy of Sciences, Kunming, Yunnan, People's Republic of China

Yunnan Key Laboratory for Fungal Diversity and Green Development, Kunming, Yunnan, People's Republic of China

CONTENTS

* Corresponding author.

5.1.1 Introduction

Yunnan Province is in the southwest part of China and covers about 4.1% of China's national territorial area; 84% of the province is mountainous. The elevation drop in Yunnan is about 6700 m, and its average elevation is around 2000 m. Three huge mountain systems (Gaoligongshan, Nushan, and Yunling) and the famous "three parallel rivers" (Lancang, Nu, and Jinsha Rivers) lie in the north of Yunnan, and the Hengduan Mountains, such as Ailaoshan, Wuliangshan, and Bangmashan, and so on, cover the south (People's Government of Yunnan Province, 2020). The climate in Yunnan is generally a subtropical plateau monsoon but shows strikingly vertical differences with cold, temperate, subtropical, and tropical zones in roughly a north–south divide (Jiang, 1980a,b, Duan et al., 2011). Because of the complex topography and geography, highly variable climate, and other abiotic and biotic factors (Sun et al., 2017), Yunnan becomes an ideal area for the growth and reproduction of plants, and has been called the "Kingdom of Plants," and the region of Hengduan Mountains, partially located in Yunnan, is one of the world's top-36 hotpots of biodiversity (Xing and Ree 2017, Boufford and Dijk 2000, Myers et al., 2000). Accordingly, Yunnan is considered a favorable region for fungi in the forests where a high species diversity of fungi exists (Feng and Yang, 2018, Ying et al., 1994, Yang, 2005, Feng et al., 2012, Han et al., 2018, Wu et al., 2016), thus, earning the reputation of "Kingdom of fungi."

Focusing on this area with high biodiversity, herein, we intend to summarize the species diversity of medicinal mushrooms in Yunnan for better exploration and utilization of them in future. In addition, we introduce some highly valued or commonly used medicinal mushrooms in Yunnan, and finally propose some measurements for the sustainable utilization of medicinal fungi. The aim is to introduce the resources and diversity of medicinal fungi of Yunnan to a wider audience.

5.1.2 Brief History of the Discovery of Medicinal Fungi in China and Yunnan Province

Fungi are inextricably linked to human life, especially in China, the most well-known aspects of which are their taste (Li and Song, 2002), toxicity (Chen et al.,

2016, Mao, 2006), and medicinal values (Liu, 1984, Ying et al., 1987, Dai et al., 2009). Fungi, especially mushrooms, have been used in traditional Chinese medicine for over 3000 years (Chang, 2006). The first herbal classic in the world, *Shen Nong's Herbal Classic* (*Shen Nong Ben Cao Jing*), published thousands of years ago, recorded some medicinal fungi such as *Ganoderma lingzhi* Sheng H. Wu et al., *Omphalia lapidescens* (Horan.) E. Cohn & J. Schröt., *Polyporus umbellatus* (Pers.) Fr., and *Wolfiporia cocos* (Schwein.) Ryvarden & Gilb. Another important book, *Compendium of Materia Medica*, completed in 1578 by Shi-Zhen Li, evaluated the properties, effects, and toxicities of over 20 medicinal fungi and ranked them accordingly. For example, *Lycoperdon* spp. were put into the lower-grade group and had an acrid taste without toxicity. They can be used for curing malignant sores and scabies. *Ganoderma* spp. were assigned to the top-grade medicines, which can prolong peoples' lives.

In recent decades, Liu (1978) recorded 117 medicinal fungi in China and increased the number of Chinese medicinal fungi to 272 in 1984 (Liu, 1984). The authoritative book *Chinese Materia Medica* edited in 1999 by the State Administration of Traditional Chinese Medicine included 135 medicinal fungi. Liu and Zhen (1996a, b, 1997, 1998, 1999, 2001) reported 267 medicinal fungi in China. Wu et al. (2012, 2013b) summarized 150 medicinal Ascomycetes species in China. Wu et al. (2013a) updated and documented 835 medicinal fungi of China. Wu et al. (2019) collated 692 medicinal fungi in China based on their previous data (Dai and Yang 2008, Dai et al., 2009) and recent research progresses.

In the region of Yunnan, a local famous herbalist of traditional Chinese medicine, Mao Lan, published a famous medical book *Materia Medica of South Yunnan* in 1436. It is the first book which systematically recorded the traditional Chinese medicine utilized in Yunnan. Since 1959, the group of *Southern Yunnan Materia Medica* completed and reorganized the book several times in which they made necessary collation and supplemented the contents of each medicine's taxonomic attribution, and morphology, habitat, medicinal part, and necessary annotation. In Mao Lan's book, 458 commonly used medicinal herbs in Yunnan were recorded including 12 fungal species, such as *Cantharellus cibarius* Fr., *Cryptoporus sinensis* Sheng H. Wu & M. Zang, *Ganoderma lingzhi* Sheng H. Wu & M. Zang, *Neoboletus brunneissimus* (W.F. Chiu) Gelardi et al., *Ramaria formosa* (Pers.) Quél., and *Russula virescens* (Schaeff.) Fr., and so on. In the last few years, the research team led by Ji-Kai Liu from the Kunming Institute of Botany, Chinese Academy of Sciences, had investigated the chemical components and biological activities of over 200 fungal species, most of which were from Yunnan (Gao et al., 2001, Yang et al., 2004, Ye et al., 2005, Liu et al., 2007, Tang et al., 2008, Wei et al., 2015, Zhu et al., 2018, Liu et al., 2020).

Regarding diversity, Ying et al. (1994) surveyed the diversity and resources of economic macrofungi in southwestern China including Yunnan, Guizhou, and Sichuan Provinces, and documented 236 medicinal fungi. Wang et al. (2004) reported eight common medicinal macrofungi traded in the wild mushroom markets of Yunnan. Zhao et al. (2006) investigated the resources and utilization of medicinal fungi on the Laojun Mountain, located in Lijiang in the northwest of Yunnan, and estimated over 200 medicinal species there. Wu et al. (2013a) reported over half of 835 Chinese medicinal fungi distributed throughout Yunnan.

5.1.3 The Resources and Diversity of Medicinal Mushrooms in Yunnan

In 2016 and 2017, the Department of Ecology and Environment of Yunnan Province funded the Kunming Institute of Botany, Chinese Academy of Sciences to compile a checklist of biological species in Yunnan (Gao and Sun, 2016). The checklist included 19,333 species of higher plants, over 50% of China's total higher plants (Gao and Sun, 2016), and 2753 species of macrofungi, accounting for 57.4% of China's reported fungal species (Yang et al., 2016). The threat levels to all the recorded species were evaluated as well.

By combining the latest checklist of macrofungi in Yunnan (Yang et al., 2016) and the documents of medicinal macrofungi in China (Wu et al., 2019, Wu et al., 2013a, Dai and Yang, 2008, Dai et al., 2009, Liu, 1984), we now have an opportunity to update our knowledge of the diversity and resources of the medicinal mushrooms in Yunnan. After a careful examination, we summarized, in total, 417 common medicinal species in Yunnan (Table 5.1.1), which belong to 70 families, 187 genera, and account for 65.9% of the total in China. Among them, some species are under threat, probably due to over-harvesting and habitat destruction and other global change factors. For example, 12 medicinal species were evaluated as vulnerable (VU), such as *Engleromyces sinensis*, *Ophiocordyceps sinensis*, *Sanghuangporus sanghuang*, and *Shiraia bambusicola*, etc. Additionally, 15 are at the nearly threatened (NT) level, such as *Amanita caojizong*, *Catathelasma laorentou*, *Ganoderma tropicum*, *Lanmaoa asiatica*, and *Russula griseocarnosa*, etc. It is worth noting that many unknown medicinal fungal species are still awaiting discovery from nature. For example, *Ophiocordyceps highlandensis* was described in 2015, and *O. lanpingensis* in 2013. Both species contain bioactive components (Cai et al., 2017, Zhou et al., 2020).

5.1.4 Some Representative Medicinal Mushrooms in Yunnan

Among the medicinal mushrooms in Yunnan, there are some species of concern due to their important medicinal properties or the threatened situation of survival and are presented herein to better understand them and further protect them.

5.1.4.1 *Cordyceps Militaris* (L.) Fr. 蛹蟲草

Ascoma 3~5 cm tall, orange, usually non-branched. The fertile part club-shaped, 1~2 cm long, 3~5 cm in diameter, surface coarse. The sterile stipe 2.5~4 cm long, 2~4 mm in diameter, subcylindrical, solid. Inner tissue whitish to pale orange. Ascospores smooth, hyaline, long-filiform, 1 μm in diameter, and often septate when mature (Figure 5.1.1 A, B).

Found growing out of dead underground pupae of *Lepidoptera* in the forests.

Note: *Cordyceps militaris* is widely cultivated in China for food and medicinal purposes.

TABLE 5.1.1
The List of Common Medicinal Species in Yunnan, China

Family Names	Medicinal Species	Medicinal Uses (Ref. Wu et al., 2019)
AGARICACEAE	*Agaricus bisporus*	Promoting digestion, lowering blood pressure, antibacteria, anti-tumor, anti-oxidant, immunomodulation
AGARICACEAE	*Agaricus campestris*	Treating anemia, dermatophytosis and hypopepsia, antibacteria, anti-tumor
AGARICACEAE	*Agaricus crocopeplus*	Anti-oxidant
AGARICACEAE	*Agaricus dulcidulus*	Anti-tumor, promoting digestion
AGARICACEAE	*Agaricus micromegethus*	Anti-tumor, promoting digestion
AGARICACEAE	*Agaricus subrufescens*	Anti-tumor, anti-oxidant
AGARICACEAE	*Bovista plumbea*	Hemostasis, detumescence, detoxification
AGARICACEAE	*Bovista pusilla*	Detumescence, hemostasis, detoxification, clearing the lung, relieving sore throat
AGARICACEAE	*Bovistella sinensis*	Hemostasis, detumescence, disinfecting, clearing the lung
AGARICACEAE	*Calvatia candida*	Relieving fever, anti-oxidant
AGARICACEAE	*Calvatia craniiformis*	Anti-inflammatory, detumescence, analgesic, antifungus
AGARICACEAE	*Calvatia cyathiformis*	Detumescence, hemostasis, detoxification
AGARICACEAE	*Calvatia gigantea*	Detumescence, analgesic, clearing the lung, detoxification, treating dermatomycosis, anti-tumor, analgesic, antibacteria, anti-oxidant, anti-inflammation, hepatoprotection
AGARICACEAE	*Calvatia lilacina*	Hemostasis, detumescence, detoxification, anti-cancer
AGARICACEAE	*Coprinus comatus*	Promoting digestion, treating hemorrhoids and diabetes, anti-tumor, antifungus, anti-oxidant, antiproliferation, HIV-1 reverse transcriptase inhibitor, antidiabetic activities
AGARICACEAE	*Coprinus sterquilinus*	Promoting digestion, eliminating phlegm, detoxification, detumescence, anti-tumor
AGARICACEAE	*Cyathus hookeri*	Anti-inflammation
AGARICACEAE	*Cyathus stercoreus*	Treating gastropathy
AGARICACEAE	*Cyathus striatus*	Antibacteria, treating gastropathy
AGARICACEAE	*Leucoagaricus leucothites*	Antimicrobial, anti-oxidant

(*Continued*)

TABLE 5.1.1 (CONTINUED)
The List of Common Medicinal Species in Yunnan, China

Family Names	Medicinal Species	Medicinal Uses (Ref. Wu et al., 2019)
AGARICACEAE	*Macrolepiota detersa*	Anti-oxidant
AGARICACEAE	*Macrolepiota dolichaula*	Antimicrobial
AGARICACEAE	*Macrolepiota procera*	Promoting digestion, anti-cancer, antimicrobial, anti-oxidant
AGARICACEAE	*Nidula niveotomentosa*	Antifungus
AGARICACEAE	*Phaeolepiota aurea*	Anti-tumor
AGARICACEAE	*Tulostoma brumale*	Hemostasis
ALBATRELLACEAE	*Albatrellus ellisii*	Anti-oxidant
ALBATRELLACEAE	*Albatrellus ovinus*	Reducing effects of Alzheimer, anti-oxidant, relief of heat pain on hyperalgesic skin
AMANITACEAE	*Amanita avellaneosquamosa*	Treating lumbago and skelalgia, and limb numbness
AMANITACEAE	*Amanita caojizong*†	Anti-tumor
AMANITACEAE	*Amanita exitialis*	Anti-tumor
AMANITACEAE	*Amanita farinosa*	Antiproliferation
AMANITACEAE	*Amanita griseofolia*	Antieczematic activity
AMANITACEAE	*Amanita hemibapha*	Anti-tumor
AMANITACEAE	*Amanita ochracea*	Hemagglutination
AMANITACEAE	*Amanita pallidorosea*	Antifungus
AURICULARIACEAE	*Auricularia cornea*	Promoting blood circulation, treating hemorrhoids, analgesic, anti-tumor, antibacteria, antihypercholesterolemic activity, anti-oxidant, immunoenhancement
AURICULARIACEAE	*Auricularia delicata*	Replenishing the blood, moistening the lung, hemostasis
AURICULARIACEAE	*Auricularia heimuer*	Antiulcer, replenishing the blood, moistening the lung, hemostasis, lowering blood glucose, anti-oxidant, hepatoprotection, immunomodulation
AURICULARIALES	*Pseudohydnum gelatinosum*	Anti-tumor, anti-oxidant
AURISCALPIACEAE	*Lentinellus cochleatus*	Anti-tumor

(*Continued*)

TABLE 5.1.1 (CONTINUED)

The List of Common Medicinal Species in Yunnan, China

Family Names	Medicinal Species	Medicinal Uses (Ref. Wu et al., 2019)
BANKERACEAE	*Boletopsis leucomelaena*	Treating asthma, rheumatism and psoriasis
BANKERACEAE	*Hydnellum concrescens*	Antivirus
BANKERACEAE	*Sarcodon imbricatus*	Lowering cholesterol, anti-oxidant, anti-tumor, immunoenhancement
BANKERACEAE	*Sarcodon scabrosus*	Antibacteria, antimicrobial, anti-tumor
BOLETACEAE	*Boletus bainiugan*	Invigorating spleen and resolving food stagnation, invigorating kidney
BOLETACEAE	*Boletus sinoedulis*	Treating lumbago and skelalgia, and deadlimb, anti-tumor
BOLETACEAE	*Boletus violaceofuscus*	Anti-tumor
BOLETACEAE	*Butyriboletus roseoflavus*†	Promoting digestion, anti-tumor
BOLETACEAE	*Butyriboletus subsplendidus*	Anti-oxidant
BOLETACEAE	*Cyanoboletus pulverulentus*	Anti-tumor
BOLETACEAE	*Hortiboletus rubellus*	Anti-tumor
BOLETACEAE	*Hourangia nigromaculatus*	Anti-tumor
BOLETACEAE	*Lanmaoa asiatica*†	Anti-tumor, immunomodulation
BOLETACEAE	*Leccinellum crocipodium*	Antiproliferation
BOLETACEAE	*Leccinum aurantiacum*	Anti-oxidant
BOLETACEAE	*Neoboletus brunneissimus*†	Relieving fever, anti-tumor, lowering blood glucose, immunomodulation
BOLETACEAE	*Pulveroboletus ravenelii*	Treating lumbago and skelalgia, limb numbness, improving the system of meridians and collaterals, anti-cancer
BOLETACEAE	*Retiboletus kauffmanii*	Anti-oxidant
BOLETACEAE	*Rubroboletus satanas*	Anti-tumor, anti-oxidant
BOLETACEAE	*Strobilomyces strobilaceus*	Anti-tumor
BOLETACEAE	*Suillellus luridus*	Antihyperglycemic activity, anti-oxidant
BOLETACEAE	*Xerocomus subtomentosus*	Anti-oxidant

(*Continued*)

TABLE 5.1.1 (CONTINUED)

The List of Common Medicinal Species in Yunnan, China

Family Names	Medicinal Species	Medicinal Uses (Ref. Wu et al., 2019)
BONDARZEWIACEAE	*Bondarzewia submesenterica*	Detoxification
BONDARZEWIACEAE	*Heterobasidion parviporum*	Antibacteria
BULGARIACEAE	*Bulgaria inquinans*	Reducing blood viscosity, anti-tumor, anti-cancer, antimicrobial, anti-oxidant, anti-malarial, eliminating blood stasis, immunomodulation, relieving itching
CALOSTOMATACEAE	*Calostoma japonicum*	Anti-tumor
CANTHARELLACEAE	*Cantharellus cibarius*†	Improving eyesight, promoting digestion, treating the respiratory and gastrointestinal tract infection, anti-tumor, antimicrobial, anti-oxidant, antihyperlipidemic activity, anti- inflammation, neuroprotective
CANTHARELLACEAE	*Cantharellus minor*	Improving eyesight, improving the lung, invigorating the stomach
CANTHARELLACEAE	*Craterellus aureus*	Antibacteria
CANTHARELLACEAE	*Craterellus cornucopioides*	Anticomplement, anti-tumor, anti-inflammation
CANTHARELLACEAE	*Craterellus tubaeformis*	Antibacteria, anti-oxidant
CLAVARIADELPHACEAE	*Clavariadelphus truncatus*	Antibacteria
CLAVICIPITACEAE	*Metacordyceps liangshanensis*	Tranquilizing, invigorating kidney, treating pulmonary diseases, anti-tumor
CORDYCIPITACEAE	*Cordyceps memorabilis*	Inhibiting the coagulation of platelet
CORDYCIPITACEAE	*Cordyceps militaris*	Hemostasis and resolving phlegm, anti-tumor, antibacteria, invigorating kidney, treating bronchitis, alleviation of non-alcoholic fatty liver, antifungus, anti-oxidant, anti-fatigue, anti-inflammation, hepatoprotection, hypnotic activity, improving immunity, protection of alcohol-induced acute liver injury
CORDYCIPITACEAE	*Cordyceps polyarthra*	Reinforcing insufficiency, invigorating the lung and kidney
CORDYCIPITACEAE	*Cordyceps takaomontana*	Anti-tumor, improving immunity, antibacteria, antidepression, anti-aging, lowering serum lipids, lowering blood glucose
CORTICIACEAE	*Laeticorticium roseum*	Antibiotics
CORTINARIACEAE	*Cortinarius bovinus*	Anti-tumor

(Continued)

TABLE 5.1.1 (CONTINUED)

The List of Common Medicinal Species in Yunnan, China

Family Names	Medicinal Species	Medicinal Uses (Ref. Wu et al., 2019)
CORTINARIACEAE	*Cortinarius cinnamomeus*	Anti-tumor
CORTINARIACEAE	*Cortinarius collinitus*	Anti-tumor
CORTINARIACEAE	*Cortinarius glutinosus*	Anti-tumor
CORTINARIACEAE	*Cortinarius hemitrichus*	Anti-tumor
CORTINARIACEAE	*Cortinarius latus*	Anti-tumor
CORTINARIACEAE	*Cortinarius pholideus*	Anti-tumor
CORTINARIACEAE	*Cortinarius salor*	Anti-tumor
CORTINARIACEAE	*Cortinarius sanguineus*	Anti-tumor
CORTINARIACEAE	*Cortinarius torvus*	Anti-tumor
CORTINARIACEAE	*Cortinarius turmalis*	Anti-tumor
CORTINARIACEAE	*Cortinarius violaceus*	Anti-tumor, anti-oxidant
DACRYMYCETACEAE	*Calocera viscosa*	Anti-oxidant
DIPLOCYSTIDIACEAE	*Astraeus hygrometricus*	Hemostasis, treating chilblain, antifungus, anti-oxidant, anti-tumor, anti-inflammation, hepatoprotection
ENTOLOMATACEAE	*Entoloma clypeatum*	Anti-tumor
ENTOLOMATACEAE	*Entoloma murrayi*	Anti-tumor
ENTOLOMATACEAE	*Entoloma rhodopolium*	Anti-tumor
ENTOLOMATACEAE	*Entoloma salmoneum*	Anti-tumor
ENTOLOMATACEAE	*Entoloma sinuatum*	Anti-tumor
FISTULINACEAE	*Fistulina subhepatica*	Anti-tumor, treating gastrointestinal diseases
FOMITOPSIDACEAE	*Antrodia subserpens*	Anti-tumor
FOMITOPSIDACEAE	*Antrodia xantha*	Antibacteria, anti-tumor

(*Continued*)

TABLE 5.1.1 (CONTINUED)
The List of Common Medicinal Species in Yunnan, China

Family Names	Medicinal Species	Medicinal Uses (Ref. Wu et al., 2019)
FOMITOPSIDACEAE	*Calcipostia guttulata*	Anti-tumor
FOMITOPSIDACEAE	*Daedalea dickinsii*	Anti-tumor, anti-oxidant
FOMITOPSIDACEAE	*Fomitopsis betulina*	Antibacteria, anti-tumor, anti-oxidant
FOMITOPSIDACEAE	*Fomitopsis pinicola*	Dispelling wind-evil, eliminating dampness, anti-tumor, antifungus, anti-oxidant, immunomodulation, neuroprotective activities
FOMITOPSIDACEAE	*Ischnoderma resinosum*	Anti-tumor
FOMITOPSIDACEAE	*Laricifomes officinalis*	Inducing diuresis, treating gastropathy, detumescence, anti-tumor, anti-oxidant, antivirus
FOMITOPSIDACEAE	*Osteina obducta*	Anti-tumor
FOMITOPSIDACEAE	*Phaeolus schweinitzii*	Anti-tumor, antimicrobial, anti-oxidant
FOMITOPSIDACEAE	*Postia lactea*	Anti-tumor
GANODERMATACEAE	*Amauroderma rude*	Anti-inflammation, eliminating blood stasis, anti-cancer, anti-oxidant, anti-tumor, immunomodulation
GANODERMATACEAE	*Amauroderma rugosum*	Anti-inflammation, inducing diuresis, improving the stomach function, anti-tumor, antimicrobial, anti-oxidant, immunomodulation
GANODERMATACEAE	*Ganoderma applanatum*	Anti-tumor, antiviral, lowering blood glucose, improving immunity, antibacteria, antimicrobial, anti-oxidant
GANODERMATACEAE	*Ganoderma australe*	Anti-tumor, immunomodulation
GANODERMATACEAE	*Ganoderma capense*	Relieving cough, anti-oxidant, anti-glycated and antiradical activities
GANODERMATACEAE	*Ganoderma leucocontextum*	Anti-cancer, antihyperlipidemic and antidiabetic activities, anti-obesity
GANODERMATACEAE	*Ganoderma lingzhi*	Anti-cancer, antihypertension, anti-oxidant, antidiabetic activity, hepatoprotection, immunomodulation, stimulating neural stem cell proliferation, treatment of nephropathy
GANODERMATACEAE	*Ganoderma lobatum*	Anti-tumor, immunomodulation

(*Continued*)

TABLE 5.1.1 (CONTINUED)

The List of Common Medicinal Species in Yunnan, China

Family Names	Medicinal Species	Medicinal Uses (Ref. Wu et al., 2019)
GANODERMATACEAE	*Ganoderma lucidum*	Anti-tumor, lowering blood pressure, improving immunity, antithrombotic, anti-oxidant, biofortified with essential elements (Se, Cu, and Zn), immunomodulation, preventing radiation-induced DNA damage and apoptosis
GANODERMATACEAE	*Ganoderma mutabile*	Anti-tumor
GANODERMATACEAE	*Ganoderma resinaceum*	Anti-tumor
GANODERMATACEAE	*Ganoderma sinense*†	Anti-tumor, anti-inflammatory, inducing diuresis, invigorating the stomach, anti-oxidant, antiproliferation, anti-HIV-1, immunomodulation, healing effect of radius fracture
GANODERMATACEAE	*Ganoderma subresinosum*	Anti-oxidant
GANODERMATACEAE	*Ganoderma tropicum*†	Treating coronary artery disease, anti-tumor
GANODERMATACEAE	*Ganoderma tsugae*†	Calming the nervousness and invigorating the liver, anti-tumor, antimicrobial, antiproliferation, enhancing memory activity immunomodulation
GEASTRACEAE	*Geastrum fimbriatum*	Anti-inflammation, hemostasis, detoxification, antibacteria, anti-oxidant, anti-tumor
GEASTRACEAE	*Geastrum rufescens*	Hemostasis
GEASTRACEAE	*Geastrum saccatum*	Hemostasis
GEASTRACEAE	*Geastrum triplex*	Hemostasis, disinfecting, clearing the lung, relieving sore throat, detoxification, antibacterial,
GEASTRACEAE	*Geastrum velutinum*	Hemostasis, detoxification
GLOEOPHYLLACEAE	*Gloeophyllum sepiarium*	Anti-tumor
GLOEOPHYLLACEAE	*Gloeophyllum trabeum*	Anti-tumor
GOMPHACEAE	*Gomphus clavatus*	Anti-cancer, anti-oxidant
GOMPHACEAE	*Ramaria apiculata*	Anti-tumor
GOMPHACEAE	*Ramaria aurea*	Anti-tumor
GOMPHACEAE	*Ramaria botrytis*	Antibacteria, anti-oxidant, anti-tumor
GOMPHACEAE	*Ramaria botrytoides*	Anti-oxidant
GOMPHACEAE	*Ramaria flava*	Anti-tumor, antibacteria, anti-oxidant

(Continued)

TABLE 5.1.1 (CONTINUED)

The List of Common Medicinal Species in Yunnan, China

Family Names	Medicinal Species	Medicinal Uses (Ref. Wu et al., 2019)
GOMPHACEAE	*Ramaria formosa*	Anti-tumor, neutrophil elastase inhibitor
GOMPHACEAE	*Ramaria hemirubella*	Anti-tumor
GYROPORACEAE	*Gyroporus castaneus*	Anti-tumor
HERICIACEAE	*Hericium coralloides*	Treating gastric ulcer and neurasthenia, promoting digestion
HERICIACEAE	*Hericium yumthangense**	Anti-tumor, antithrombotic, anti-aging, lowering blood glucose and cholesterol, improving immunity, antimicrobial, anti-oxidant, anti- inflammation, a-glucosidase inhibitor, hepatoprotection, immunomodulation, improvement of mild cognitive impairment, repair effects on gastric mucosal injury
HYDNACEAE	*Hydnum repandum*	Anti-tumor
HYDNANGIACEAE	*Laccaria laccata*	Anti-tumor, anti-oxidant
HYGROPHORACEAE	*Ampulloclitocybe clavipes*	Anti-tumor
HYGROPHORACEAE	*Hygrocybe cantharellus*	Antifungus
HYGROPHORACEAE	*Hygrocybe conica*	Anti-oxidant
HYGROPHORACEAE	*Hygrophorus lucorum*	Antimicrobial, anti-oxidant, anti-tumor, immunomodulation
HYGROPHOROPSIDACEAE	*Hygrophoropsis aurantiaca*	Anti-oxidant, antiproliferation
HYMENOCHAETACEAE	*Fomitiporia bannaensis*	Treating coronary artery diseases
HYMENOCHAETACEAE	*Fomitiporia hartigii*	Anti-tumor, anti-oxidant
HYMENOCHAETACEAE	*Fomitiporia punctata*	Treating coronary artery diseases, anti-oxidant, anti-tumor, antivirus
HYMENOCHAETACEAE	*Fomitiporia robusta*	Anti-tumor, antibacteria, anti-oxidant
HYMENOCHAETACEAE	*Fulvifomes mcgregorii*	Anti-tumor, improving immunity
HYMENOCHAETACEAE	*Fuscoporia torulosa*	Detoxification, treating anemia, antibacteria, anti-oxidant
HYMENOCHAETACEAE	*Inonotus cuticularis*	Hemostasis, anti-tumor

(*Continued*)

TABLE 5.1.1 (CONTINUED)

The List of Common Medicinal Species in Yunnan, China

Family Names	Medicinal Species	Medicinal Uses (Ref. Wu et al., 2019)
HYMENOCHAETACEAE	*Inonotus hispidus*	Improving digestion, hemostasis, anti-tumor, activating the circulation to remove blood stasis, antibacteria, anti-oxidant, antihyperglycemic activity, immunomodulation, treatment of candidiasis
HYMENOCHAETACEAE	*Onnia flavida*	Anti-tumor
HYMENOCHAETACEAE	*Phellinopsis conchata*	Promoting blood circulation, detoxification, improving immunity, anti-tumor
HYMENOCHAETACEAE	*Phellinus igniarius*	Hemostasis, anti-tumor antibacteria, anti-oxidant, anti-fatigue, anti-inflammatory, immunomodulation, hepatoprotection, lowering blood glucose
HYMENOCHAETACEAE	*Phellinus lundellii*	Anti-tumor, improving immunity
HYMENOCHAETACEAE	*Phellinus monticola*	Hemostasis, anti-tumor
HYMENOCHAETACEAE	*Phellinus parmastoi*	anti-tumor, improving immunity
HYMENOCHAETACEAE	*Phellinus piceicola*	Anti-tumor, hemostasis
HYMENOCHAETACEAE	*Phellinus tremulae*	Anti-tumor, improving immunity
HYMENOCHAETACEAE	*Porodaedalea himalayensis*	Anti-tumor, improving immunity
HYMENOCHAETACEAE	*Porodaedalea yunnanensis*	Anti-tumor, improving immunity
HYMENOCHAETACEAE	*Pyrrhoderma adamantinum*	Treating gastropathy
HYMENOCHAETACEAE	*Pyrrhoderma lamaoense*	Anti-tumor
HYMENOCHAETACEAE	*Sanghuangporus sanghuang**	Antimicrobial, anti-oxidant, anti-tumor, anti-inflammation, antiproliferation
HYMENOCHAETACEAE	*Sanghuangporus vaninii*	Anti-oxidant, antiproliferation, anti-inflammation
HYMENOCHAETACEAE	*Xanthoporia radiata*	Anti-tumor
HYMENOCHAETALES	*Trichaptum abietinum*	Anti-tumor
HYMENOCHAETALES	*Trichaptum biforme*	Antibacteria, antifungus, anti-tumor, anti-oxidant
HYMENOCHAETALES	*Trichaptum byssogenum*	Anti-tumor
INOCYBACEAE	*Inocybe rimosa*	Anti-tumor, antieczematic

(*Continued*)

TABLE 5.1.1 (CONTINUED)

The List of Common Medicinal Species in Yunnan, China

Family Names	Medicinal Species	Medicinal Uses (Ref. Wu et al., 2019)
LYCOPERDACEAE	*Lycoperdon asperum*	Hemostasis, antibacteria
LYCOPERDACEAE	*Lycoperdon fuscum*	Hemostasis
LYCOPERDACEAE	*Lycoperdon pedicellatum*	Treating pulmonary diseases
LYCOPERDACEAE	*Lycoperdon perlatum*	Detumescence, hemostasis, antibacteria, clearing the lung, relieving sore throat, detoxification, antimicrobial, anti-oxidant
LYCOPERDACEAE	*Lycoperdon pyriforme*	Anti-tumor, antibacteria, hemostasis, clearing the lung, relieving sore throat, detoxification
LYCOPERDACEAE	*Lycoperdon umbrinum*	Anti-inflammatory, hemostasis, antibacteria
LYOPHYLLACEAE	*Hypsizygus marmoreus**	Hemagglutination inhibition, antifungus, antihypertension, anti-oxidant, antiproliferation, anti-HBV, immunoenhancement
LYOPHYLLACEAE	*Hypsizygus ulmarius*	Anti-tumor, antidiabetic activity, anti-oxidant, anti-tumor, anti-inflammation, hepatoprotection, lowering serum lipids, protecting vascular endothelium cells
LYOPHYLLACEAE	*Lyophyllum decastes*	Anti-tumor, antimicrobial, lowering blood glucose
LYOPHYLLACEAE	*Lyophyllum semitale*	Anti-tumor
LYOPHYLLACEAE	*Tephrocybe anthracophila*	Anti-tumor
LYOPHYLLACEAE	*Termitomyces aurantiacus*	Anti-oxidant, lowering blood glucose
LYOPHYLLACEAE	*Termitomyces clypeatus*	Antimicrobial, anti-oxidant
LYOPHYLLACEAE	*Termitomyces eurrhizus*	Improving the stomach, treating hemorrhoids, anti-tumor, anti-ulcerogenic activity
LYOPHYLLACEAE	*Termitomyces microcarpus*	Hypocholesterolemic activity
MARASMIACEAE	*Marasmius androsaceus*	Treating arthritis, anti-tumor, analgesic effect on neuropathic pain
MARASMIACEAE	*Marasmius oreades*	Treating lumbago and skelalgia, limb numbness, anti-tumor
MARASMIACEAE	*Xeromphalina campanella*	Anti-tumor
MERIPILACEAE	*Rigidoporus ulmarius*	Anti-tumor
MERULIACEAE	*Abortiporus biennis*	Anti-tumor, immunomodulation

(Continued)

TABLE 5.1.1 (CONTINUED)

The List of Common Medicinal Species in Yunnan, China

Family Names	Medicinal Species	Medicinal Uses (Ref. Wu et al., 2019)
MERULIACEAE	*Bjerkandera adusta*	Anti-tumor
MERULIACEAE	*Bjerkandera fumosa*	Anti-tumor, anti-oxidant, immunomodulation
MERULIACEAE	*Flavodon flavus*	Anti-tumor, anti-oxidant
MERULIACEAE	*Irpex lacteus*	Treating oliguria, edema and lumbago, lowering blood pressure, antiproliferation, anti-inflammation, immunomodulation, prevention and treatment of chronic glomerulonephritis
MERULIACEAE	*Mycoleptodonoides aitchisonii*	Antidiabetic activity, anti-oxidant, immune stimulation
MERULIACEAE	*Oxyporus corticola*	Antibacteria, anti-tumor
MERULIACEAE	*Phlebia tremellosa*	Anti-tumor, antibacteria
MORCHELLACEAE	*Morchella sextelata*	Anti-oxidant, neuroprotective
MYCENACEAE	*Mycena galericulata*	Anti-tumor
MYCENACEAE	*Mycena haematopus*	Anti-tumor
MYCENACEAE	*Mycena pura*	Anti-tumor
MYCENACEAE	*Panellus stipticus*	Hemostasis, anti-tumor
MYCENACEAE	*Sarcomyxa edulis*	Improving immunity, anti-tumor, antibacteria
NAEMATELIACEAE	*Naematelia aurantialba*	Treating breathless, dissipating phlegm, treating tracheitis, anti- hypertensive, anti-oxidant, lowering blood glucose, immunostimulant
OMPHALOTACEAE	*Gymnopus confluens*	Antibacteria, antifungus
OMPHALOTACEAE	*Lentinula edodes*	Improving immunity, lowering cholesterol and blood pressure, anti-tumor, anti-oxidant
OMPHALOTACEAE	*Marasmiellus ramealis*	Antibacteria, anti-tumor
OMPHALOTACEAE	*Omphalotus japonicus*	Anti-tumor
OMPHALOTACEAE	*Omphalotus olearius*	Antibiotics, anti-oxidant
OPHIOCORDYCIPITACEAE	*Ophiocordyceps crassispora**	Tranquilizing
OPHIOCORDYCIPITACEAE	*Ophiocordyceps gracilis*†	Tranquilizing, invigorating kidney, treating pulmonary diseases, anti-tumor

(*Continued*)

TABLE 5.1.1 (CONTINUED)

The List of Common Medicinal Species in Yunnan, China

Family Names	Medicinal Species	Medicinal Uses (Ref. Wu et al., 2019)
OPHIOCORDYCIPITACEAE	*Ophiocordyceps highlandensis*	Tranquilizing, invigorating kidney, treating pulmonary diseases, anti-tumor, immunomodulation, invigorating kidney
OPHIOCORDYCIPITACEAE	*Ophiocordyceps nutans*	Invigorating the lung and kidney
OPHIOCORDYCIPITACEAE	*Ophiocordyceps sinensis**	Tranquilizing, invigorating kidney, treating pulmonary diseases, anti-tumor, immunomodulation, invigorating kidney
PAXILLACEAE	*Paxillus involutus*	Treating lumbago and skelalgia, limb numbness, anti-oxidant
PHALLACEAE	*Phallus echinovolvatus*	Antimicrobial, anti-oxidant, anti-tumor
PHALLACEAE	*Dictyophora rubrovolvata*	Antiglycation, anti-tumor, anti-oxidant, anti-fatigue, anti-hypoxia, hepatoprotection
PHALLACEAE	*Lysurus mokusin*	Anti-tumor
PHALLACEAE	*Phallus dongsun*	Analgesic, promoting blood circulation, treating rheumatism, clearing the lung
PHALLACEAE	*Phallus rubicundus*	Detumescence
PHALLACEAE	*Phallus tenuis*	Relieving fever, detoxification, detumescence
PHANEROCHAETACEAE	*Climacodon septentrionalis*	Anti-oxidant
PHYSALACRIACEAE	*Armillaria mellea*	Improving immunity, treating insomnia, anti-tumor, anti-cancer, antiedema, antimicrobial, anti-oxidant, anti-inflammation, anti-neuroinflammation
PHYSALACRIACEAE	*Armillaria ostoyae*	Tranquilizing, improving immunity, treating neurasthenia, insomnia and limb numbness
PHYSALACRIACEAE	*Desarmillaria tabescens*	Treating hepatopathy, anti-tumor
PHYSALACRIACEAE	*Flammulina filiformis*	Lowering blood pressure, lowering cholesterol, anti-tumor, antibacteria, antimicrobial, anti-oxidant
PHYSALACRIACEAE	*Flammulina rossica*	Anti-tumor, immunomodulation
PHYSALACRIACEAE	*Oudemansiella submucida*	Antifungus, anti-tumor
PLEUROTACEAE	*Hohenbuehelia grisea*	Antibacteria, anti-tumor
PLEUROTACEAE	*Hohenbuehelia petaloides*	Anti-tumor

(Continued)

TABLE 5.1.1 (CONTINUED)

The List of Common Medicinal Species in Yunnan, China

Family Names	Medicinal Species	Medicinal Uses (Ref. Wu et al., 2019)
PLEUROTACEAE	*Pleurotus citrinopileatus*	Improving immunity, lowering serum lipids, anti-tumor, antidiabetic activity, antimicrobial, anti-oxidant, anti-inflammation
PLEUROTACEAE	*Pleurotus cornucopiae*	Anti-tumor, anti-oxidant, hepatoprotection, HIV-1 reverse transcriptase inhibitor
PLEUROTACEAE	*Pleurotus cystidiosus*	Anti-cancer, antifungus, anti-oxidant, antiproliferation, HIV-1 reverse transcriptase inhibitor, hypoglycemic activity
PLEUROTACEAE	*Pleurotus djamor*	Anti-oxidant, anti-tumor, hepatoprotection
PLEUROTACEAE	*Pleurotus flabellatus*	Antimicrobial
PLEUROTACEAE	*Pleurotus giganteus*	Antifungus, anti-oxidant
PLEUROTACEAE	*Pleurotus ostreatus*	Treating lumbago and skelalgia, limb numbness, improving the system of meridians and collaterals, anti-tumor, anti-oxidant
PLEUROTACEAE	*Pleurotus pulmonarius*	Anti-tumor, anticholinesterase, anticoagulant, antidiabetic activity, antimicrobial, antinociception, anti-oxidant, anti- inflammation, immunomodulation
PLUTEACEAE	*Volvariella bombycina*	Anti-oxidant
PLUTEACEAE	*Volvariella volvacea*	Treating scurvy, anti-tumor, anti-tumorlowering blood glucose
POLYPORACEAE	*Cerrena unicolor*	Treating chronic bronchitis, anti-tumor, anti-cancer, antimicrobial, antivirus, anti-oxidant, immunomodulation
POLYPORACEAE	*Cerrena zonata*	Antibacteria, anti-tumor
POLYPORACEAE	*Cryptoporus sinensis*	Treating asthma and tracheitis, antibacteria and anti-inflammatory, anti-oxidant, anti-tumor
POLYPORACEAE	*Daedaleopsis tricolor*	Anti-tumor, anti-oxidant
POLYPORACEAE	*Earliella scabrosa*	Promoting blood circulation, relieving itching, antibacteria, antifungus, anti-oxidant
POLYPORACEAE	*Fomes fomentarius*	Eliminating blood stasis, anti-tumor, antibacteria, anti-infection, immunomodulation
POLYPORACEAE	*Funalia trogii*	Anti-cancer, anti-oxidant, anti-tumor

(*Continued*)

TABLE 5.1.1 (CONTINUED)

The List of Common Medicinal Species in Yunnan, China

Family Names	Medicinal Species	Medicinal Uses (Ref. Wu et al., 2019)
POLYPORACEAE	*Grifola frondosa*	Treating hepatopathy and diabetes, anti-hypertensive, anti-tumor, antidiabetic activity, anti-oxidant, antivirus, hypolipidemic, immunomodulation
POLYPORACEAE	*Hexagonia apiaria*	Tonifying intestines, promoting digestion, anti-inflammation
POLYPORACEAE	*Hexagonia tenuis*	Antitrypanosomal activity
POLYPORACEAE	*Lentinus sajor-caju*	Antihypertension, anti-oxidant
POLYPORACEAE	*Lentinus squarrosulus*	Anti-oxidant, antiulcer, improving immunity
POLYPORACEAE	*Lentinus tigrinus*	Lowering blood glucose, antimicrobial, anti-oxidant
POLYPORACEAE	*Lenzites betulinus*	Dispelling cold, relaxing tendons, anti-cancer, antimicrobial, anti-oxidant
POLYPORACEAE	*Meripilus giganteus*	Antibacteria, anti-oxidant
POLYPORACEAE	*Neolentinus adhaerens*	Anti-tumor
POLYPORACEAE	*Panus conchatus*	Treating lumbago and skelalgia, limb numbness, anti-tumor
POLYPORACEAE	*Perenniporia fraxinea*	Anti-tumor
POLYPORACEAE	*Perenniporia subacida*	Anti-tumor, anti-oxidant, anti-tumor, relieving itching
POLYPORACEAE	*Polyporus arcularius*	Anti-tumor, antimicrobial
POLYPORACEAE	*Polyporus elegans*	Tonifying meridians and collaterals
POLYPORACEAE	*Polyporus squamosus*	Improving immunity, anti-tumor, antimicrobial, anti-oxidant
POLYPORACEAE	*Polyporus umbellatus*†	Inducing diuresis, treating hepatopathy, anti-tumor, antimicrobial, immunostimulant, prevention of early renal injury, reducing hepatitis B infection
POLYPORACEAE	*Polyporus varius*	Dispelling wind-evil, tonifying meridians and collaterals
POLYPORACEAE	*Pycnoporus sanguineus*	Antibacteria, anti-tumor, hemostasis, treating rheumatism, relieving itching, anti-oxidant
POLYPORACEAE	*Trametes elegans*	Dispelling pathogenic wind, relieving itching, antimicrobial
POLYPORACEAE	*Trametes gibbosa*	Anti-tumor, antimicrobial, anti-oxidant, antivirus, hypoglycemic activity
POLYPORACEAE	*Trametes hirsuta*	Treating rheumatism, relieving cough, anti-tumor, antimicrobial

(*Continued*)

TABLE 5.1.1 (CONTINUED)

The List of Common Medicinal Species in Yunnan, China

Family Names	Medicinal Species	Medicinal Uses (Ref. Wu et al., 2019)
POLYPORACEAE	*Trametes orientalis*	Treating pulmonary diseases, anti-tumor, anti-oxidant, hepatoprotection, immunomodulation
POLYPORACEAE	*Trametes pubescens*	Anti-tumor
POLYPORACEAE	*Trametes versicolor*	Relieving fever, treating hepatopathy, anti-inflammatory, anti-tumor, anti-oxidant, antivirus
POLYPORACEAE	*Wolfiporia cocos*	Relieving cough, inducing diuresis, calming the nervousness, relieving fever, anti-tumor, adjustment of intestinal bacterial flora, antihyperlipidemic activity, anti-oxidant, anti-hepatitis B virus, anti-inflammation, anti-metastasis, anti-tyrosinase, hypoglycemic activity, improvement of cardiac function, improvement of learning and memory abilities, improvement of liver fibrosis, prevention of diabetic nephropathy, sedative and hypnotic activities
PSATHYRELLACEAE	*Coprinellus micaceus*	Anti-tumor, antimicrobial, anti-oxidant, antidiabetic activity, anti-acetylcholinesterase, anti-inflammation, anti-tyrosinase, α-glucosidase inhibitory
PSATHYRELLACEAE	*Coprinopsis atramentaria*	Promoting digestion, eliminating phlegm, detoxification, detumescence, anti-tumor, antibacteria, antifungus
PSATHYRELLACEAE	*Coprinopsis cinerea*	Anti-tumor
PSATHYRELLACEAE	*Coprinopsis lagopus*	Anti-tumor
PSATHYRELLACEAE	*Parasola plicatilis*	Anti-tumor
PSATHYRELLACEAE	*Psathyrella candolleana*	Antibacteria
PYRONEMATACEAE	*Aleuria aurantia*	Immunomodulation
RHIZOPOGONACEAE	*Rhizopogon luteolus*	Antibacteria, anti-cancer, anti-oxidant
RHIZOPOGONACEAE	*Rhizopogon piceus*	Hemostasis
RHIZOPOGONACEAE	*Rhizopogon roseolus*	Anti-tumor, anti-oxidant
RUSSULACEAE	*Lactarius chichuensis*	Anti-tumor
RUSSULACEAE	*Lactarius controversus*	Anti-cancer, antibiotic, antimicrobial, anti-oxidant
RUSSULACEAE	*Lactarius deliciosus*	Anti-tumor, anti-cancer, antimicrobial, anti-oxidant, immunostimulant

(Continued)

TABLE 5.1.1 (CONTINUED)

The List of Common Medicinal Species in Yunnan, China

Family Names	Medicinal Species	Medicinal Uses (Ref. Wu et al., 2019)
RUSSULACEAE	*Lactarius hatsudake*[†]	Anti-tumor, anti-oxidant, antivirus, lowering serum lipids
RUSSULACEAE	*Lactarius lignyotus*	Anti-tumor
RUSSULACEAE	*Lactarius pallidus*	Anti-tumor
RUSSULACEAE	*Lactarius pubescens*	Anti-oxidant
RUSSULACEAE	*Lactarius quietus*	Anti-oxidant
RUSSULACEAE	*Lactarius rufus*	Antinociceptive, anti-inflammation
RUSSULACEAE	*Lactarius subzonarius*	Anti-tumor
RUSSULACEAE	*Lactarius zonarius*	Treating lumbago and skelalgia, limb numbness
RUSSULACEAE	*Lactifluus hygrophoroides*	Anti-tumor
RUSSULACEAE	*Lactifluus piperatus*	Treating lumbago and skelalgia, limb numbness, anti-tumor
RUSSULACEAE	*Lactifluus vellereus*	Treating lumbago and skelalgia, limb numbness, anti-tumor, antifungus, antimicrobial, anti-oxidant immunosuppressive
RUSSULACEAE	*Lactifluus volemus*	Anti-tumor, anti-oxidant
RUSSULACEAE	*Russula adusta*	Anti-tumor
RUSSULACEAE	*Russula alutacea*	Tonifying meridians and collaterals, anti-tumor, anti-oxidant
RUSSULACEAE	*Russula crustosa*	Anti-tumor
RUSSULACEAE	*Russula cyanoxantha*	Anti-tumor, antimicrobial, anti-oxidant
RUSSULACEAE	*Russula delica*	Anti-tumor, antimicrobial, anti-oxidant, HIV-1 reverse transcriptase inhibitor
RUSSULACEAE	*Russula densifolia*	Treating lumbago and skelalgia, deadlimb, anti-tumor
RUSSULACEAE	*Russula emetica*	Anti-tumor, anti-oxidant, anti-tyrosinase, hyperglycemic inhibitor
RUSSULACEAE	*Russula foetens*	Treating lumbago and skelalgia, limb numbness, anti-tumor
RUSSULACEAE	*Russula fragilis*	Antibacteria, anti-oxidant

(*Continued*)

TABLE 5.1.1 (CONTINUED)

The List of Common Medicinal Species in Yunnan, China

Family Names	Medicinal Species	Medicinal Uses (Ref. Wu et al., 2019)
RUSSULACEAE	*Russula griseocarnosa*[†]	Treating anemia, anti-tumor, anti-oxidant
RUSSULACEAE	*Russula integra*	Treating lumbago and skelalgia, limb numbness
RUSSULACEAE	*Russula laurocerasi*	Antimicrobial, anti-oxidant
RUSSULACEAE	*Russula lilacea*	Anti-tumor
RUSSULACEAE	*Russula nigricans*	Treating lumbago and skelalgia, limb numbness, anti-tumor
RUSSULACEAE	*Russula sanguinea*	Anti-tumor
RUSSULACEAE	*Russula senecis*	Anti-tumor, anti-oxidant, immunostimulant
RUSSULACEAE	*Russula sororia*	Anti-tumor
RUSSULACEAE	*Russula subnigricans*	Anti-tumor
RUSSULACEAE	*Russula vesca*	Promoting digestion, anti-tumor, antimicrobial, anti-oxidant
RUSSULACEAE	*Russula virescens*	Improving eyesight, anti-tumor, anti-oxidant
SCHIZOPHYLLACEAE	*Schizophyllum commune*	Treating neurasthenia, anti-inflammatory, anti-tumor, antimicrobial, anti-oxidant, anti-aging
SCLERODERMATACEAE	*Pisolithus arhizus*	Antifungus, anti-oxidant
SCLERODERMATACEAE	*Scleroderma areolatum*	Anti-inflammation, hemostasis, anti-oxidant
SCLERODERMATACEAE	*Scleroderma bovista*	Anti-inflammation, hemostasis
SCLERODERMATACEAE	*Scleroderma cepa*	Detoxification, detumescence, hemostasis
SCLERODERMATACEAE	*Scleroderma citrinum*	Anti-inflammation
SCLERODERMATACEAE	*Scleroderma flavidum*	Anti-inflammation
SCLERODERMATACEAE	*Scleroderma polyrhizum*	Detumescence, hemostasis, anti-inflammation
SCLERODERMATACEAE	*Scleroderma verrucosum*	Hemostasis
SHIRAIACEAE	*Shiraia bambusicola**	Relieving cough, tonifying meridians and collaterals, invigorating qi, replenishing the blood, promoting menstruation, anti-cancer, antidiabetic activity, antimicrobial, anti-oxidant

(*Continued*)

TABLE 5.1.1 (CONTINUED)
The List of Common Medicinal Species in Yunnan, China

Family Names	Medicinal Species	Medicinal Uses (Ref. Wu et al., 2019)
SPARASSIDACEAE	*Sparassis latifolia*	Antibacteria, anti-cancer, antimicrobial, anti-oxidant, antiproliferation, anti-angiogenesis, anti-inflammation, anti- metastasis, immunostimulant, improving the impaired healing of diabetic wounds
STEREACEAE	*Aleurodiscus mirabilis*	Antibacteria
STEREACEAE	*Stereum gausapatum*	Anti-tumor
STEREACEAE	*Stereum hirsutum*	Anti-tumor, antimicrobial, immunosuppressant
STEREACEAE	*Xylobolus annosus*	Anti-tumor
STEREACEAE	*Xylobolus frustulatus*	Anti-tumor
STEREACEAE	*Xylobolus princeps*	Anti-tumor
STROPHARIACEAE	*Agrocybe chaxingu*	Immunopotentiator (Zhang et al., 2013)
STROPHARIACEAE	*Cyclocybe salicaceicola*	Inducing diuresis, invigorating the spleen, improving immunity, anti-tumor, antidiarrheal, antifungus, anti-oxidant, anti-nematode
STROPHARIACEAE	*Gymnopilus aeruginosus*	Anti-tumor
STROPHARIACEAE	*Gymnopilus liquiritiae*	Anti-tumor
STROPHARIACEAE	*Gymnopilus spectabilis*	Anti-oxidant
STROPHARIACEAE	*Hemipholiota populnea*	Anti-tumor
STROPHARIACEAE	*Hypholoma capnoides*	Anti-oxidant
STROPHARIACEAE	*Hypholoma fasciculare*	Anti-tumor
STROPHARIACEAE	*Pholiota adiposa*	Antibacteria, improving immunity, anti-oxidant, antiproliferation, anti-tumor
STROPHARIACEAE	*Pholiota aurivella*	Antimicrobial
STROPHARIACEAE	*Pholiota flammans*	Anti-tumor
STROPHARIACEAE	*Pholiota highlandensis*	Anti-tumor
STROPHARIACEAE	*Pholiota lenta*	Anti-tumor
STROPHARIACEAE	*Pholiota lubrica*	Anti-tumor

(*Continued*)

TABLE 5.1.1 (CONTINUED)

The List of Common Medicinal Species in Yunnan, China

Family Names	Medicinal Species	Medicinal Uses (Ref. Wu et al., 2019)
STROPHARIACEAE	*Pholiota microspora*	Antibacteria, anti-tumor, anti-oxidant
STROPHARIACEAE	*Pholiota spumosa*	Anti-tumor
STROPHARIACEAE	*Pholiota squarrosa*	Anti-tumor, immunomodulation
STROPHARIACEAE	*Pholiota terrestris*	Anti-tumor
STROPHARIACEAE	*Stropharia rugosoannulata*	Anti-tumor, antibacteria, anti-oxidant, anti-fatigue, lowering blood glucose, hypoglycemic activity,
SUILLACEAE	*Suillus bovinus*	Anti-tumor, anti-oxidant
SUILLACEAE	*Suillus cavipes*	Treating lumbago and skelalgia, and deadlimb
SUILLACEAE	*Suillus granulatus*	Treating Kaschin–Beck's disease, anti-tumor, anti-oxidant
SUILLACEAE	*Suillus grevillei*	Treating lumbago and skelalgia, limb numbness, anti-tumor, anti-oxidant
SUILLACEAE	*Suillus luteus*	Treating Kaschin–Beck's disease, anti-tumor, antidiabetic and anti-oxidant activities
SUILLACEAE	*Suillus subaureus*	Anti-oxidant
SUILLACEAE	*Suillus viscidus*	Anti-tumor
TAPINELLACEAE	*Pseudomerulius aureus*	Anti-tumor
TAPINELLACEAE	*Pseudomerulius curtisii*	Antibacteria, anti-oxidant
TAPINELLACEAE	*Tapinella atrotomentosa*	Antibacteria
THELEPHORACEAE	*Polyozellus multiplex*	Antivirus, anti-angiogenesis
THELEPHORACEAE	*Thelephora aurantiotincta*	Anti-cancer
THELEPHORACEAE	*Thelephora ganbajun**	Anti-oxidant, anti-oxidant, antiproliferation, lowering serum lipids
THELEPHORACEAE	*Thelephora vialis*	Treating lumbago and skelalgia, limb numbness
TREMELLACEAE	*Phaeotremella foliacea*	Treating gynecopathy
TREMELLACEAE	*Tremella fuciformis*	Invigorating the kidney, moistening the lung, relieving fever, anti-oxidant, anti-tumor, anti-inflammation, immunomodulation, protection of alcohol-induced liver injury

(Continued)

TABLE 5.1.1 (CONTINUED)

The List of Common Medicinal Species in Yunnan, China

Family Names	Medicinal Species	Medicinal Uses (Ref. Wu et al., 2019)
TREMELLACEAE	*Tremella mesenterica*	Treating neurasthenia and breathless, anti-hypertensive, anti-oxidant, treating asthma
TRICHOLOMATACEAE	*Calocybe gambosa*	Invigorating qi, thermolysis, antifungus, anti-tumor, immunomodulation
TRICHOLOMATACEAE	*Catathelasma laorentou*†	Antibacteria, anti-oxidant, anticoagulant activity, antihyperglycemic, antihyperlipidemic and antidiabetic activities
TRICHOLOMATACEAE	*Clitocybe candida*	Antibacteria
TRICHOLOMATACEAE	*Clitocybe fragrans*	Anti-tumor
TRICHOLOMATACEAE	*Clitocybe maxima*	Anti-oxidant
TRICHOLOMATACEAE	*Clitocybe nebularis*	Antibacteria, anti-tumor, antiproliferation
TRICHOLOMATACEAE	*Clitocybe odora*	Anti-tumor, antifungus, anti-oxidant
TRICHOLOMATACEAE	*Infundibulicybe geotropa*	Antibacteria, anti-oxidant
TRICHOLOMATACEAE	*Lepista irina*	Anti-tumor, anti-oxidant
TRICHOLOMATACEAE	*Lepista luscina*	Anti-tumor, antibacteria
TRICHOLOMATACEAE	*Lepista nuda*	Antibacteria, anti-tumor, anti-oxidant
TRICHOLOMATACEAE	*Lepista sordida*	Calming the nervousness, invigorating the liver, antibacteria, anti-oxidant, anti-tumor, immunomodulation, treating laryngeal cancer
TRICHOLOMATACEAE	*Leucopaxillus giganteus*	Invigorating qi, relieving fever, treating cold and flu, tuberculosis, antimicrobial, anti-oxidant
TRICHOLOMATACEAE	*Omphalia lapidescens*	Anti-tumor, anthelmintic action, improving immunity, relieving fever, anti-oxidant, antiproliferation
TRICHOLOMATACEAE	*Pseudoclitocybe cyathiformis*	Anti-tumor, antifungus
TRICHOLOMATACEAE	*Tricholoma acerbum*	Anti-tumor, anti-oxidant
TRICHOLOMATACEAE	*Tricholoma albobrunneum*	Anti-tumor
TRICHOLOMATACEAE	*Tricholoma album*	Anti-tumor
TRICHOLOMATACEAE	*Tricholoma bakamatsutake*	Anti-tumor

(*Continued*)

TABLE 5.1.1 (CONTINUED)

The List of Common Medicinal Species in Yunnan, China

Family Names	Medicinal Species	Medicinal Uses (Ref. Wu et al., 2019)
TRICHOLOMATACEAE	*Tricholoma flavovirens*	Anti-tumor
TRICHOLOMATACEAE	*Tricholoma fulvum*	Anti-tumor
TRICHOLOMATACEAE	*Tricholoma imbricatum*	Anti-oxidant, cholinesterase inhibitor
TRICHOLOMATACEAE	*Tricholoma matsutake**	Invigorating the stomach, treating bronchitis, anti-tumor, angiotensin converting enzyme inhibitor, antihypertension, antimicrobial, anti-oxidant, immunomodulation
TRICHOLOMATACEAE	*Tricholoma robustum*	anti-tumor
TRICHOLOMATACEAE	*Tricholoma saponaceum*	Antibacteria
TRICHOLOMATACEAE	*Tricholoma sulphureum*	Anti-tumor, antimicrobial, anti-oxidant
TRICHOLOMATACEAE	*Tricholoma vaccinum*	Anti-tumor
TRICHOLOMATACEAE	*Tricholoma virgatum*	Anti-tumor
TRICHOLOMATACEAE	*Tricholomopsis bambusina*	Anti-tumor
TUBERACEAE	*Tuber huidongense**	Anti-oxidant
TUBERACEAE	*Tuber indicum**	Anti-tumor, anti-oxidant, immunomodulation
TUBERACEAE	*Tuber latisporum*†	Anti-oxidant
TUBERACEAE	*Tuber liyuanum*	Anti-oxidant
TUBERACEAE	*Tuber panzhihuanense**	Immunomodulation
TUBERACEAE	*Tuber pseudohimalayense*	Anti-oxidant
TUBERACEAE	*Tuber subglobosum*†	Anti-oxidant
XYLARIACEAE	*Daldinia concentrica*	Treating infantile convulsion, anti-oxidant
XYLARIACEAE	*Engleromyces sinensis**	Anti-inflammation, antibacteria
XYLARIACEAE	*Xylaria carpophila*	Anti-tumor
XYLARIACEAE	*Xylaria nigripes*	Inducing diuresis, invigorating the kidney, improving immunity; anti-oxidant, anti-depression, hepatoprotection, hypoglycemic activity, mitigation of spatial memory impairment.

Note: * shows the species in the vulnerable (VU) category, † shows the species in the near-threatened (NT) category. The evaluation results were based on "Yunnan Province Biological Species Red List (2017)" (Gao and Sun, 2017).

FIGURE 5.1.1 (A–B) *Cordyceps militaris* (L.) Fr. (photos by Zhu-Liang Yang); (C–D) *Cryptoporus sinensis* Sheng H. Wu & M. Zang (photos by Qi Zhao); (E–F) *Engleromyces sinensis* M.A. Whalley et al., (left: photo by Qi Zhao; right: photo by Zhu-Liang Yang).

5.1.4.2 *Cryptoporus Sinensis* Sheng H. Wu & M. Zang 中華隱孔菌

Basidioma annual, stipitate or nearly sessile, without special flavor when fresh, soft woody-textured, becoming harder when dry. Pileus broadly globose, projecting up to 2 cm, 3 cm wide, up to 1 cm thick at base; surface milky white to egg-shell colored, becoming yellowish brown to reddish brown, glabrous; margin obtuse, lighter than the central part, forming a marginal veil covering the entire hymenophoral surface, and remaining a small hole at base. Hymenophoral surface grayish brown when dry, non-glossy; pores round to nearly round, 3~5/mm, edge thick, even; tubes cream, hard woody-textured, up to 3 mm long. Context cream, becoming woody-textured when dry. Basidiospores 8.3~9.5 × 3.8~4.2 μm, cylindrical, colorless, thick-walled, smooth, inamyloid, weakly cyanophilous (Figure 5.1.1 C, D).

Found on the dead trunks, fallen or rotten woods of coniferous trees, especially pines.

Note: *Cryptoporus sinensis* was recorded in *Materia Medica of South Yunnan* by Mao Lan. It was previously misidentified as *C. volvatus*. Wu and Zang (2000) delimited *C. sinensis* by its smaller basidiospores and its habitat with warmer climate.

5.1.4.3 *Engleromyces Sinensis* M.A. Whalley et al. 中華肉球菌

Ascoma 5~8 cm in diameter, globose to subglobose, surface verrucose, orange yellow, yellowish brown to light brown, internal flesh white to light wood colored. Perithecia 700~800 × 450~600 μm, globose, ovoid to vase shaped. Asco 130~150 × 15~20 μm, cylindrical to clavate, with eight ascospores in a single row, funnel to "T" shaped, with an apical apparatus (a region at the ascus tip that forms the spore-shooting mechanism) staining blue in Melzer's reagent. Ascospores 15~20 × 11~15 μm, broadly ellipsoid to ovoid, dark brown to black, without germ pores (Figure 5.1.1 E, F).

Found on the stem branches of bamboos in subalpine forests.

Note: *Engleromyces sinensis* was previously considered as *E. goetzii*, and Whalley et al. (2010) recognized it as a new taxon, which differs from *E. goetzii* by the overall size of the stromata, ascospore shape and dimensions, and the unique funnel or T-shaped apical apparatus.

5.1.4.4 *Ganoderma Leucocontextum* T.H. Li et al. 白肉靈芝

Basidioma annual, stipitate, soft woody-textured when fresh, becoming harder when dry. Pileus projecting up to 10 cm, 20 cm wide, up to 3 cm thick at base; surface paint-glossy, dark red-brown, dark purple-red or black-red-brown when mature, with concentric striations, and weakly radiate wrinkles; margin white to light yellow, becoming yellow to reddish brown. Hymenophoral surface white to cream when fresh, turning light brown to brown, pores round, 4~6/mm; tubes ochraceous to light grayish brown or grayish brown, up to 8 mm long. Context white, becoming cream when dry, soft woody-textured to woody-textured, with a thin brown shell near to the cuticular layer with 2.2 cm in thickness. Stipe cylindrical or slightly tabular, lateral to eccentrical, sometimes sessile, dark red-brown to dark purple-brown, glossy. Basidiospores 9.5~12.5 × 7~9 μm, ellipsoid with a truncate apex, light brown, double-walled, exospore smooth, endospore echinulate, inamyloid, cyanophilous (Figure 5.1.2 A, B).

Found on the rotten woods of *Cyclobalanopsis* trees.

Note: *Ganoderma leucocontextum* is a recently recognized and cultivated species with potential as a high-value medicine (Li et al., 2014, Wang et al., 2015, Chen et al., 2018). Morphologically, *Ganoderma leucocontextum* is similar to *G. lingzhi*, *G. lucidum*, and *G. sichuanense*, but it differs from the latter by its lighter context and smaller basidiospores (Li et al., 2014).

FIGURE 5.1.2 (A–B) *Ganoderma leucocontextum* T.H. Li et al. (photos by Qi Zhao); (C–D) *Ganoderma lingzhi* Sheng H. Wu et al. (photos by Gang Wu); (E) *Ganoderma tropicum* (Jungh.) Bres. (photo by Geng-Shen Wang); (F) *Hericium yumthangense* K. Das et al. (photo by Zhu-Liang Yang); (G) *Hypocrella bambusae* (Berk. & Broome) Sacc. (photo by Jian-Wei Liu); (H) *Naematelia aurantialba* (Bandoni & M. Zang) Millanes & Wedin (photo by Qi Zhao).

5.1.4.5 *Ganoderma Lingzhi* Sheng H. Wu et al. 靈芝

Basidioma annual, soft woody-textured when fresh, becoming harder when dry. Pileus applanate, operculiform, projecting up to 12 cm, 16 cm wide, up to 2.6 cm thick at base; surface colorful, light yellow, light yellowish brown to yellow-brown initially, becoming yellow-brown to reddish brown; edge obtuse or sharp, sometimes slightly incurved. Hymenophoral surface white when young, becoming sulfur yellow when mature, becoming brown to dark brown when touched, becoming light yellow when dry; pores roundish or angular, 5~6/mm, dissepiments thin, entire; sterile margin distinctive, 4 mm wide; tubes brown, woody-textured, darker in color than context, up to 1.7 cm thick. Context woody colored to light brown, double-layered, the upper layer lighter in color than the lower part, soft woody textured, up to 1 cm thick. Stipe lateral to eccentrical, tabular to subcylindrical, orange-yellow to light yellow-brown when young, becoming reddish brown to purplish black, up to 22 cm long, 3.5 cm in diameter. Basidiospores 9~10.7 × 5.8~7 μm, ellipsoid, with a truncate apex, light brown, double-walled, endospores echinulate, inamyloid, cyanophilous (Figure 5.1.2 C, D).

Found on the dying, fallen, and rotten woods of broad-leaved trees.

Note: *Ganoderma lingzhi* is one of the most famous medicinal mushrooms, which has been utilized in Chinese traditional medicine for over 2000 years. It is widely cultivated in China. The epithet of this species, *lingzhi*, just referred to the Chinese name *Lingzhi*. *Ganoderma lingzhi* was previously identified as *G. lucidum*. Cao et al. (2012) found it represented a new taxon and is different from *G. lucidum* by its whitish to sulfur yellow hymenophoral surface, and the blackish brown zones in the context when mature. Another similar species, *G. sichuanense*, was also considered as the Chinese *Lingzhi* (Wang et al., 2012). However, the ITS sequence from its type specimen is different from those of the commonly cultivated *Lingzhi*. Moreover, on morphology, *G. sichuanense* differs in the sessile basidiocarps and smaller basidiospores (7.4~9.2×5~6.6 μm). Therefore, *G. sichuanense* represents an independent species distinct from *G. lingzhi* (Cao et al., 2012).

5.1.4.6 *Ganoderma Tropicum* (Jungh.) Bres. 熱帶靈芝

Basidioma annual, sessile or with a short lateral stipe, woody-textured when dry. Pileus semicircular to circular, projecting up to 12 cm, 16 cm wide, up to 2.5 cm thick at base; surface yellowish brown to purplish brown, covered with a crust, glossy; margin thin, obtuse, lighter than the center. Hymenophoral surface dirty white to grayish brown; pores roundish, 3~4/mm, dissepiments thick, entire; tubes light brown, with multiple but indistinct layers, up to 15 mm long. Sterile margin distinctive. Context yellowish brown, 1 cm thick. Stipe cylindrical, concolorous with the pileus, up to 3 cm long, 1.5 cm in diameter. Basidiospores 8.8~10.5 × 6.1~7.8 μm, ovoid, truncate at the apex, brown, double-walled, exospores colorless, smooth, endospores echinulate, inamyloid, cyanophilous (Figure 5.1.2 E).

Found on the trunk, and on fallen or rotten woods of broad-leaved trees.

5.1.4.7 *Hericium Yumthangense* K. Das et al. 高山猴頭菌

Basidioma subhemispherical, fleshy, sessile, 8~12 cm high, with soft spines hanging from a tough, hidden base that is attached to the trunk of the tree; spines 0.6~1.5 cm long, 0.1~0.2 cm wide, white to ivory-white when young, yellowish to brownish when mature. Context white, unchanging when damage. Basidiospores 5~6.5 × 4~5.5 μm, subglobose to broadly ellipsoid, amyloid, surface nearly smooth (Figure 5.1.2 F).

Found on the rotten wood in subalpine coniferous forests.

Note: *Hericium yumthangense* was previously misidentified as *H. erinaceus* but was delimited as a new taxon by Das et al. (2013).

5.1.4.8 *Hypocrella Bambusae* (Berk. & Broome) Sacc. 竹亞肉座菌

Ascoma nearly hemispherical, 0.5~1.8 cm in diameter, pinkish red to light salmon when fresh, comparatively soft, becoming grayish brown to dark red-brown when dry, comparatively hard; surface covered with irregularly coronoid warts; context pinkish red to dark salmon. Perithecia single-rowed, wall transparent, colorless, 620~700 × 500~600 μm. Asco slender, 330~400 × 16~20 μm; Ascospores worm-like, right-handed-rotated, 260~300 × 7.5~8 μm, colorless to pale yellow, broken into several parts when mature, each part 18~28 μm long; paraphyses slightly longer than asco, inflated at the apex (Figure 5.1.2 G).

Found surrounding the branches of *Sinarundinaria* spp.

5.1.4.9 *Naematelia Aurantialba* (Bandoni & M. Zang) Millanes & Wedin 金耳

Basidioma up to 4.5 cm tall, 6~17 cm long, and 3.5~11 cm wide, consisting of one to many flat but contorted lobes, many-lobed basidiomes mostly with lobes irregularly or completely connate through most of their length, the short free tips flat, contorted, giving the surface a strongly wrinkled appearance; lobes with a white fleshy interior zone of host hyphae, the outer portions gelatinous, orange, drying brownish, horny, the basidiocarp not shrinking greatly upon drying. Basidiospores 11~14 × 9~12.5 μm, subglobose to very broadly ellipsoid (Figure 5.1.2 H).

Parasitic species of *Stereum* spp. on *Quercus* and *Betula* spp.

5.1.4.10 *Neoboletus Brunneissimus* (W.F. Chiu) Gelardi et al. 茶褐新牛肝菌

Pileus 2~8 cm in diameter, surface yellow-brown to pale brown or brown; context yellowish, becoming blue when damage. Hymenophore adnate, surface yellow-brown, pale brownish to brown with red tinge, staining blue when damage; tubes yellow, staining blue when damage. Stipe 4~9 cm long, 0.5~1.3 cm wide, yellow to yellow-brown; surface covered with brown granular-like squamules; staining blue when damage; basal mycelium yellow. Basidiospores 9~14 × 4~6 μm, subfusiform to oblong, surface smooth (Figure 5.1.3 A).

Found on soil in the subtropical broad-leaved forests or mixed forests.

Note: *Neoboletus brunneissimus* was recorded in *Materia Medica of South Yunnan* by Mao Lan, and is one of the most commonly traded boletes in the wild mushroom markets, which is probably the reason for its near-threatened situation.

5.1.4.11 *Omphalia Lapidescens* (Horan.) E. Cohn & J. Schröt. 雷丸

Sclerotia small, irregularly globose, nubby, 0.5~5 cm, surface yellowish brown, brown, blackish brown to black, subglabrous to finely wrinkled, sometimes with a rhizomorph at the depressed place, hard when dry; inner context white to waxy yellow, slightly sticky (Figure 5.1.3 B).

Growing in the soil 10~20 cm deep in the bamboo forests.

Note: there is a popular fungal medicine used in Chinese traditional medicine, which is called *Leiwan* in Chinese. However, its scientific name has been controversial for a long time. Currently, the name *Omphalia lapidescens* is more acceptable (Chen and Bao, 2012).

FIGURE 5.1.3 (A) *Neoboletus brunneissimus* (W.F. Chiu) Gelardi et al. (photo by Yan-Chun Li); (B) *Omphalia lapidescens* (Horan.) E. Cohn & J. Schröt (photo by Gang Wu); (C) *Ophiocordyceps sinensis* (Berk.) G.H. Sung et al. (photo by Qi Zhao); (D–G) *Polyporus umbellatus* (Pers.) Fr. (D and E: photos by Geng-Sheng Wang; F and G: photos by Jian-Wei Liu).

5.1.4.12 *Ophiocordyceps Sinensis* (Berk.) G.H. Sung et al. 冬蟲夏草

Ascoma 5~10 cm, brown to yellowish brown, context white. The sterile upper part 3~6 mm in diameter, subcylindrical, dark brown, surface verrucose, with a sterile acute apex. The sterile stipe slenderer, 3~6 mm in diameter. Perithecia 300~400 × 120~250 μm, ovoid to ellipsoid, buried or sub-buried, occasionally epigeous. Asco 250~450 × 8~12 μm, with a thickened cap which is penetrated by a narrow pore. Ascospores 180~350 × 5~6.5 μm, linear, colorless, with septate, unbroken (Figure 5.1.3 C).

Grown out from the head of pupae of *Lepidoptera*. Distributed in the subalpine areas at an elevation of 3000~5000 m.

Note: *Ophiocordyceps sinensis* is one of the most famous fungal medicines. In 1757, the famous herbalist Yi-Luo Wu recorded it as medicine for the first time in the medical book *New Compilation of Materia Medica*, Volume 1. This medicinal fungus was initially exported abroad in the middle period of the Ming Dynasty. Due to its high medicinal value, *O. sinensis* has been over-harvested and its resources are being depleted year by year.

5.1.4.13 *Polyporus Umbellatus* (Pers.) Fr. 豬苓

Basidioma annual, with a central stipe growing from the underground sclerotium, branched into numerous caps with central stipes, fleshy to leathered. Pileus subglobose to infundibuliform, 4 cm in diameter, up to 0.4 cm thick; surface grayish brown, covered with concolorous minute scales, wrinkled when dry; margin concolorous with the central part, wavy, incurved when dry. Hymenophoral surface white to cream, pore irregular, 2~3/mm, edge thin, even, or slightly split; tubes concolorous with pores, up to 1.5 mm long, decurrent to the upper part of the stipe. Context white to cream, up to 2.5 mm thick. Stipe multiply branched, cream, up to 7 cm long, 2.5 cm in diameter at base. Basidiospores 9~12 × 3.5~4.3 μm, cylindrical to boat-shaped, colorless, thin-walled, smooth, inamyloid, non-cyanophilous (Figure 5.1.3 D, E, F, G).

Found on the trunk of the broad-leaved trees. Can be cultivated.

5.1.4.14 *Sanghuangporus Sanghuang* (Sheng H. Wu et al.) Sheng H. Wu et al.桑黃

Basidioma perennial, sessile, with sour taste when fresh, woody textured. Pileus hippocrepiform, projecting up to 5 cm, 7 cm wide, up to 4 cm thick at base; surface yellowish brown to grayish brown, with distinctive circular grooves and zones; margin obtuse, vivid yellow. Hymenophoral surface yellow to brown; pores round to angular, 8~9/mm, edge thin, even, with distinctive sterile edge, up to 3 mm wide; tubes brown, up to 5 mm long. Context yellow, with circular zones, up to 3.5 cm thick. Basidiospores 3.6~4.6 × 3~3.5 μm, broadly ellipsoid, yellow, thick-walled, smooth, inamyloid, non-cyanophilous (Figure 5.1.4 A).

Found on the trees of *Morus*. Can be cultivated.

5.1.4.15 *Shiraia Bambusicola* Henn. 竹黄

Ascoma 3~5 cm long, 1~3 cm wide, verrucose to kidney-shaped, surface pinkish red, salmon to light salmon, turning bluish green in potassium hydroxide. Context red. Perithecia subglobose, buried in the ascoma. Asco 350~400 × 20~30 μm, enclosed with 6~8 ascospores. Ascospores 60~80 × 15~25 μm, subfusoid, with septa, colorless or light yellow. Paraphysis linear, erected, 1~2 μm in diameter (Figure 5.1.4 B).

FIGURE 5.1.4 (A) *Sanghuangporus sanghuang* (Sheng H. Wu et al.) Sheng H. Wu et al. (photo by Qi Zhao); (B) *Shiraia bambusicola* Henn. (photo by Qi Zhao); (C) *Thelephora ganbajun* M. Zang (photo by Zhu-Liang Yang); (D) *Tremella fuciformis* Berk. (photo by Zhu-Liang Yang); (E–F) *Wolfiporia cocos* (Schwein.) Ryvarden & Gilb. (photos by Gang Wu); (G–H) *Xylaria nigripes* (Klotzsch) Cooke (photos by Wen-Fei Lin).

Found on the branches of bamboos.

5.1.4.16 *Thelephora Ganbajun* M. Zang 乾巴糙孢革菌(乾巴菌)

Basidioma 2~5 cm tall, 4~10 cm in diameter, leathery, multi-branched from base; branches flat, spathulate to petal-like; upper surface (abhymenial surface) blackish to dark grey, with irregularly striate; lower surface (hymenial surface) blackish to greyish black, smooth to slightly rugose, sometimes warty; context grey to greyish, unchanging when injured. Basidiospores 5~7.5 × 4.5~6.5 μm, ellipsoid to ovoid, angular, warty, yellowish brown (Figure 5.1.4 C).

On soil in subtropical coniferous or mixed forests.

Note: *Thelephora ganbajun* is a favorite edible fungus in Yunnan, especially in the capital city, Kunming. Probably due to the over-harvest, it is becoming a vulnerable and endangered species.

5.1.4.17 *Tremella Fuciformis* Berk. 銀耳

Basidioma composed of many flat branches; branched foliaceous to petaloid, white or cream, becoming cream to yellowish when mature or dry. Basidia longitudinally cruciate-septate, 10~13 × 9~10 μm, ovoid to subglobose; basidiospores 6~8.5 × 5~6 μm, broadly ellipsoid to ellipsoid, smooth. Hyphae with clamp connections (Figure 5.1.4 D).

Parasitic on species of Xylariales on rotten wood in tropical to subtropical forests. Can be cultivated.

5.1.4.18 *Wolfiporia Cocos* (Schwein.) Ryvarden & Gilb. 茯苓

Basidioma annual, appressed, leathered, up to 10 cm long, 8 cm wide, up to 2 mm thick at the center. Hymenophoral surface white when fresh, becoming cream when dry; pores round to roundish or angular, 0.5~2/mm, edge thin, split, with distinctive sterile edge. Tubes concolorous with pores or lighter, up to 1.5 mm long. Context cream, up to 0.5 mm thick. Basidiospores 6.5~8.1 × 3.5~4.3 μm, cylindrical, thin-walled, smooth, inamyloid, non-cyanophilous. Sclerotia globose to subglobose, up to 30 cm in diameter; surface dirty brownish; context white (Figure 5.1.4 E–F).

The medicinal part is the sclerotia occurring under the rotten wood of pines. Can be cultivated.

5.1.4.19 *Xylaria Nigripes* (Klotzsch) Cooke 烏靈參

Ascoma 6~12 cm tall for the part above the ground, 4~8 mm in diameter, usually non-branched, clavate, obtuse on the top, brown to blackish, leathered when fresh, becoming hard wood-textured to wood-textured. Fertile part coarsely. Sterile part subglabrous to slightly wrinkled. Sclerotia in the empty termitaria ovoid, globose to subglobose, 2~5 cm in diameter; surface black to dark brown; context white to whitish. Ascospores 4~5 × 2~3 μm, nearly ellipsoid to subglobose, black, thick-walled, inamyloid (Figure 5.1.3 G–H).

On soil in broad-leaved forests, usually associated with the underground empty termitaria without termites. Parts of sclerotia are used as medicine.

5.1.5 The Sustainable Utilization of Medicinal Mushrooms in Yunnan

In the utilization of the medicinal mushrooms in Yunnan, like edible mushrooms (Yu et al., 2002), some common problems have occurred, such as (1) insufficient development efforts and basic researches, (2) low technology content of products, and (3) uncoordinated development and protection resulting in serious resource damage. Therefore, we need to adopt some measures to address the above problems and can refer to those recommendations for the sustainable utilization of edible mushrooms (Yang, 2002; Yu et al., 2002, Zang et al., 2005).

5.1.5.1 Comprehensively Develop and Utilize the Resources of Medicinal Mushrooms Based on Resource Advantages

The wild mushroom production season in Yunnan is distinctive, mostly from June to September every year. In the peak season, due to the lack of "fresh-keeping" processing equipment, many wild mushrooms including medicinal ones cannot be processed in a timely manner, resulting in a backlog and waste of resources. At present, most wild medicinal mushrooms in Yunnan are sold on the original product market. Since deep processing and high value-added products are rare, it is imperative to give consideration to the advantages of resources, adapt to the industrial deep processing, and increase the added value of products of wild medicinal mushrooms. To achieve this goal, we should first strengthen the research on post-harvest physiology of medicinal mushrooms to provide a theoretical basis for their preservation, storage, and further processing. Meanwhile, it is necessary to develop the deep fermentation of medicinal mushrooms and determine the best process conditions to lay the foundation for industrial production. In addition, we can develop mushroom foods with both edible and medicinal properties that meet the trend of food development and have market potential.

5.1.5.2 Strengthen the Basic Research and Develop New Techniques of Artificial Cultivation of Medicinal Mushrooms

We should strengthen the scientific investigation of medicinal mushrooms to uncover their species diversity, geographic distribution, resources, and ecological environments. Moreover, we can pay more attention to basic biological research on the classification, physiology, ecology, chemistry, and molecular biology of valuable medicinal mushrooms to provide a basis for the development of resources. While giving consideration to the advantages of medicinal mushroom resources in Yunnan, we should vigorously develop the artificial cultivation of medicinal mushrooms with existing and new techniques, and strengthen the research on intensive cultivation to improve the quality of medicinal mushrooms, and the degree of mechanization.

5.1.5.3 Maintain Ecological Balance and Sustainable Development

Resource protection is the foundation of industrial development. Therefore, based on the traditional use of fungal resources by various nationalities in Yunnan, we should increase the message of resource protection, improve protection awareness of mushroom farmers, and implement the closing off of mountain areas and afforestation. On the one hand, we should strengthen the *in situ* protection of wild medicinal mushrooms, establish protected areas, promote regular and quantitative continuous utilization, and strengthen the research on artificial promotion of wild medicinal fungi. Conversely, we must avoid quick successes and instant benefits, and resolutely end destructive predatory collection to reduce the pressure on wild resources. At the same time, the forestry department of the government should adopt a combination of thinning and nursery afforestation (shaving-head-cutting is strictly prohibited), and this approach can lead to gradually implementing seedling mycorrhization or inoculation of mycelium in afforestation to maintain the amount of underground mycelium, which is conducive to the survival and reproduction of underground mycelium.

REFERENCES

Boufford D, Dijk P, 2000. South-Central China. In: Mittermeier R, Myers N, Mittermeier C (eds) *Hotspots: Earth's Biologically Richest and Most Endangered Terrestrial Ecoregions*. Cemex, Mexico, pp. 338–351.

Cai J, Wang YB, Sun D, et al., 2017. Bioactive component profiles in stroma and sclerotium of *Ophiocordyceps highlandensis*. *Acta Edulis Fungi*, 24: 63–66.

Cao Y, Wu SH, Dai YC, 2012. Species clarification of the prize medicinal *Ganoderma* mushroom "Lingzhi". *Fung Divers*, 56: 49–62.

Chang ST, 2006. The need for scientific validation of culinary-medicinal mushroom products. *Int J Med Mushrooms*, 8: 187–195.

Chen H, Zhang J, Ren J, et al., 2018. Triterpenes and meroterpenes with neuroprotective effects from *Ganoderma leucocontextum*. *Chem Biodivers*, 15: e1700567.

Chen HZ, Bao HY, 2012. Bentsaological research of fungal medicine *Omphalia lapidescens*. *J Fung Res*, 10: 57–62.

Chen ZH, Yang ZL, Bau T, et al., 2016. *Poisonous Mushrooms: Recognition and Poisoning Treatment*. Science Press, Beijing.

Dai YC, Yang ZL, 2008. A revised checklist of medicinal fungi in China. *Mycosystema*, 27: 801–824.

Dai YC, Yang ZL, Cui BK, et al., 2009. Species diversity and utilization of medicinal mushrooms and fungi in China. *Int J Med Mushrooms*, 11: 287–302.

Das K, Stalpers JA, Stielow JB, 2013. Two new species of hydnoid-fungi from India. *IMA Fungus*, 4: 359–369.

Duan X, Tao Y, Duan CC, 2011. A fine mesh climate division and the selection of representative climate stations in Yunnan Province. *Trans Atmos Sci*, 34: 336–342.

Feng B, Xu J, Wu G, et al., 2012. DNA sequence analyses reveal abundant diversity, endemism and evidence for Asian origin of the porcini mushrooms. *PloS One*, 7: e37567.

Feng B, Yang ZL, 2018. Studies on diversity of higher fungi in Yunnan, southwestern China: A review. *Plant Divers*, 40: 165–171.

Gao J, Hu L, Liu J, 2001. A novel sterol from Chinese truffles *Tuber indicum*. *Steroids*, 66: 771–775.

Gao ZW, Sun H, 2016. *List of Biological Species Yunnan*. Yunnan Science and Technology Press, Kunming.

Gao ZW, Sun H, 2017. *Yunnan Province Biological Species Red List (2017). Macrofungi*. Yunnan Science and Technology Press, Kunming.

Han LH, Feng B, Wu G, et al., 2018. African origin and global distribution patterns: Evidence inferred from phylogenetic and biogeographical analyses of ectomycorrhizal fungal genus *Strobilomyces*. *J Biogeogr*, 45: 201–212.

Jiang HQ, 1980a. Distributional features and zonal regularity of vegetation in Yunnan. *Acta Bot Yunnan*, 2: 22–32.

Jiang HQ, 1980b. Distributional features and zonal regularity of vegetation in Yunnan (continued). *Acta Bot Yunnan*, 2: 142–151.

Li TH, Hu HP, Deng WQ, et al., 2014. *Ganoderma leucocontextum*, a new member of the *G. lucidum* complex from southwestern China. *Mycoscience*, 56: 81–85.

Li T, Song B, 2002. Species and distributions of Chinese edible boletes. *Shi Yong Jun Xue Bao*, 9: 22–30.

Liu B, 1978. *The Chinese Medical Fungi*. The second edition. Shanxi People's Press, Taiyuan.

Liu B, 1984. *The Chinese Medical Fungi*. Shanxi People's Press, Taiyuan.

Liu DZ, Wang F, Yang LM, et al., 2007. A new cadinane sesquiterpene with significant Anti-HIV-1 activity from the cultures of the Basidiomycete *Tyromyces chioneus*. *J Antibiot Res*, 60: 332–334.

Liu YP, Dai Q, Wang WX, et al., 2020. Psathyrins: Antibacterial diterpenoids from *Psathyrella candolleana*. *J Nat Prod*, 83: 1725–1729.

Liu ZN, Zhen SF, 1996a. The current knowledge and resources of medicinal fungi in China. *Edib Fung China*, 15: 20–22.

Liu ZN, Zhen SF, 1996b. The current knowledge and resources of medicinal fungi in China (Continued). *Edib Fung China*, 15: 29–31.

Liu ZN, Zhen SF, 1997. The current knowledge and resources of medicinal fungi in China (Continued). *Edib Fung China*, 16: 30–31.

Liu ZN, Zhen SF, 1998. The current knowledge and resources of medicinal fungi in China (Continued). *Edib Fung China*, 17: 22–24.

Liu ZN, Zhen SF, 1999. The current knowledge and resources of medicinal fungi in China (Continued). *Edib Fung China*, 18: 17–19.

Liu ZN, Zhen SF, 2001. The current knowledge and resources of medicinal fungi in China (Continued). *Edib Fung China*, 20: 22–24.

Mao XL, 2006. Poisonous mushrooms and their toxins in China. *Mycosystema*, 25: 345–363.

Myers N, Mittermeier RA, Mittermeier CG, et al., 2000. Biodiversity hotspots for conservation priorities. *Nature*, 403: 853–858.

People's Government of Yunnan Province 2020. Nature overview. Available at http://www.yn.gov.cn/yngk/gk/201904/t20190403_96255.html.

Sun H, Zhang J, Deng T, et al., 2017. Origins and evolution of plant diversity in the Hengduan Mountains, China. *Plant Divers*, 39: 161–166.

Tang JG, Wang YH, Wang RR, et al., 2008. Synthesis of analogues of Flazin, in particular, Flazinamide, as promising anti-HIV agents. *Chem Biodivers*, 5: 447–460.

Wang K, Bao L, Xiong W, et al., 2015. Lanostane triterpenes from the Tibetan medicinal mushroom *Ganoderma leucocontextum* and their Inhibitory effects on HMG-CoA reductase and α-glucosidase. *J Nat Prod*, 78: 1977–1989.

Wang XC, Xi RJ, Li Y, et al., 2012. The species identity of the widely cultivated *Ganoderma, 'G. lucidum'* (Ling-zhi), in China. *PloS One*, 7: e40857–e40857.
Wei K, Wang GQ, Bai X, et al., 2015. Structure-based optimization and biological evaluation of pancreatic lipase inhibitors as novel potential antiobesity agents. *Nat Prod Bioprospect*, 5: 129–157.
Whalley MA, Khalil AMA, Wei TZ, et al., 2010. A new species of *Engleromyces* from China, a second species in the genus. *Mycotaxon*, 112: 317–323.
Wu F, Zhou LW, Yang ZL, et al., 2019. Resource diversity of Chinese macrofungi: Edible, medicinal and poisonous species. *Fung Divers*, 98: 1–76.
Wu G, Li YC, Zhu XT, et al., 2016. One hundred noteworthy boletes from China. *Fung Divers*, 81: 25–188.
Wu SH, Zang M, 2000. *Cryptoporus sinensis* sp. nov., a new polypore found in China. *Mycotaxon*, 74: 415–422.
Wu XL, Mao XL, Bau T, et al., 2013a. *Medicinal Fungi of China*. Science Press, Beijing.
Wu XL, Mao XL, Song B, et al., 2012. Study on species diversity and chemical composition of medicinal Ascomycota in China. *Guizhou Sci*, 30: 1–20.
Wu XL, Mao XL, Song B, et al., 2013b. Study on species diversity and chemical composition of medicinal Ascomycota in China II. *Guizhou Sci*, 31: 1–22.
Xing Y, Ree RH, 2017. Uplift-driven diversification in the Hengduan Mountains, a temperate biodiversity hotspot. *PNAS*, 114: E3444.
Yang WM, Liu JK, Hu L, et al., 2004. Antioxidant properties of natural p-Terphenyl derivatives from the mushroom *Thelephora ganbajun. Z Naturforsch C J Biosci*, 59: 359–362.
Yang ZL, 2002. On wild mushroom resources and their utilization in Yunnan Province, Southwest China. *J Nat Resourc*, 17: 463–469.
Yang ZL, 2005. Diversity and biogeography of higher fungi in China. In: Xu JP (ed) *Evolutionary Genetics of Fungi*. Horizon Bioscience, Norfolk, pp 35–62.
Yang ZL, Ge ZW, Li YC, et al., 2016. Macrofungi. In: Gao Z-W, Sun H (eds) *List of Biological Species Yunnan*. Yunnan Science and Technology Press, Kunming, pp. 3–63.
Ye M, Liu JK, Lu Z, et al., 2005. Grifolin, a potential antitumor natural product from the mushroom *Albatrellus confluens*, inhibits tumor cell growth by inducing apoptosis in vitro. *FEBS Lett*, 579: 3437–3443.
Ying JZ, Mao XL, Ma QM, et al., 1987. *Icons of Medicinal Fungi from China*. Science Press, Beijing.
Ying JZ, Zang M, Zong YC, et al., 1994. *Economic Macrofungi from Southwestern China*. Science Press, Beijing.
Yu FQ, Wang XH, Liu PG, 2002. Prospects of exploitation and utilization on edible fungi resource in Yunnan. *Chin Wild Plant Resourc*, 4: 21–25.
Zang M, Li XJ, Zhou KY, 2005. Biodiversity of edible fungi in Yunnan and its resource protection. *Edib Fung China*, 24: 3–6.
Zhang X, Cao F, Sun Z, et al., 2013. Sulfation of *Agrocybe chaxingu* polysaccharides can enhance the immune response in broiler chicks. *J Appl Poult Res*, 22: 778–791.
Zhou S, Zhou Y, Yu J, et al., 2020. *Ophiocordyceps lanpingensis* polysaccharides attenuate pulmonary fibrosis in mice. *Biomed Pharmacother*, 126: 110058.
Zhu JJ, Yang HX, Li ZH, et al., 2018. Anti-inflammatory lupane triterpenoids from *Menyanthes trifoliata. J Asian Nat Prod Res*, 21: 597–602.

5.2

Medicinal Mushrooms and Fungi from Yunnan Province, Part 2: Chemistry and Bioactivity

Ji-Kai Liu
School of Pharmaceutical Sciences, South-Central University for Nationalities, No. 182 Minzu Road, Wuhan 430074, People's Republic of China

CONTENTS

5.2.1 Introduction

Fungi, as an independent kingdom, are critical for decomposing dead organic matters and recycling nutrients. They also provide many direct benefits to humans, from edible and medicinal mushrooms to medicinal drugs and antibiotics. Fungi that produce spore-bearing structures visible to the naked eye are often referred to as macrofungi (such as mushrooms). Those not producing spore-bearing structures are often referred to as microfungi: they are too small to be seen without a microscope. Fungi are immensely diverse. The latest best estimate suggests that the total number of fungal species on Earth is between 2.2 and 3.8 million, but only 144,000 species have been named and classified around the world to date (Royal Botanic Gardens

Kew, 2018). Yunnan Province, in the southwest of China, is one of the areas with the richest and most diverse bioresources including medicinal mushrooms and fungi in the world based on its unique geo-environment, diverse geomorphology, and three-dimensional differentiation of climate. Fungi as a bioresource produce a large and diverse variety of secondary metabolites (Liu, 2002). We have been interested in biologically active substances present in untapped and diverse sources of medicinal mushrooms and fungi in Yunnan. In recent years, several hundred new natural products and bioactive compounds were found in selected fungi based on our knowledge of the collection of fruiting bodies, strain preservation, fermentation, biological screening, and chemical investigation (Chen and Liu, 2017). The isolation, structural elucidation, and biological activity of pigments, the novel terpenoids, nitrogen-containing compounds, and other compounds will be discussed in this chapter.

5.2.2 Pancreatic Lipase Inhibitors: Vibralactone and Their Derivatives from *Boreostereum vibrans* (syn. *Stereum vibrans*)

Vibralactone (**1**), a well-studied molecule featuring a β-lactone group, displaying pancreatic lipase inhibition, was isolated from the culture broth of *Boreostereum vibrans* (syn. *Stereum vibrans*) (Figure 5.2.1) (Liu et al., 2006). The biosynthetic pathway of vibralactone has been elucidated. The prenylated 4-hydroxybenzoid acid (**2**) was reduced to a prenylated 4-(hydroxymethyl)phenol (**4**), then the cleavage of the benzene ring led to the production of the key intermediate 1,5-*seco*-vibralactone (**3**), which further underwent a 1,5-C-C bond formation to yield **1** (Figure 5.2.2A) (Zhao et al., 2013). Vibralactone derivatives have been used as a chemical probe to study the structure and activity of ClpP1P2 (Zeiler et al., 2011).

An in-depth study mainly by large-scale fermentation of *B. vibrans* resulted in the isolation of vibralactone derivatives, vibralactones B-Q (**6**) (Jiang et al., 2008, Jiang et al., 2010, Wang et al., 2012, Wang et al., 2014, Chen et al., 2014), 1,5-*seco*-vibralactone (**3**) (Jiang et al., 2008), and 10-lactyl vibralactone G (**5**) (Jiang et al., 2010). Recently, a series of oximes and polyoxime esters with a vibralactone backbone, namely vibralactoximes A-P (**7–10**), show pancreatic lipase inhibitory activity and are more potent than vibralactone (Figure 5.2.1). Moreover, most of these compounds also exhibit significant anti-tumor activities against five human cancer cell lines (HL-60, SMMC-7721, A-549, MCF-7, and SW480) (Chen et al., 2016).

A thorough analysis of the secondary metabolome of *B. vibrans* led to unravelling the divergent vibralactone biosynthetic pathways. Yang et al. proposed that prenylated 4-(hydroxymethyl)phenol (**3**) was the key intermediate for the generation of 20 analogues with different scaffolds, and this hypothesis was further confirmed by feeding experiments with 3-allyl-4-hydroxybenzylalcohol to obtain the corresponding derivatives with allyl moieties rather than isoprenyl moieties (Yang et al., 2016). In general, the secondary metabolome of *B. vibrans* was mainly involved in producing five classes with the skeletons A–E, as depicted in Figure 5.2.2B. The isoprenyl moiety of those skeletons can be considered as "conservative," while the benzene ring was presented in various forms. In particular, compound **3** was positioned at a "crossroads" to enable various biosynthetic procedures. One pathway led to oxygenation and splitting

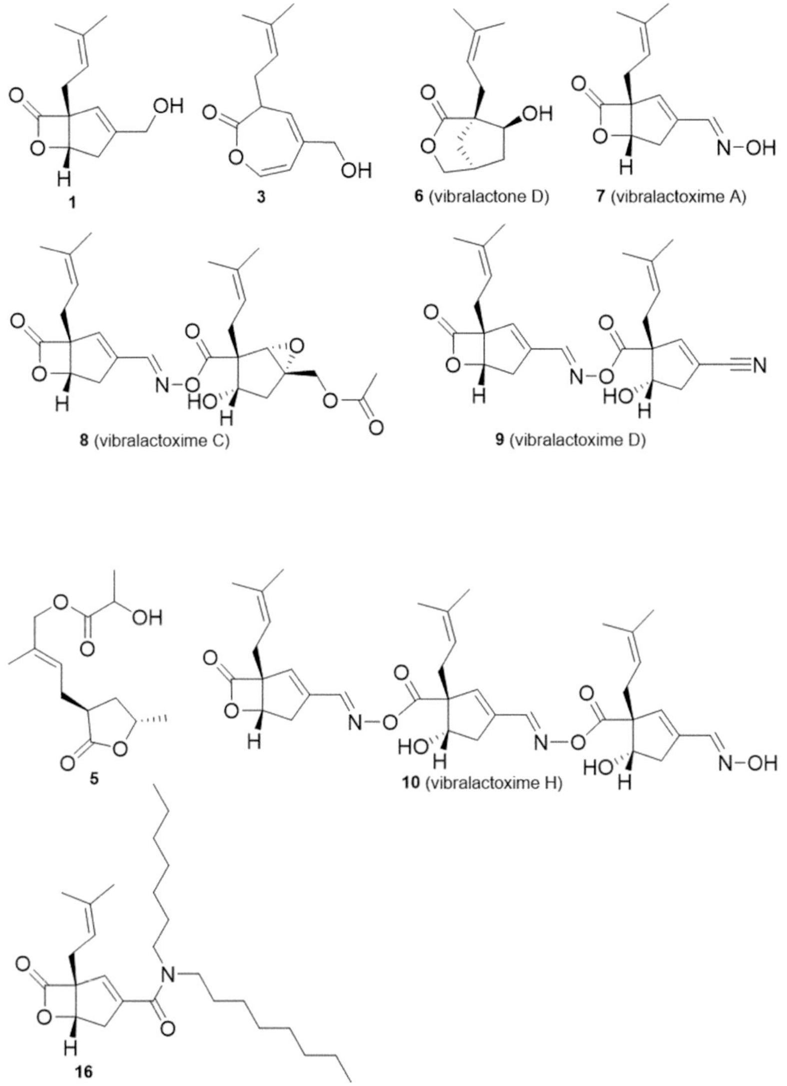

FIGURE 5.2.1 Prenylated benzene derivatives.

of the benzene ring to give vibralactone J (**11**) with the scaffold type A, while a carbon–carbon formation reaction yielded vibralactone and its derivatives, representing the scaffold type B. Thirdly, an oxygenation and reduction on the benzene ring of **3** followed by a key ring contraction reaction led to vibralactone I (**12**), representing scaffold type C. On the other hand, oxygenation of the hydroxy alcohol led to 3-prenyl-3-hydroxybenzoaldehyde (**13**): this intermediate underwent a C_2 extension with pyruvate to give vibranether, namely the skeleton D. Moreover, **13** was further oxygenated to 3-prenyl-3-hydroxybenzoic acid (**14**), and the decarboxylation of **14**

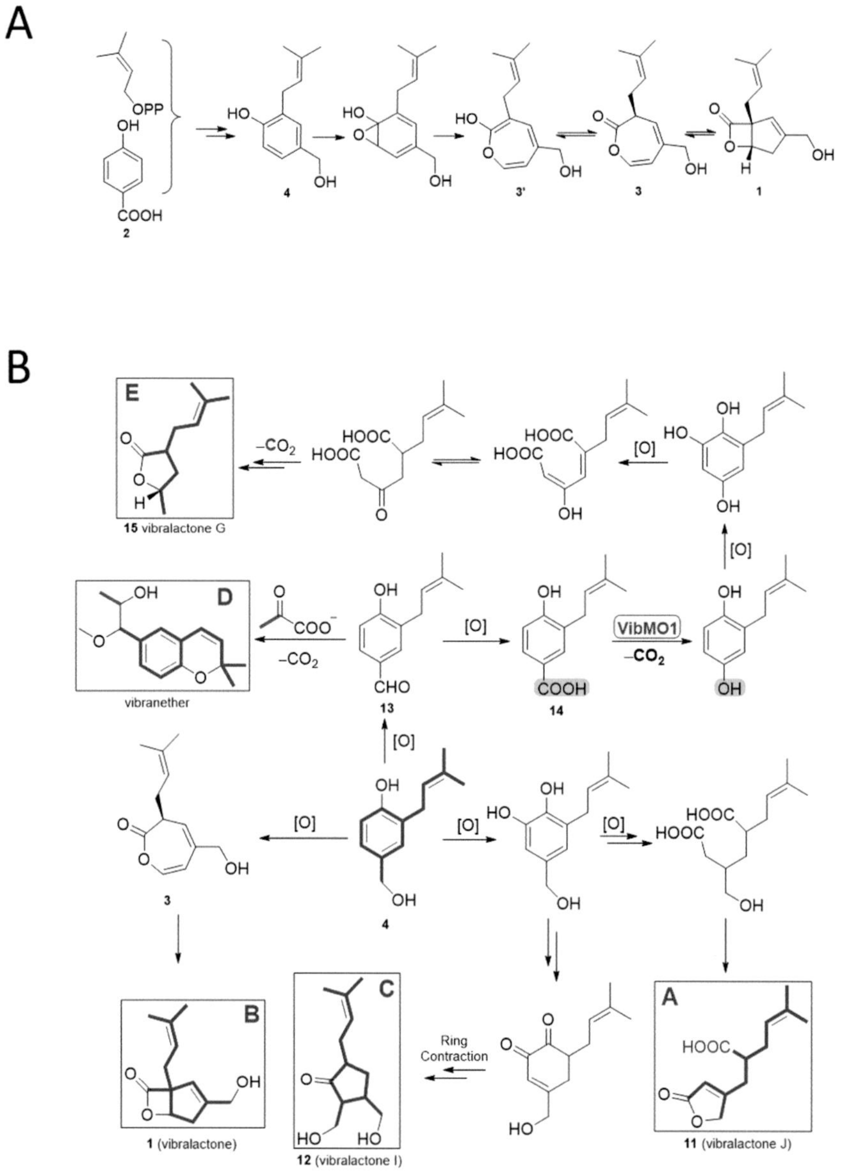

FIGURE 5.2.2 (A) The biosynthetic pathway of vibralactone (**1**); (B) Proposed divergent biosynthetic pathways for five classes of secondary metabolites from *B. vibrans*.

followed by cascade oxygenation/decarboxylation reactions yielded vibralactone G (**15**), representing skeleton E.

It was noteworthy that Yang et al. also identified a FAD-dependent monooxygenase (VibMO1) that converted prenyl-4-hydroxybenzoate into prenylhydroquinone. Heterologous expression of VibMO1 confirmed this function. This finding provided pioneering information for the determination of enzymes essential for similar conversion steps in other organisms.

Very recently, we reported the discovery of the cyclase VibC which belongs to the α/β-hydrolase superfamily and is involved in the vibralactone biosynthesis. Biochemical and crystal studies suggested that VibC may catalyze an aldol or an electrocyclic reaction initiated by the Ser-His-Asp catalytic triad. For the aldol and pericyclic chemistry in living cells, VibC is a unique hydrolase performing the carbocycle formation of an oxepinone to a fused bicyclic β-lactone (Figure 5.2.3). This reaction represents a naturally occurring new enzyme reaction in both aldol and hydrolase (bio)chemistry that will guide future exploitation of these enzymes in synthetic biology for chemical diversity expansion of natural products (Feng et al., 2020).

A structure-based lead optimization of vibralactone resulted in three series of 104 analogs, among which compound **C1** (**16**) exhibited the most potent inhibition of pancreatic lipase, with an IC_{50} value of 14 nM. This activity is more than 3000-fold higher than that of vibralactone. The effect of **16** on obesity was investigated using high-fat diet (HFD)-induced C57BL/6 J obese mice. Treatment with **16** at a dose of 100 mg/kg significantly decreased HFD-induced obesity, primarily through the improvement of metabolic parameters, such as triglyceride levels (Wei et al., 2015).

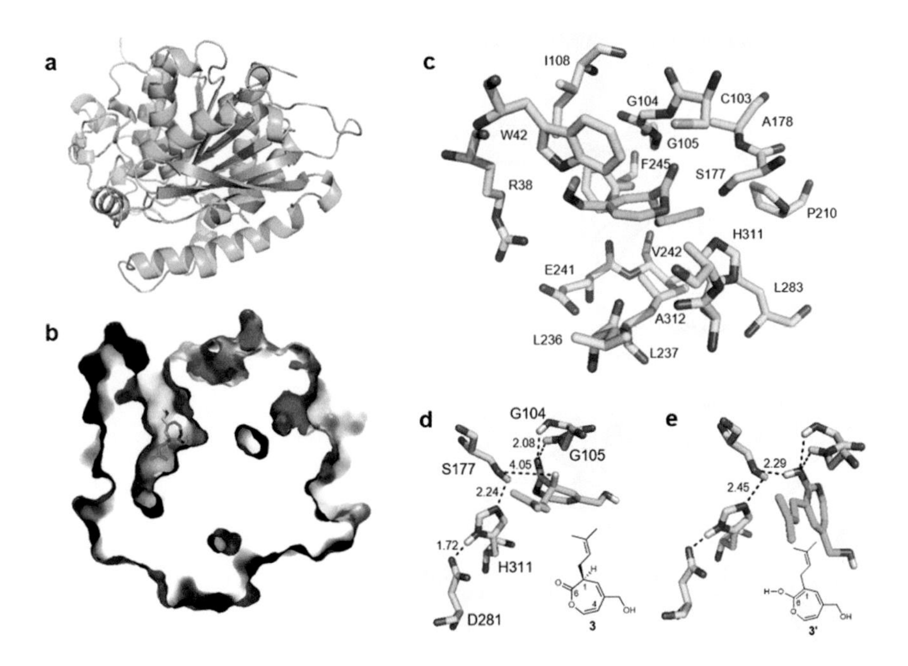

FIGURE 5.2.3 Crystal structure and catalytic sites of VibC. (a) Crystal structure of VibC (PDB code: 6KD0); (b–c) Pocket-shaped regions on VibC surface (b) and residues close to the substrate (c); (d–e) Molecular docking for distances (Å) of natural substrate 3 (in cyan, d) and its enol isomer 3′ (in green, e) to the catalytic triad residues (Ser177-His311-Asp281) and oxyanion hole (Gly104-Gly105).

5.2.3 Toxic Non-Protein Amino Acids and *Tricholoma* Triterpenoids

The mushroom *Trogia venenata* Zhu L. Yang was a previously undescribed species from Yunnan Province, southwest China (Figure 5.2.4A). Epidemiological studies implicated that intake of this mushroom was responsible for the sudden unexpected death (SUD) of more than 300 people over the past 30 years. We have isolated and characterized three toxic non-protein amino acids from the fruiting bodies of this mushroom, namely 2R-amino-4S-hydroxy-5-hexynoic acid (**17**), 2R-amino-5-hexynoic

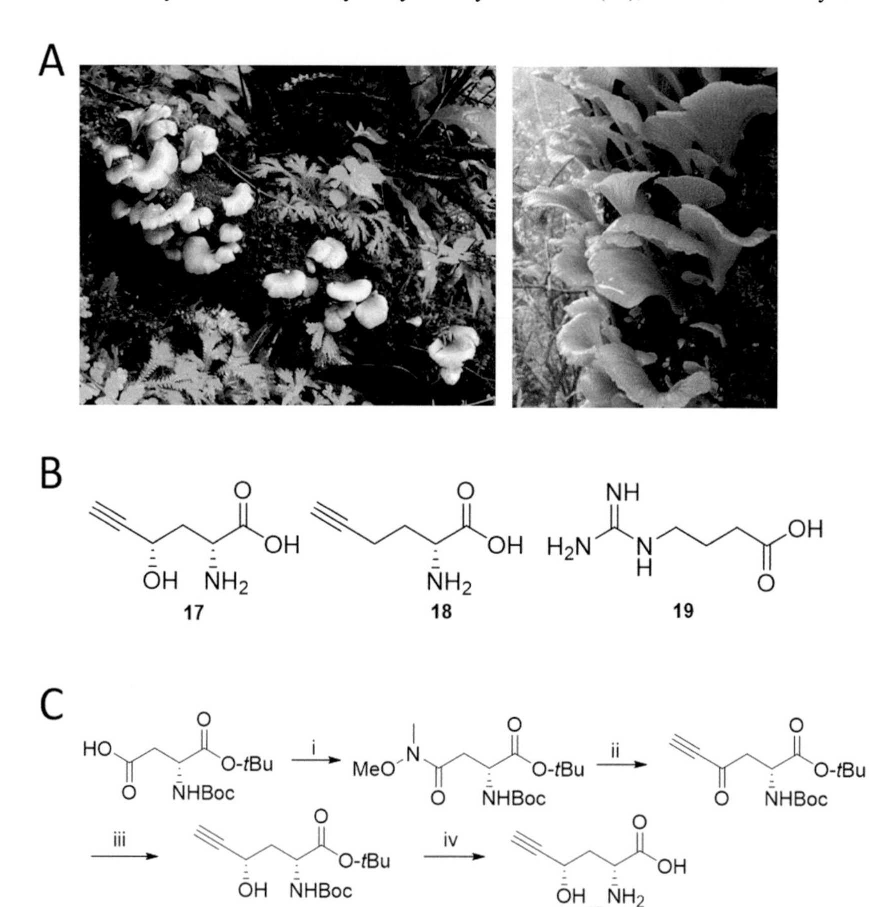

FIGURE 5.2.4 (A) The toxic mushroom *Trogia venenata* Zhu L. Yang (photo by J.K. Liu); (B) Structures of non-protein amino acid toxins (**17, 18** and **19**) from mushroom *T. venenata*; (C) Total synthesis of compound **17**. ***Reagents and conditions***: (i) Et_3N, $CH_3ONHCH_3{\cdot}HCl$, $BOP{\cdot}PF_6$, CH_2Cl_2; (ii) HC≡CMgBr (5 equiv), Et_2O, –78°C, 78% yield; (iii) (*S*)-B-methyl Corey-Bakshi-Shibata (CBS) catalyst (2 equiv), BH_3-SMe_2 (2 equiv), toluene, 61% yield; (iv) CF_3CO_2H, 99% yield. BOP = benzotriazol-1-yl-oxy-tris-(dimethylamino)-phosphonium.

acid (**18**), and γ-guanidinobutyric acid (**19**), guided by oral toxicity tests in mice (Figure 5.2.4B). The absolute configuration of **17** was determined as 2R, 4S by both matrix-mode and optical rotation computations based on DFT methods, further confirmed by total synthesis (Figure 5.2.4C). Both **17** and **18** were lethal for ICR mice with LD_{50} values of 71 and 84 mg/kg, respectively. The total content of **17** and **18** in fruiting bodies was 0.2%, which equated to a LD_{50} value for humans (60 kg) ingestion of about 400 g of dried fruiting bodies. It was noteworthy that **17** was also detected from the blood in the heart of a victim of Yunnan SUD, which provided solid evidence for the long-time SUD in Yunnan Province (Zhou et al., 2012).

Repeated ingestion of the wild mushroom *Tricholoma equestre* caused rhabdomyolysis in France (Bedry et al., 2001). The mushroom *T. terreum* was a co-occurred species of *T. equestre* in southwestern France (Figure 5.2.5A). Both the crude extracts

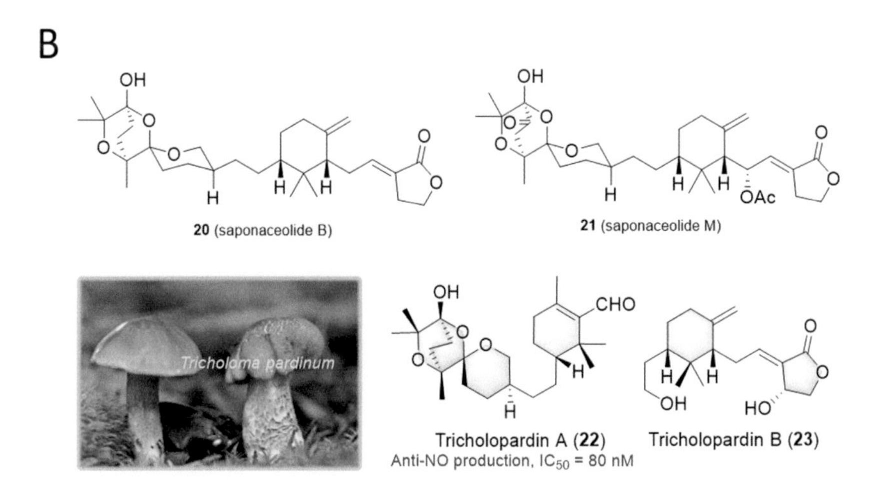

FIGURE 5.2.5 (A) The mushroom *Tricholoma terreum* (photo by T. Feng); (B) The mushroom *Tricholoma pardinum* (photo by T. Feng) and structures of saponaceolides A (**20**), M (**21**), and tricholopardins A (**22**) and B (**23**).

($CHCl_3$/MeOH, 1:1) of these two kinds of mushroom are toxic to mice, while only the non-polar fraction (ethyl acetate layer) was toxic when the *T. terreum* extract was partitioned between water and ethyl acetate. Further chemical investigation on the secondary metabolites of *T. terreum* led to the isolation of 15 new triterpenoids, namely terreolides A–F and saponaceolides H–P. Acute toxicity and the serum creatine kinase (CK) assays in mice of these compounds revealed that saponaceolide B (**20**) and M (**21**) were the toxic principles with the LD_{50} values of 88.3 and 63.7 mg/kg, respectively, and caused a 1.52- to 1.65-fold increase in serum CK levels relative to mice that received water or 1%-Tween-80 (Figure 5.2.5B). This research disclosed a hitherto unknown poisonous European mushroom *T. terreum* (Yin et al., 2014). Four new meroterpenoids, terreumols A–D, with a rare ten-membered ring system, were isolated from the fruiting bodies of *Tricholoma terreum*. Their structures with absolute stereochemistry were determined by comprehensive spectroscopic methods, as well as single-crystal X-ray diffractions. They were evaluated for their cytotoxicities against five human cancer cell lines. All of them exhibited inhibitory effects, with IC_{50} values comparable to those of cisplatin (Yin et al., 2013).

Two new terpenoids, tricholopardins A (**22**) and B (**23**), were isolated from the fruiting bodies of the basidiomycetes *Tricholoma pardinum*. Their structures were elucidated by spectroscopic methods, as well as electronic circular dichroism and optical rotatory dispersion calculations. Tricholopardin A (**22**) potently inhibited nitric oxide production in lipopolysaccharide-induced RAW264.7 macrophages with an IC_{50} of 0.08 μM. Its anti-inflammatory effects on three inflammatory mediators were also evaluated (Feng et al., 2015). From the same genus, matsutakone, a novel sterol with an unprecedented polycyclic ring system, together with a new norsteroid matsutoic acid were isolated from the fruiting bodies of *Tricholoma matsutake*. Their structures and absolute configurations were assigned by extensive spectroscopic analyses and computational methods. Bioassay results showed that matsutakone exhibited inhibitory activities against acetylcholinesterase (IC_{50} = 20.9 μM) (Zhao et al., 2017).

5.2.4 Antroalbol H Enhances Cellular Glucose Uptake in Studies Initiated from the Mushroom *Antrodiella albocinnamomea*

Hypoglycemic drugs such as metformin increase the uptake and utilization of glucose by peripheral tissue to achieve hypoglycemic activity, and the AMP-activated protein kinase (AMPK) signaling pathway is an important mechanism of this pharmacological action. Liver kinase B1 (LKB1) acts as a kinase upstream of AMPK and plays an important regulatory role in glucose metabolism. In recent years, as a tumor suppressor, the anti-tumor activity of LKB1 has been widely investigated, but its hypoglycemic activity has seldom been reported. In addition, specific activators of LKB1 derived from the natural products are still limited. We confirmed that antroalbol H (**24**) isolated from fermentation of the medicinal mushroom *Antrodiella albocinnamomea* (Figure 5.2.6) increased the glucose uptake in L6 myotubes and 3T3-L1 adipocytes. By means of AMPK signaling pathway-specific activators and inhibitors, we demonstrated that the effect of antroalbol H on cellular glucose uptake is closely related to AMPKα phosphorylation. Moreover, antroalbol H induced the phosphorylation of LKB1 specifically at the T189 residue, which was further supported by in-point mutation experiments. Moreover, antroalbol H altered the subcellular localization of LKB1

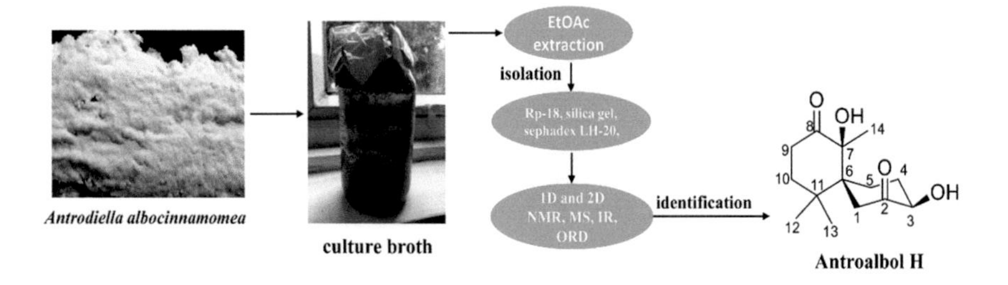

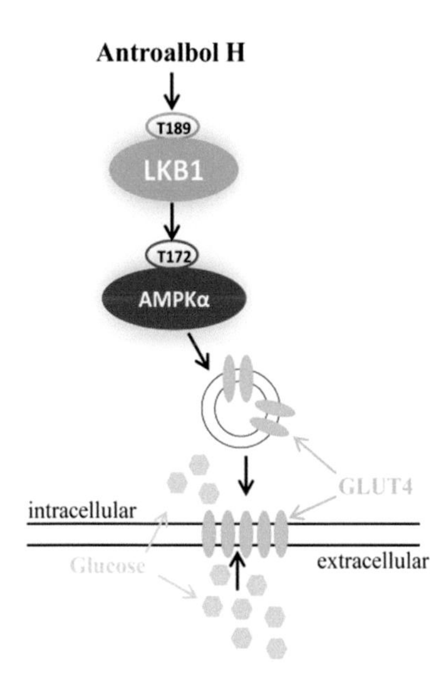

FIGURE 5.2.6 Summary of the role of antroalbol H in AMPK signaling pathway. Photo of *Antrodiella albocinnamomea* by J.K. Liu.

in cells, indicating that nuclear translocation regulation is also present. In addition, cellular immunofluorescence and western blot analysis confirmed that antroalbol H can promote GLUT4 translocation on the plasma membrane (Figure 5.2.6). In summary, our results indicated that, under certain circumstances, antroalbol H is a rare activator of the T189 phosphorylation residue of LKB1 and this residue is revealed as a potential target for regulating cellular glucose uptake (Wang et al., 2019).

5.2.5 Anti-Cancer Agents: Grifolin, Neoalbaconol, and Albaconol in Studies of the Mushroom *Albatrellus confluens*

Grifolin (**25**), a farnesyl phenolic compound (Figure 5.2.7A), is a secondary metabolite isolated from the fresh fruiting bodies of the mushroom *Albatrellus confluens*. For

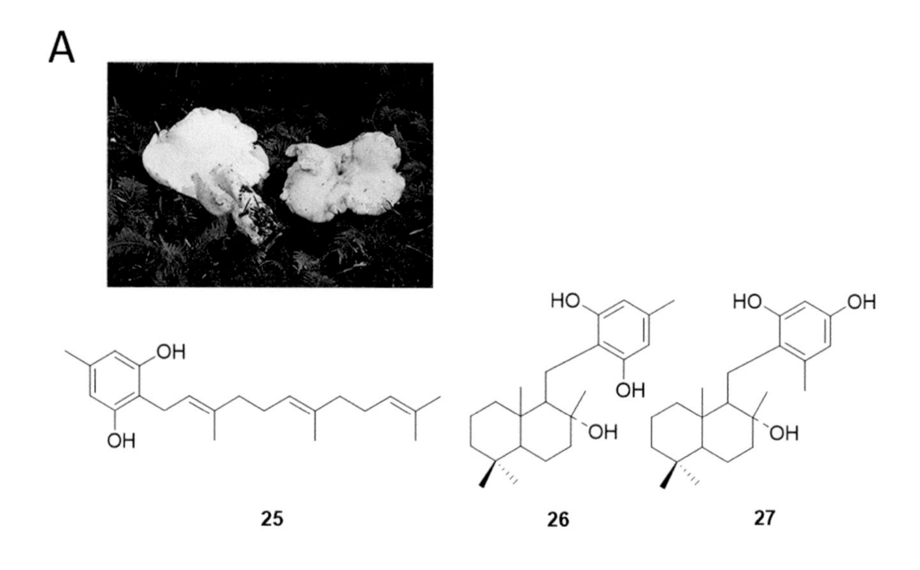

B

28

29 R=H
31 R=OH

30 R=H
32 R=OH

33 R=H
35 R=OH

34 R=H
36 R=OH

FIGURE 5.2.7 (A) *Albatrellus confluens* (photo by J.K. Liu) and structures of compounds **25**, **26**, **27**; (B) Structures of conflamides A–I.

the first time, we described a novel activity of grifolin, namely its ability to inhibit the growth of tumor cells by the induction of apoptosis. Grifolin strongly inhibited the growth of tumor cell lines: CNE1, HeLa, MCF7, SW480, K562, Raji, and B95-8. Analysis of acridine orange (AO)/ethidium bromide (EB) staining and flow cytometry showed that grifolin possessed apoptosis induction activity to CNE1, HeLa, MCF7, and SW480. Furthermore, the cytochrome c release from mitochondria was detected by confocal microscopy in CNE1 cells after a 12 h treatment with grifolin. The increase of caspase-8, 9, 3 activities revealed that caspase was a key mediator of the apoptotic pathway induced by grifolin, and the under-expression of Bcl-2 and up-regulation of Bax resulted in the increase of Bax: Bcl-2 ratio, suggesting that the Bcl-2 family is involved in the control of apoptosis (Ye et al., 2005).

We found that the effect of grifolin on the human nasopharyngeal carcinoma cell line CNE1 occurs primarily via the ERK1/2 pathway. At high doses, both the ERK1/2 and the ERK5 pathways may be involved in the inhibition. Because inhibition of the ERK1/2 or the ERK5 pathway has been associated with cell-cycle arrest and growth inhibition, we evaluated the cell cycle distribution after grifolin treatment. We found that grifolin significantly caused cell-cycle arrest in the G1 phase. To investigate the underlying mechanisms, G1-related proteins were assayed by western blotting. Following grifolin treatment, a concomitant inhibition of cyclin D1, cyclin E, CDK4 expression, and subsequent reduction in pRB phosphorylation occurred. Meanwhile, grifolin treatment also resulted in a significant upregulation of CKI (p19INK4D). These results suggested that the inhibition of the ERK1/2 or the ERK5 pathway is responsible for at least part of the induction of cell-cycle arrest in G1 phase by grifolin (Ye et al., 2007).

We also found that grifolin induced dephosphorylation of DAPK1 (Ser308) to activate DAPK1 and subsequent phosphorylation of its potential downstream effector p21 (Thr145) in nasopharyngeal carcinoma cell CNE1. Inhibition of DAPK1 by introducing siRNA targeting DAPK1 reversed the grifolin-induced phosphorylation of p21. Furthermore, we confirmed that grifolin increased the half-life of p21 and promoted its stability. Flow cytometry analysis demonstrated that DAPK1 was involved in grifolin-induced G1 phase arrest in CNE1 cells. The similar effects induced by grifolin and underlying mechanism were identified in another nasopharyngeal carcinoma cell HONE1. In addition, we observed that grifolin promoted the protein–protein interaction of DAPK1 and ERK1/2 to prevent ERK1/2 nucleolus translocation. Our findings indicated that DAPK1 plays a crucial role in the induction of cell-cycle arrest at the G1 phase by grifolin (Luo et al., 2011a, b).

We observed that grifolin inhibited tumor cells' adhesion and migration. Moreover, grifolin reduced reactive oxygen species (ROS) production and caused cellular ATP depletion in high-metastatic tumor cells. PGC1α (peroxisome proliferator-activated receptor γ, coactivator 1α) encodes a transcriptional co-activator involved in mitochondrial biogenesis and respiration and plays a critical role in the maintenance of energy homeostasis. Interestingly, grifolin suppressed the mRNA as well as protein level of PGC1α. We further identified that MMP2 and CD44 expressions were PGC1α inducible. PGC1α can bind with metastatic-associated transcription factors: Fra-1 and LSF and the protein-protein interaction was attenuated by grifolin treatment. Overall, these findings suggest that grifolin decreased ROS generation and intracellular ATP to suppress tumor cell adhesion/migration via impeding the interplay between PGC1α and Fra-1/LSF-MMP2/CD44 axes (Luo et al., 2015, 2016). We demonstrated that grifolin attenuates glycolytic flux and recovery of mitochondrial OXPHOS function by inhibiting DNMT1 expression and activity, as well as its mitochondrial retention in NPC cells (Luo et al., 2018). Thus, grifolin holds the promise of being an interesting anti-tumor agent that deserves further laboratory and *in vivo* exploration.

We have reported that neoalbaconol (**26**), a novel small molecular compound isolated from the fungus, *Albatrellus confluens*, could target 3-phosphoinositide-dependent protein kinase 1 (PDK1) and inhibit its downstream phosphoinositide-3 kinase (PI3-K)/Akt-hexokinase 2 (HK2) pathway, which eventually resulted in energy depletion. By targeting PDK1, NA reduced the consumption of glucose and ATP generation, activated autophagy, and caused apoptotic and necroptotic death of cancer cells through an independent pathway (Deng et al., 2013).

We determined that neoalbaconol-induced cell death is partly dependent on tumor necrosis factor α (TNFα) feed-forward signaling. More importantly, it abolished the ubiquitination of RIPK1 by down-regulating E3 ubiquitin ligases, cellular inhibitors of apoptosis protein 1/2 (cIAP1/2) and TNFα receptor-associated factors (TRAFs). The suppression of RIPK1 ubiquitination induced the activation of the non-canonical nuclear factor-κB (NF-κB) pathway and stimulated the transcription of TNFα. Moreover, we also found that neoalbaconol caused RIPK3-mediated reactive oxygen species (ROS) production and contributed to cell death. Taken together, these results suggested that two distinct mechanisms are involved in neoalbaconol-induced necroptosis and include RIPK1/NF-κB-dependent expression of TNFα and RIPK3-dependent generation of ROS (Yu et al., 2015).

We determined whether neoalbaconol could attenuate angiogenesis and how it occurs. Data suggest that neoalbaconol could inhibit the proliferation of breast cancer cells and induce apoptosis (Yu et al., 2017). Also, neoalbaconol suppressed vascular endothelial growth factor (VEGF)-induced human umbilical vascular endothelial cells (HUVECs) proliferation, migration, invasion, and capillary-like tube formation *in vitro* and reduced tumor angiogenesis *in vivo*. VEGF receptor activation and the downstream signal transduction cascades activation were inhibited by neoalbaconol. Additionally, neoalbaconol blocked EGFR-mediated VEGF production. EGFR overexpression reversed the neoalbaconol-induced VEGF reduction, confirming the importance of the EGFR inhibition in anti-angiogenesis of neoalbaconol. Furthermore, neoalbaconol inhibited tumor growth and tumor angiogenesis in a breast cancer xenograft model *in vivo* (Yu et al., 2017). Taken together, these results indicated that neoalbaconol could inhibit tumor angiogenesis and growth through direct suppression effects on vascular endothelial cells and the reduction of proangiogenic factors in cancer cells.

Albaconol (**27**) is another compound, a prenylated resorcinol, isolated from the fruiting bodies of the inedible mushroom *Albatrellus confluens*. Our studies showed that albaconol can inhibit tumor cell growth and dendritic cell maturation (Liu et al., 2008). We investigated the effects of albaconol on the proliferation and LPS-induced proinflammatory cytokine production of macrophages. Albaconol, when used at a dose higher than 1.0 μg/mL, inhibited the proliferation of RAW264.7 cells in a dose- and time-dependent manner and induced cellular apoptosis when used at high dosage ($\geq$7.5 μg/mL). Furthermore, we found that albaconol used at a lower dosage without apoptosis induction could significantly inhibit LPS-induced TNF-alpha, IL-6, IL-1beta, and NO production in RAW264.7 cells. The inhibition of NF-kappaB activation and enhancement of SOCS1 expression in LPS-stimulated macrophages by albaconol may contribute to the above immunosuppressive or anti-inflammatory activities of albaconol (Yu et al., 2008a, b). Thus, our results suggest that albaconol may be a potential immunosuppressive and anti-inflammatory drug.

Several prenylphenols from basidiocarps of European and Chinese *Albatrellus* spp., namely grifolin, neogrifolin, confluentin, scutigeral, and albaconol, were investigated in test models for vanilloid receptor modulation. The isolation of these compounds from *A. confluens* and structure elucidation of the novel natural product confluentin are described. The effects of scutigeral and neogrifolin on vanilloid receptors were studied by means of an electrophysiological methodology on rat dorsal root ganglion neurons as well as on recombinant cell lines expressing the rat VR1 receptor. Concurrently, the effects of these compounds on a reporter cell line expressing the human vanilloid receptor VR1 were measured. In contrast to previously reported studies, these results suggested that fungal prenylphenols act as weak antagonists (activity in the micromole range), rather than exhibiting agonistic activities (Hellwig et al., 2003).

Eight novel heterocyclic compounds conflamides B-I (**29–36**) with an unprecedented skeleton and their precursor conflamide A (**28**) were isolated from the mushroom *Albatrellus confluens*. Their absolute configurations were determined by use of NMR studies, a total synthesis, and calculated ECD spectra (Figure 5.2.7B). Conflamides D and E exhibited potent inhibition against LPS-induced B lymphocyte cell proliferation with IC_{50} values 1.48 and 5.71 μM, respectively (Zhang et al., 2018).

5.2.6 Novel Natural Products with New Skeletons from Studies on the Fungus *Xylaria curta* E10

Two unique cytochalasans, curtachalasins A (**37**) and B (**38**), were purified from the endophytic fungus *Xylaria curta* E10 harbored in the plant *Solanum tuberosum* (Figure 5.2.8A). Their structures were determined by extensive spectroscopic methods, X-ray crystallographic analysis, and electronic circular dichroism calculations. These two compounds feature an unprecedented pyrolidine/perhydroanthracene (5/6/6/6 tetracyclic skeleton) fused ring system (Wang et al., 2018).

Curtachalasins C–E (**39–41**) were identified from the endophytic fungus *Xylaria curta*, which have an unprecedent bridged 6/6/6/6 ring system. Residual dipolar coupling (RDC) analysis associating with density functional theory (DFT) calculations were utilized to determine the relative configuration of non-crystallizable **40**, which presented challenges to structure assignment based on regular NOEs or J-coupling analysis alone. The absolute configurations of **39–41** were determined by X-ray diffraction and electronic circular dichroism (ECD) calculations. Remarkably, curtachalasin C (**39**) showed significant resistance reversal activity against fluconazole-resistant *Candida albicans* (Wang et al., 2019a). The novel architecture and biological profile of this type of compound may provide clues to develop new antifungal strategies.

A cytochalasan, xylarichalasin A (**42**), was obtained from the endophytic fungus *Xylaria cf. curta* harbored in *Solanum tuberosum*. Its structure was elucidated by comprehensive spectroscopic methods including HRESIMS, 1D/2D NMR, and residual dipolar coupling analysis as well as quantum chemistry calculations including DFT GIAO 13C NMR and ECD calculation. It has an unprecedented 6/7/5/6/6/6 fused polycyclic structure. In bioassay, xylarichalasin A showed cytotoxicity against human cancer cell lines with IC_{50} value ranging from 6.3 to 17.3 μM (Wang et al., 2019b).

Two highly conjugated alkaloids xylariadines A (**43**) and B (**44**) were obtained as racemates from the fungus *Xylaria longipes*. They were resolved into optically pure enantiomers, respectively. Their structures were determined by extensive

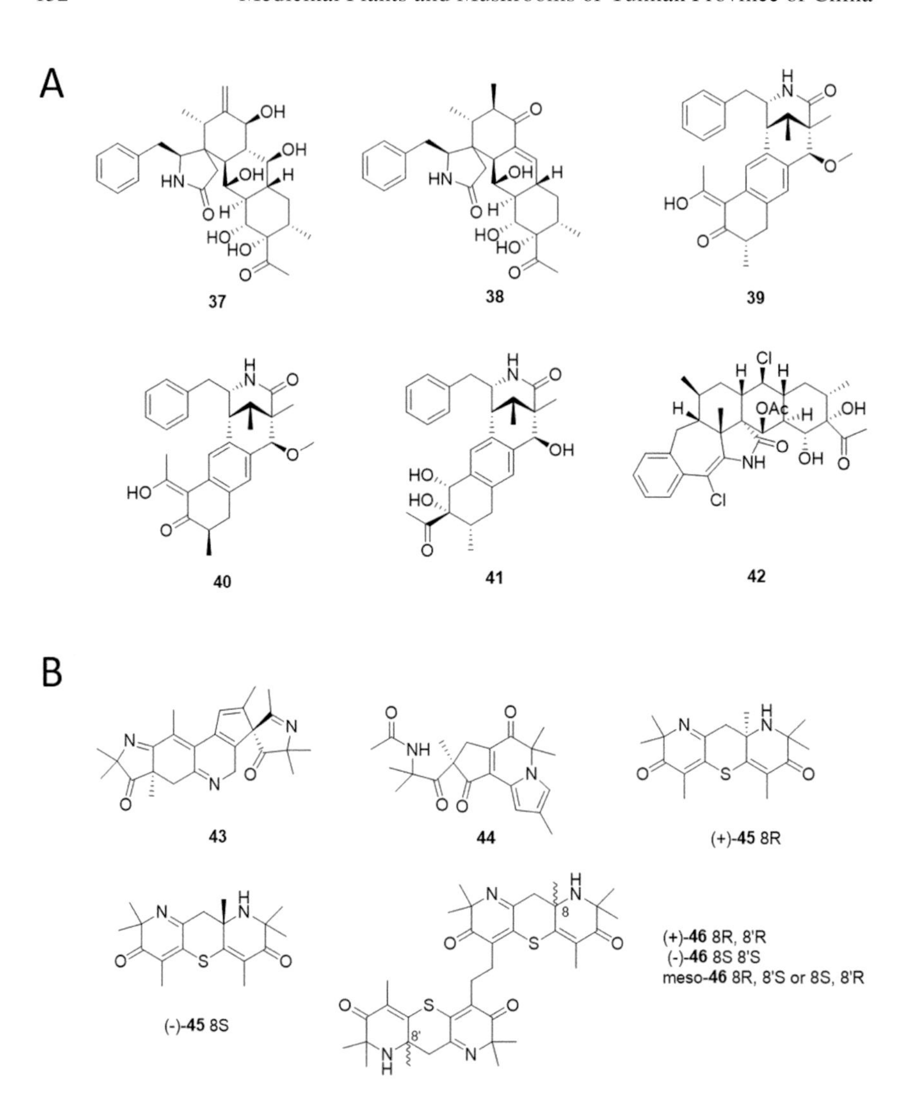

FIGURE 5.2.8 (A) Structures of **37–42**; (B) Structures of **43–46**.

spectroscopic analyses, X-ray diffraction, and ECD calculations. Compound **43** possesses a 5/6/6/5/5 fused ring system with unique 2-azaspiro[4.4]nonane substructure. The hypotheses of biosynthesis pathway for **43** and **44** were proposed (Li et al., 2019a) (Figure 5.2.8B).

Structurally unique thiopyranodipyridine alkaloids xylaridines C (**45**) and D (**46**) were isolated from the fungus *Xylaria longipes*. Their structures were established by comprehensive spectroscopic analysis combined with single-crystal X-ray diffraction and electron-capture detection (ECD) calculations. Compound **45** possesses two piperidine moieties fused with a central thiopyran ring, while compound **46** is a dimer of **45**. Compound **45** was resolved into optically pure enantiomers (+)-**45** and (–)-**45**

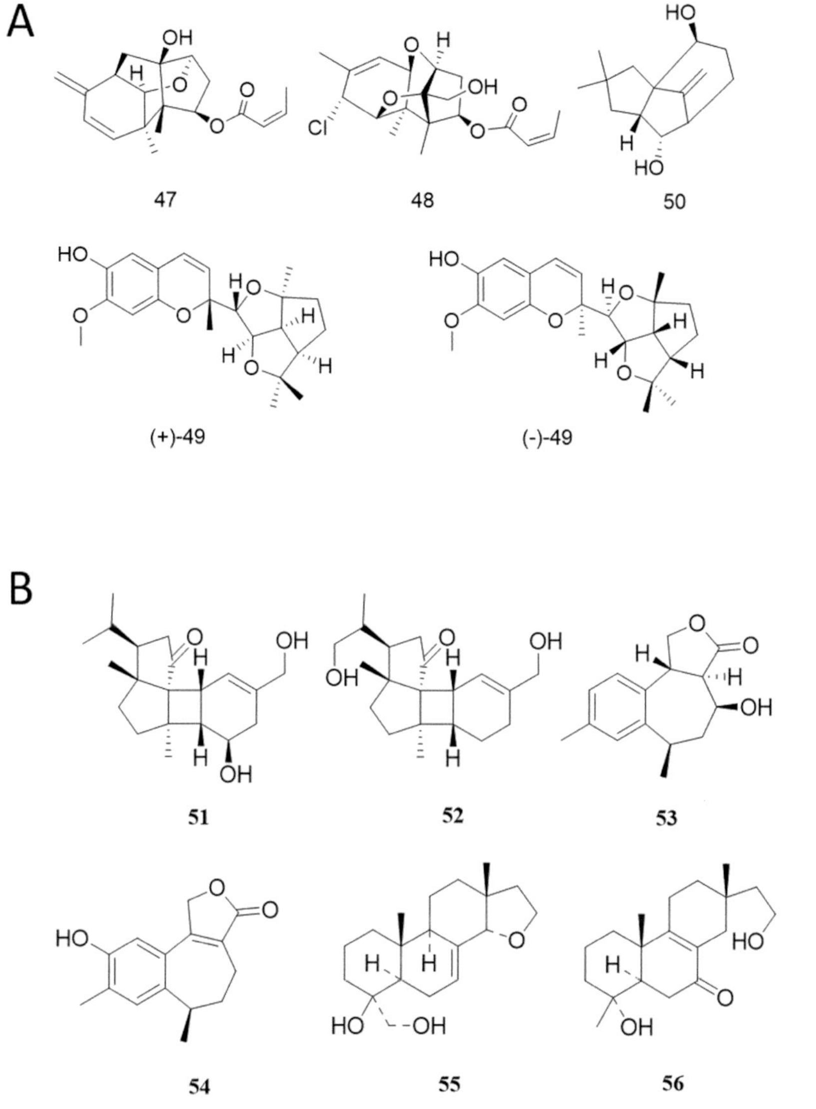

FIGURE 5.2.9 (A) Structures of **47–50**; (B) Structures of **51–56**.

by chiral HPLC. Moreover, compound **46** was resolved into an optically pure compound (+)-**46** and a mixture of (–)-**46** and meso-**46**. Plausible biosynthetic pathways of compounds **45** and **46** are proposed (Li et al., 2019b).

Two trichothecene sesquiterpenoids, trichothecrotocins A (**47**) and B (**48**), and a merosesquiterpenoid racemate, (±)-trichothecrotocin C (**49**), were obtained from potato endophytic fungus *Trichothecium crotocinigenum* by a bio-assay guided isolation (Figure 5.2.9A). The racemate **49** was resolved into pure enantiomers by a chiral separation. The structures of these compounds were elucidated by extensive spectroscopic methods, and ECD calculations, as well as the single-crystal X-ray diffraction. Compound **47** was a tetracyclic trichothecene derivative that featured a 6/5/5 bridged carbon backbone. Compound **48** was a Cl-containing trichothecene derivative featuring a bridged 6/6/5/5 ring system. Compound **49** contained a novel 6/6–5/5/5 fused ring system. Compounds **47–49** showed anti-phytopathogenic activities with MIC values of 8–128 μg/mL (Yang et al., 2018).

Antroalbocin A (**50**), a sesquiterpenoid possessing a bridged tricyclic system, was isolated from cultures of the basidiomycete *Antrodiella albocinnamomea*. The structure was elucidated by extensive spectroscopic methods with the absolute configuration determination by single-crystal X-ray diffraction. The plausible biosynthetic pathway for **50** was proposed. Compound **50** inhibited *Staphylococcus aureus* with an MIC of 30 μg/mL (Li et al., 2018).

Two undescribed tetracyclic diterpenoids, with unusual skeletons, psathyrins A (**51**) and B (**52**), have been characterized from cultures of the basidiomycete *Psathyrella candolleana*. Their structures with absolute configurations were established by means of spectroscopic methods including single-crystal X-ray diffraction, as well as ECD calculations. They possess a 5/5/4/6-fused ring system that represents a new carbon skeleton. Compounds **51** and **52** show inhibition of bacteria: *Staphylococcus aureus* and *Salmonella enterica* (Liu et al., 2020) (Figure 5.2.9B).

Eight previously undescribed sesquiterpenoids, tremutins A–H, together with three known ones, were isolated from cultures of the basidiomycetes *Irpex lacteus*. Structures of the new compounds with absolute configurations were elucidated based on extensive spectroscopic methods, as well as single-crystal X-ray diffractions and ECD methods. Compounds **53** and **54** possess an unusual 6/7-fused ring system that might be derived from a tremulane framework. Compound **53** inhibits the lipopolysaccharide (LPS)-induced proliferation of B lymphocyte cells with an IC_{50} value of 22.4 μM, while **54** inhibits concanavalin A (Con A)-induced T cell proliferation and LPS-induced B lymphocyte cell proliferation with IC_{50} values of 16.7 and 13.6 μM, respectively (Wang et al., 2020).

Eighteen new nor-isopimarane diterpenes, xylarinorditerpenes A–R, along with two previously reported compounds, 14α,16-epoxy-18-norisopimar-7-en-4α-ol and the labdane-type diterpene agatadiol, were isolated from cultures of the fungicolous fungus *Xylaria longipes* HFG1018 isolated from the wood-rotting basidiomycete *Fomitopsis betulinus*. Their structure elucidation and relative configuration assignments were accomplished by the interpretation of spectroscopic data and through computational methods. The absolute configurations of some compounds were determined by single-crystal X-ray diffraction (**55–56**). They possess an 18- or 19-nor-isopimarane skeleton, or 18,19-dinor-isopimarane skeleton. Some of them showed immunosuppressive activity but were devoid of cytotoxicity against the cell proliferation by concanavalin A-induced T lymphocytes and lipopolysaccharide-induced B lymphocytes, with IC_{50} values varying from 1.0 to 27.2 μM and from 16.1 to 51.8 μM, respectively (Chen et al., 2020).

Irpexolidal (**57**), a triterpenoid with an unprecedented carbon skeleton, along with its biogenetic-related compound irpexolide A, were isolated from the fruiting bodies of the medicinal fungus *Irpex lacteus* (Figure 5.2.10A). Irpexolidal features a 6/5/6/5/6/5-fused polycyclic skeletal system which arises from the eburicane-type triterpene by a 6,7-seco-6,8-cyclo pattern. The structures were established by means

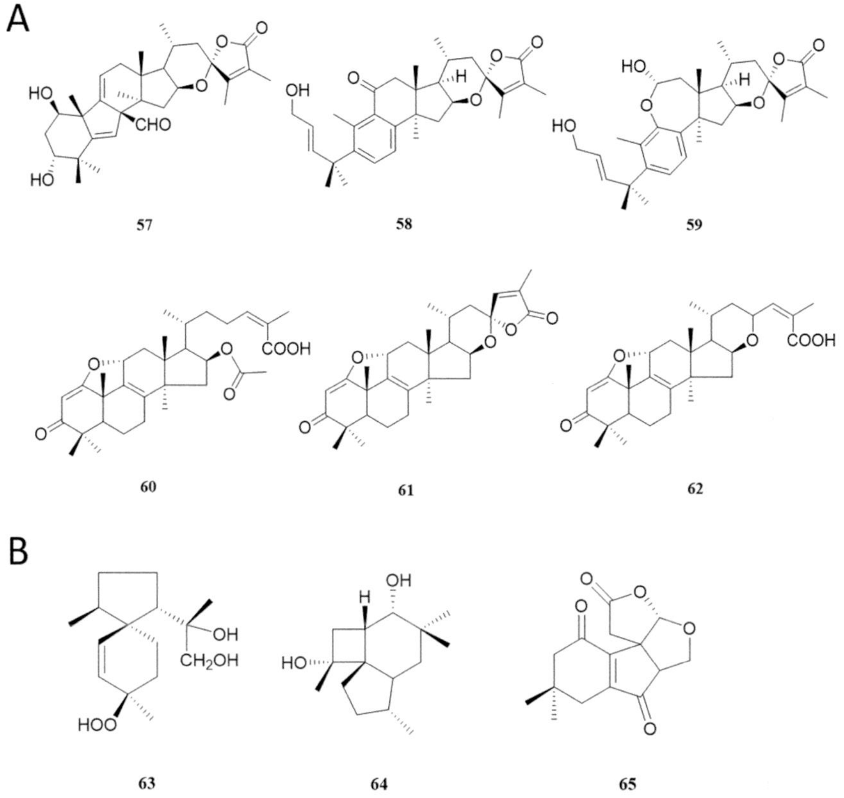

FIGURE 5.2.10 (A) Structures of **57–62**; (B) The Structure of Cordycepol C (**63**) from *Cordyceps ophioglossoides* and new skeleton sesquiterpenes **64** from *Tremella foliacea* and **65** from *Conocybe siliginea.*

of extensive spectroscopic techniques, ECD calculation, and DP4+ probability based on GIAO NMR chemical shift calculations. The plausible biosynthetic pathways for the compounds were proposed. Their biological activities were evaluated (Tang et al., 2019). Five new triterpenoids, irpeksins A–E, were isolated from fruiting bodies of the same fungus *Irpex lacteus*. The structures as well as absolute configurations of the new compounds were established via extensive spectroscopic analysis, computational methods, and Cotton effects. Four of them are featured by a scaffold of 1,10-seco- and ring B aromatic eburicane (24-methyllanostane) (**58**), and another one (**59**) is characterized by a scaffold of 1,10–9,11-diseco- and ring B aromatic eburicane, which represents unprecedented cleavage patterns in the lanostane family. The compounds showed significant inhibitory activity against NO production in LPS activated RAW 264.7 macrophage cells with IC_{50} values varying from 2.2 to 19.6 μM (Tang et al., 2018).

The mushroom *Macrolepiota procera*, also called “parasol mushroom” due to its large fruiting body resembling a parasol, is widespread in temperate regions. In Europe, *M. procera* is a highly sought-after and popular item due to its large-size fruiting bodies, frequent seasonal accessibility, and versatility in the kitchen. However, no reports have addressed the secondary metabolites of this kind of famous edible mushroom so far. As our continuous research aiming at discovery drug leads from edible mushroom, a chemical investigation on the constituents of the Poland-origin parasol mushroom *M. procera* was carried out. The isolation, structure elucidation, and biological evaluation of 12 lanostane triterpenoids, namely lepiotaprocerins A–L, from the fruiting bodies of *M. procera* were reported by us (Chen et al., 2018). Figure 5.2.10A shows three of them (**60–62**).

Cordycepol C (**63**), a novel sesquiterpene isolated from the cultured mycelia of *Cordyceps ophioglossoides*, contains a hydroperoxy group and is cytotoxic to HepG2 cells. So far, no sesquiterpenes have been found in the genus *Cordyceps* and it would be interesting to investigate the anti-tumor efficacy as well as the mechanism of action of this unusual sesquiterpene. In this study, we showed that cordycepol C induced apoptosis of the HepG2 cells without affecting the normal liver cell line L-02. Cordycepol C caused poly(ADP-ribose) polymerase-1 (PARP-1) cleavage and triggered the loss of mitochondrial membrane potential ($\Delta\psi m$) in HepG2 cells in a time- and dose-dependent manner, resulting in the nuclear translocation of apoptosis-inducing factor (AIF) and endonuclease G (Endo G). We also found that cordycepol C induced the expression of Bax protein, followed by its translocation from the cytosol to mitochondria in both wild type and p53 knockdown HepG2 cells. However, cordycepol C could not cause cleavages of procaspase-3, -8, and -9. Caspase activities were not increased and Z-VAD-fmk, a caspase inhibitor, could not prevent the apoptosis induced by cordycepol C. These findings indicate that cordycepol C induces caspase-independent apoptosis in HepG2 cells through a p53-independent and Bax-mediated mitochondrial pathway, leading to the nuclear translocation of AIF and Endo G. Our study provides the molecular mechanism by which cordycepol C induces apoptosis in hepatocellular carcinoma cells and indicates the potential use of cordycepol C as an anti-tumor agent (Sun et al., 2014).

Trefolane A (**64**), an unprecedented skeleton with a 5/6/4 tricyclic ring system, was isolated from cultures of the basidiomycete *Tremella foliacea* (Figure 5.2.10B). The structure was elucidated by means of spectroscopic methods and further confirmed by single-crystal X-ray diffraction analysis. A possible biogenesis for trefolane A was also proposed. Humulene is formed from farnesyl pyrophosphate by an enzymatic cyclization reaction. Biogenetically, it is suggested that most sesquiterpenoids derived from mushrooms, subdivision Basidiomycotina, were started from humulene with three pathways. One route leads to caryophyllane, and another pathway ends in the irregular sesquiterpenes of the tremulane type. The third pathway is the most important one. It produces the tricyclic sesquiterpene protoilludane which is at the biosynthetic crossroad for many sesquiterpene classes. We suggest that trefolane A (**64**) started from humulene with the fourth pathway, in which two new carbon bonds were formed between C-7 and C-4, and C-7 and C-11. Considering the present pathway of **64**, a new type of sesquiterpene with a 3/7/5 ring system may be discovered in future (Ding et al., 2012).

Conosilane A (**65**), a novel sesquiterpene with an unprecedented carbon skeleton, was isolated from the cultures of the basidiomycete *Conocybe siliginea*. Its structure was elucidated by extensive spectroscopic methods, and the absolute configuration was determined by single-crystal X-ray diffraction analysis. We suggest that conosilane A starts from the humulane skeleton with a new biogenetic pathway, in which two new carbon bonds are formed between C-4 and C-10, C-6 and C-9, via methyl migration and cleavage of bond C-6/C-7 to construct a 6/5 carbon ring system. Then after an accompanying oxidation and dehydration, a tetrahydrofuran ring C and a five-membered lactone ring D are built. Conosilane A is a non-isoprenoid sesquiterpenoid that represents a new skeleton type in the family of sesquiterpenoids. It was found to exhibit moderate inhibitory activities against both human and mouse 11β-HSD1 at a concentration of 10 μg/mL, with inhibitory rates of 53.3% and 70.0%, respectively (Yang et al., 2012).

Two novel fomannosane-type sesquiterpenoids, agrocybins H (**66**) and I (**67**), together with a known compound illudosin, were isolated from the culture broth of the mushroom *Agrocybe salicacola* (Figure 5.2.11A). Their structures were elucidated by extensive spectroscopic analysis. The relative stereochemistry of **66** was determined using single-crystal X-ray crystallographic diffraction (Li et al., 2012).

Seven new illudin-type sesquiterpenoids, agrocybins A–G, along with three known analogues, have been isolated from the culture broth of the fungus *Agrocybe salicacola* (Liu et al., 2011). Their structures were elucidated based on extensive

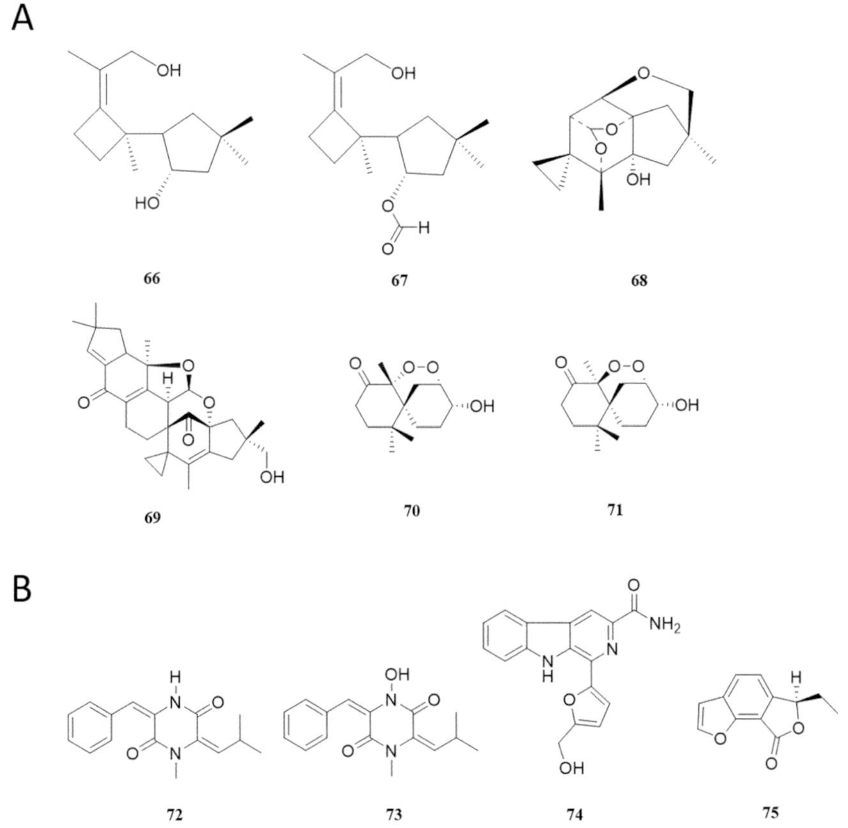

FIGURE 5.2.11 (A) Structures of **66–71**; (B) Structures of **72–75**.

spectroscopic data analysis and comparison with data reported in the literature. The relative stereo-configurations of **68** were elucidated by X-ray crystallographic diffraction analysis. Compound **68** was highly cyclized, containing seven chiral carbons, which arranged compactly into seven rings (Li et al., 2012). The isolation, structure elucidation, and relative stereochemistry assignment of a novel illudane–illudane bis-sesquiterpene, agrocybone (**69**), from the same mushroom, were reported. Agrocybone represents a structure with eight rings (including two spiro rings) and seven stereogenic carbon atoms (Zhu et al., 2010). Two novel 3-normethyl-chamigrane sesquiterpene peroxides, named steperoxide A (**70**) and B (**71**), have been isolated from basidiomycete *Steccherinum ochraceum*. This is the first report on the isolation of chamigrane sesquiterpene from mushrooms. The structures of **70** and **71** were established based on spectroscopic data and single-crystal X-ray analysis (Liu et al., 2010).

5.2.7 Miscellaneous Compounds from Various Fungal Sources

In inflammation, proinflammatory cytokines induce the formation of large amounts of nitric oxide (NO) by inducible nitric oxide synthase (iNOS), and compounds that inhibit NO production have anti-inflammatory effects. We investigated the effects of lansai C and D on NO production in lipopolysaccharide-induced RAW 264.7 cells and evaluated the mechanisms of action of the compounds Lansais C and D (**72**, **73**) (Figure 5.2.11B). They inhibited iNOS protein and mRNA expression and NO production in a dose-dependent manner. These compounds inhibited the activation of nuclear factor-kB, which is a significant transcription factor for iNOS and inhibited the activation of the signal transducer and activator of transcription-1, another important transcription factor for iNOS. The study characterizes the effects and mechanisms of lansais C and D on iNOS expression and NO production on inactivated macrophages. The results explain the pharmacological efficacy of these compounds as anti-inflammatory compounds (Taechowisan et al., 2009, 2010).

Flazin, isolated from the fruiting bodies of *Suillus granulatus*, possesses weak anti-HIV activity (EC_{50} = 2.36 μM, therapeutic index (TI) = 12). To establish a SAR study, 46 flazin analogues were synthesized, and their anti-HIV activities were evaluated *in vitro*. Among them, flazinamide (**74**) showed the most potent activity with an EC_{50} value of 0.38 μM and a TI value of 312. These results suggested that appropriate substituents at positions 3, 1', and 5' of flazin might play a crucial role in determining their anti-HIV activities, and that flazinamide can be considered as a promising, readily available anti-HIV agent (Tang et al., 2008, Wang et al., 2007).

A novel benzofuran lactone, named concentricolide (=rel-(6R)-6-ethylbenzo[2,1-b:3,4-c']difuran8(6H)-one; **75**), was isolated from the fruiting bodies of the xylariaceous ascomycete *Daldinia concentrica*. The structure of **75** was established by spectroscopic methods and X-ray crystallographic analysis. Studies showed that **75** inhibited HIV-1 induced cytopathic effects. The EC_{50} value was 0.31 μg/mL, and the TI was 247. Concentricolide (**75**) exhibited the blockage (EC_{50} = 0.83 μg/mL) on syncytium formation between HIV-1 infected cells and normal cells and was synthesized for the first time as a racemate (Qin et al., 2006, Fang and Liu, 2009).

5.2.8 Conclusions

The first systematic investigations of secondary metabolites from fungi originated after the discovery and introduction into clinical practice of penicillin. From 1940 until the early 1950s, mycelial cultures or fruiting bodies of many medicinal mushrooms and fungi were screened to produce antibiotics. The investigations resulted in the discovery of pleuromutilin, the lead compound for the semisynthetic tiamulin used in veterinary practice and recently also in humans. It is highly remarkable that a synthetic analog of illudin S, (–)-irofulven, has entered clinical trials and demonstrated activity in ovarian, gastrointestinal, and non-small cell lung cancer. As compared to the natural product, irofulven has a much better therapeutic index and pharmacological profile (Chen and Liu, 2017).

As can be deduced from the numerous new structures, interest in the mushrooms' secondary metabolism has gained momentum. The biological activities are interesting and may help to define new lead compounds offering structures not easily detected by the random screening of libraries derived from combinatorial chemical synthesis. The availability of secondary metabolites from higher fungi is facilitated by important progress in fermentation technologies and genetics, opening access to templates for chemical syntheses and providing new chemical approaches to yet unexplored biological targets. The mushrooms will continue to attract the interest of natural products chemists as a source of toxins, hallucinogens, and pigments (Chen and Liu, 2017).

Fungi should be viewed on a par with the plant and animal kingdoms, and we have only just started to scratch the surface of knowledge of this incredible and diverse group of organisms. What also becomes apparent is that when looking for nature-based solutions to some of our most critical global challenges, fungi could provide many of the answers (Royal Botanic Gardens Kew, 2018).

REFERENCES

Bedry R, Baudrimont I, Deffieux G, et al., 2001. Wild-mushroom intoxication as a cause of rhabdomyolysis. *N Engl J Med*, 345: 798–802.

Chen HP, Zhao ZZ, Yin RH, et al., 2014. Six new vibralactone derivatives from cultures of the fungus *Boreostereum vibrans*. *Nat Prod Bioprospect*, 4: 271–276.

Chen HP, Zhao ZZ, Li ZH, et al., 2016. Novel natural oximes and oxime esters with a vibralactone backbone from the basidiomycete *Boreostereum vibrans*. *Chem Open*, 5: 142–149.

Chen HP, Liu JK, 2017. Secondary metabolites from higher fungi. *Prog Chem Org Nat Prod*, 106: 1–201.

Chen HP, Zhao ZZ, Li ZH, et al., 2018. Anti-proliferative and anti-inflammatory lanostane triterpenoids from the Polish edible mushroom *Macrolepiota procera*. *J Agric Food Chem*, 66: 3146–3154.

Chen HP, Zhao ZZ, Cheng GG, et al., 2020. Immunosuppressive nor-isopimarane diterpenes from cultures of the fungicolous fungus *Xylaria longipes* HFG1018. *J Nat Prod*, 83: 401–412.

Deng Q, Yu X, Xiao L, et al., 2013. Neoalbaconol induces energy depletion and multiple cell death in cancer cells by targeting PDK1-PI3-K/Akt signaling pathway. *Cell Death Dis*, 4: e804.

Ding JH, Feng T, Li ZH, et al., 2012. Trefolane A, a sesquiterpenoid with a new skeleton from cultures of the Basidiomycete *Tremella foliacea*. *Org Lett*, 14: 4976–4978.
Fang LZ, Liu JK, 2009. First synthesis of racemic concentricode, an anti-HIV-1 agent isolated from the fungus *Daldinia concentrica*. *Hetereocycles*, 78: 2107–2113.
Feng KN, Yang YL, Xu YX, et al., 2020. A hydrolase-catalyzed cyclization forms the fused bicyclic β-lactone in vibralactone. *Angew Chem Int Ed*, 59: 7209–7213.
Feng T, He J, Ai HL, et al., 2015. Three new triterpenoids from European mushroom *Tricholoma terreum*. *Nat Prod Bioprospect*, 5: 205–208.
Hellwig V, Nopper R, Mauler F, et al., 2003. Activities of prenylphenol derivatives from fruitbodies of Albatrellus spp. on the human and rat vanilloid receptor 1 (VR1) and characterization of the novel natural product, confluentin. *Archiv der Pharmazie*, 336: 119–126.
Jiang MY, Wang F, Yang XL, et al., 2008. Derivatives of vibralactone from cultures of the basidiomycete *Boreostereum vibrans*. *Chem Pharm Bull*, 56: 1286–1288.
Jiang MY, Zhang L, Dong ZJ, et al., 2010. Vibralactones D-F from cultures of the basidiomycete *Boreostereum vibrans*. *Chem Pharm Bull*, 58: 113–116.
Li J, Wang WX, Chen HP, et al., 2019a. (±)-Xylaridines A and B, highly conjugated alkaloids from the fungus *Xylaria longipes*. *Org Lett*, 21: 1511–1514.
Li J, Wang WX, Li ZH, et al., 2019b. Xylaridines C and D, unusual thiopyranodipyridine alkaloids from the fungus *Xylaria longipes*. *Org Lett*, 21: 6145–6148.
Li W, He J, Feng T, et al., 2018. Antroalbocin A, an antibacterial sesquiterpenoid from higher fungus *Antrodiella albocinnamomea*. *Org Lett*, 20: 8019–8021.
Liu DZ, Wang F, Liao TG, et al., 2006. Vibralactone: a lipase inhibitor with an unusual fused *β*-lactone produced by cultures of the basidiomycete *Boreostereum vibrans*. *Org Lett*, 8: 5749–5752.
Liu DZ, Dong ZJ, Wang F, et al., 2010. Two novel norsesquiterpene peroxides from basidiomycete *Steccherinum ochraceum*. *Tetrahedron Lett*, 51: 3152–3153.
Liu JK, 2002. Biologically active substances from mushrooms in Yunnan, China. *Heterocycles*, 57: 157–167.
Liu LY, Zhang L, Feng T, et al., 2011. Unusual illudin-type sesquiterpenoids from cultures of *Agrocybe salicacola*. *Nat Prod Bioprospect*, 1: 87–92.
Liu LY, Li ZH, Dong ZJ, et al., 2012. Two novel fomannosane-type sesquiterpenoids from the culture of the basidiomycete *Agrocybe salicacola*. *Nat Prod Bioprospect*, 2: 130–132.
Liu Q, Shu X, Sun A, et al., 2008. Plant-derived small molecule albaconol suppresses LPS-triggered proinflammatory cytokine production and antigen presentation of dendritic cells by impairing NF-κB activation. *Int Immunopharmacol*, 8: 1103–1111.
Liu YP, Dai Q, Wang WX, et al., 2020. Psathyrins: antibacterial diterpenoids from *Psathyrella candolleana*. *J Nat Prod*, 83: 1725–1729.
Luo X, Yang L, Xiao L, et al., 2015. Grifolin directly targets ERK1/2 to epigenetically suppress cancer cell metastasis. *Oncotarget*, 6: 42704–42716.
Luo X, Hong L, Cheng C, et al., 2018. DNMT1 mediates metabolic reprogramming induced by Epstein-Barr virus latent membrane protein 1 and reversed by grifolin in nasopharyngeal carcinoma. *Cell Death Dis*, 9: 619–634.
Luo XJ, Li W, Yang LF, et al., 2011a. DAPK1 mediates the G1 phase arrest in human nasopharyngeal carcinoma cells induced by grifolin, a potential antitumor natural product. *Eur J Pharmacol*, 670: 427–434.

Luo XJ, Li LL, Deng QP, et al., 2011b. Grifolin, a potent antitumour natural product upregulates death-associated protein kinase 1 DAPK1 via p53 in nasopharyngeal carcinoma cells. *Eur J Cancer*, 47: 316–325.

Luo XJ, Li NM, Zhong JF, et al., 2016. Grifolin inhibits tumor cells adhesion and migration via suppressing interplay between PGC1α and Fra-1 / LSF- MMP2 / CD44 axes. *Oncotarget*, 7: 68708–68720.

Qin XD, Dong ZJ, Liu JK, et al., 2006. Concentricolide, an anti-HIV agent from the Ascomycete *Daldinia concentrica*. *Helv Chim Acta*, 89: 127–133.

Royal Botanic Gardens, Kew. Willis, Katherine (Ed.) In: State of the World's Fungi 2018. https://kew.iro.bl.uk/work/ns/e30de436-455d-410e-8605-8c533a0398ce

Sun YS, Lv LX, Zhao Z, et al., 2014. Cordycepol C induces caspase-independent apoptosis in human hepatocellular carcinoma HepG2 Cells. *Biol Pharm Bull*, 37: 608–617.

Taechowisan T, Wanbanjob A, Tuntiwachwuttikul P, et al., 2009. Anti-inflammatory activity of lansais from endophytic *Streptomyces* sp. SUC1 in LPS-induced RAW 264.7 cells. *Food Agric Immunol*, 20: 67–77.

Taechowisan T, Wanbanjob A, Tuntiwachwuttikul P, et al., 2010. Anti-inflammatory effects of lansai C and D cause inhibition of STAT-1 and NF-kB activations in LPS-induced RAW 264.7 cells. *Food Agric Immunol*, 21: 57–64.

Tang JG, Wang YH, Wang RR, et al., 2008. Synthesis of analogues of flazin, in particular, flazinamide, as promising anti-HIV agents. *Chem Biodiver*, 5: 447–460.

Tang Y, Zhao ZZ, Yao JN, et al., 2018. Irpeksins A–E, 1,10-seco-eburicane-type triterpenoids from the medicinal fungus *Irpex lacteus* and their anti-NO activity. *J Nat Prod*, 81: 2163–2168.

Tang Y, Zhao ZZ, Hu K, et al., 2019. Irpexolidal represents a class of triterpenoid from the fruiting bodies of the medicinal fungus *Irpex lacteus*. *J Org Chem*, 84: 1845–1852.

Wang F, Yang XY, Lu YT, et al., 2019. The natural product antroalbol H promotes phosphorylation of liverkinase B1 (LKB1) at threonine 189 and thereby enhances cellular glucose uptake. *J Biol Chem*, 294: 10415–10427.

Wang GQ, Wei K, Feng T, et al., 2012. Vibralactones G–J from cultures of the basidiomycete *Boreostereum vibrans*. *J Asian Nat Prod Res*, 14: 115–120.

Wang GQ, Wei K, Zhang L, et al., 2014. Three new vibralactone-related compounds from cultures of basidiomycete *Boreostereum vibrans*. *J Asian Nat Prod Res*, 16: 447–452.

Wang M, Du JX, Yang HX, et al., 2020. Sesquiterpenoids from cultures of the Basidiomycetes *Irpex lacteus*. *J Nat Prod*, 84: 1524–1531.

Wang WX, Li ZH, Feng T, et al., 2018. Curtachalasins A and B, two cytochalasans with a tetracyclic skeleton from the endophytic fungus *Xylaria curta* E10. *Org Lett*, 20: 7758–7761.

Wang WX, Lei XX, Ai HL, et al., 2019a. Cytochalasans from the endophytic fungus *Xylaria cf. curta* with resistance reversal activity against fluconazole-resistant *Candida albicans*. *Org Lett*, 21: 1108–1111.

Wang WX, Lei XX, Yang YL, et al., 2019b. Xylarichalasin A, a halogenated hexacyclic cytochalasan from the fungus *Xylaria cf. curta*. *Org Lett*, 21: 6957–6960.

Wang YH, Tang JG, Wang RR, et al., 2007. Flazinamide, a novel β-carboline compound with anti-HIV actions. *Biochem Biophys Res Commun*, 355: 1091–1095.

Wei K, Wang GQ, Bai X, et al., 2015. Structure-based optimization and biological evaluation of pancreatic lipase inhibitors as novel potential antiobesity agents. *Nat Prod Bioprospect*, 5: 129–157.

Yang HX, Ai HL, Feng T, et al., 2018. Trichothecrotocins A–C, antiphytopathogenic agents from potato endophytic fungus *Trichothecium crotocinigenum*. *Org Lett*, 20: 8069–8072.

Yang XY, Feng T, Li ZH, et al., 2012. Conosilane A, an unprecedented sesquiterpene from the cultures of Basidiomycete *Conocybe siliginea*. *Org Lett*, 14: 5382–5384.

Yang YL, Zhou H, Du G, et al., 2016. A monooxygenase from *Boreostereum vibrans* catalyzes oxidative decarboxylation in a divergent vibralactone biosynthesis pathway. *Angew Chem Int Ed*, 55: 5463–5466.

Ye M, Liu JK, Lu ZX, et al., 2005. Grifolin, a potential antitumor natural product from the mushroom *Albatrellus confluens*, inhibits tumor cell growth by inducing apoptosis *in vitro*. *FEBS Lett*, 579: 3437–3443.

Ye M, Luo XJ, Li LL, et al., 2007. Grifolin, a potential antitumor natural product from the mushroom *Albatrellus confluens*, induces cell-cycle arrest in G1 phase via the ERK1/2 pathway. *Cancer Lett*, 258: 199–207.

Yin X, Feng T, Li ZH, et al., 2013. Highly oxygenated meroterpenoids from fruiting bodies of the mushroom *Tricholoma terreum*. *J Nat Prod*, 76: 1365–1367.

Yin X, Feng T, Shang JH, et al., 2014. Chemical and toxicological investigations of a previously unknown poisonous European mushroom *Tricholoma terreum*. *Chem Eur J*, 20: 7001–7009.

Yu QL, Shu XL, Sun AN, et al., 2008a. Plant-derived small molecule albaconol suppresses LPS-triggered proinflammatory cytokine production and antigen presentation of dendritic cells by impairing NF-kB activation. *Int Immunopharmacol*, 8: 1103–1111.

Yu QL, Shu XL, Wang L, et al., 2008b. Albaconol, a plant-derived small molecule, inhibits macrophage function by suppressing NF-kB activation and enhancing SOCS1 expression. *Cell Mol Immunol*, 4: 271–278.

Yu X, Deng Q, Li W, et al., 2015. Neoalbaconol induces cell death through necroptosis by regulating RIPK-dependent autocrine TNFα and ROS production. *Oncotarget*, 6: 1995–2008.

Yu XF, Li W, Deng QP, et al., 2017. Neoalbaconol inhibits angiogenesis and tumor growth by suppressing EGFR-mediated VEGF production. *Mol Carcinog*, 56: 1414–1426.

Zeiler E, Braun N, Böttcher T, et al., 2011. Vibralactone as a tool to study the activity and structure of the ClpP1P2 complex from *Listeria monocytogenes*. *Angew Chem Int Ed*, 50: 11001–11004.

Zhang SB, Huang Y, He SJ, et al., 2018. Heterocyclic compounds from the mushroom *Albatrellus confluens* and their inhibitions against lipopolysaccharides-induced B lymphocyte cell proliferation. *J Org Chem*, 83: 10158–10165.

Zhao PJ, Yang YL, Du LC, et al., 2013. Elucidating the biosynthetic pathway for vibralactone: a pancreatic lipase inhibitor with a fused bicyclic *β*-lactone. *Angew Chem Int Ed*, 52: 2298–2302.

Zhao ZZ, Chen HP, Wu B, et al., 2017. Matsutakone and matsutoic acid, two (nor)steroids with unusual skeletons from the edible mushroom *Tricholoma matsutake*. *J Org Chem*, 82: 7974–7979.

Zhou ZY, Shi GQ, Fontaine R, et al., 2012. Evidence for the natural toxins from the mushroom *Trogia venenata* as a cause of sudden unexpected death in Yunnan province, China. *Angew Chem Int Ed*, 51: 2368–2370.

Zhu YC, Wang G, Yang XL, et al., 2010. Agrocybone, a novel bis-sesquiterpene with a spirodienone structure from basidiomycete *Agrocybe salicacol*. *Tetrahedron Lett*, 51: 3443–3445.

Section 3

Plants as Health Food or Supplements

6

Panax notoginseng (Burk.) F.H. Chen

Jia-Huan Shang
State Key Laboratory of Phytochemistry and Plant Resources in West China, Kunming Institute of Botany, Chinese Academy of Sciences, Kunming, Yunnan, People's Republic of China

University of Chinese Academy of Sciences, Beijing, People's Republic of China

Yi-Jun Qiao
State Key Laboratory of Phytochemistry and Plant Resources in West China, Kunming Institute of Botany, Chinese Academy of Sciences, Kunming, Yunnan, People's Republic of China

Dong Wang
State Key Laboratory of Phytochemistry and Plant Resources in West China, Kunming Institute of Botany, Chinese Academy of Sciences, Kunming, Yunnan, People's Republic of China

Yunnan Key Laboratory of Natural Medicinal Chemistry, Kunming Institute of Botany, Chinese Academy of Sciences, Kunming, Yunnan, People's Republic of China

Hong-Tao Zhu
State Key Laboratory of Phytochemistry and Plant Resources in West China, Kunming Institute of Botany, Chinese Academy of Sciences, Kunming, Yunnan, People's Republic of China

Yunnan Key Laboratory of Natural Medicinal Chemistry, Kunming Institute of Botany, Chinese Academy of Sciences, Kunming, Yunnan, People's Republic of China

Chong-Ren Yang
State Key Laboratory of Phytochemistry and Plant Resources in West China, Kunming Institute of Botany, Chinese Academy of Sciences, Kunming, Yunnan, People's Republic of China

Ying-Jun Zhang*
State Key Laboratory of Phytochemistry and Plant Resources in West China, Kunming Institute of Botany, Chinese Academy of Sciences, Kunming, Yunnan, People's Republic of China

Yunnan Key Laboratory of Natural Medicinal Chemistry, Kunming Institute of Botany, Chinese Academy of Sciences, Kunming, Yunnan, People's Republic of China

CONTENTS

6.1 Introduction

Panax notoginseng (Burk.) F. H. Chen, locally known as *Sanqi* or *Tianqi* in Chinese, is a famous and important traditional Chinese medicine belonging to the genus *Panax* of the Araliaceae family. It has been domesticated and cultivated for more than 400 years in Southwest China (Qiao et al., 2018). As one of the most valuable and distinct biological resources in Yunnan Province, *P. notoginseng* has played an important role in the sourcing of raw materials to market and in traditional Chinese medicinal preparations. Historically, the root of the herb has been used for the treatment of trauma and ischemic cardiovascular diseases. Investigations into the chemical composition and pharmacological activities of *P. notoginseng* have been carried out for nearly 40 years, leading to the isolation and identification of more than 200 constituents from its roots, stems, leaves, buds, fruits, and fruit pedicels. These compounds belong to

* Corresponding author.

different chemical types including triterpenoid saponins, cyclodipeptides, flavonoids, sterols, and polyacetylenes. Among them, *P. notoginseng* saponins (PNS) are characterized as the main type and dominant bioactive components. Clinical trials have indicated that the root of *P. notoginseng* and PNS preparations are effective when applied for cardiac cerebral vascular and other blood-related diseases.

As a local medicinal plant with the largest cultivated area in Yunnan and Guangxi Provinces, the root of *P. notoginseng* has been widely used as a tonic and hemostatic drug for promoting blood circulation, relief of bruises, and treating blood loss caused by both internal and external injuries. Research on *P. notoginseng* can be traced back to the 1930s. With the development of modern separation and purification techniques, and further studies of biological activities, more than 200 compounds in natural or transformed products have been isolated and identified from the whole plant of this herb. PNS are the major bioactive components and play a crucial role in pharmacological activities in the cardiovascular, cerebrovascular, immune, and nervous systems. Furthermore, amino acids, flavonoids, polysaccharides, and polyacetylenes found in *P. notoginseng* also possess a wide range of applications including medicines, health products, and cosmetics.

Herein, we summarize the research on *P. notoginseng* from the botany, ethnopharmacology, phytochemistry, and biological activities to its commercial applications, hoping to provide a comprehensive understanding of this important medicinal plant.

6.2 Botany

For more than 400 years of its medicinal plant history, *P. notoginseng* (Figure 6.1) was mainly named in Chinese as *Sanqi, Sanqi* ginseng, *Shanqi or Tianqi*, which were derived from its morphology and the place of origin (Yang, 2015). The name of *Sanqi* was firstly recorded as a plant with three branches and seven leaves on each branch in the Yimen Mizhi, written by Siwei Zhang (Ming Dynasty). As a precious resource

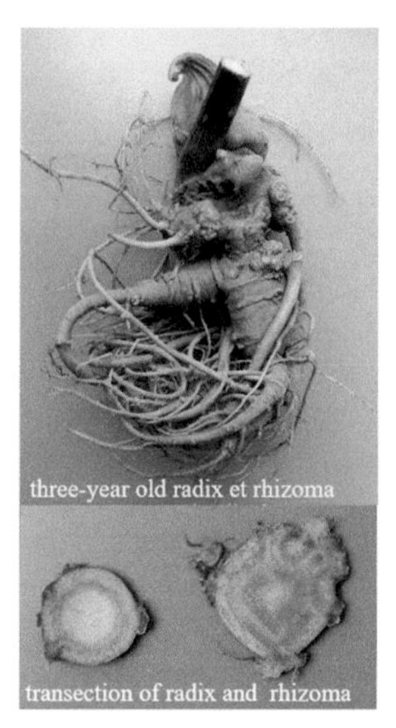

FIGURE 6.1 *P. notoginseng* and its root, flower, and fruit (drawing [middle] and photos by Dong Wang).

and a very effective folk medicine to cure blood-related diseases, *P. notoginseng* is often called *Xueqi* (blood ginseng) or *Jinbuhuan* (not to be exchanged even for gold) by local people (Shang and Lowry, 2007).

Generally, after the seeds germinate, seedlings are cultured in nursery beds with 80% shade for one year and then transplanted to the field, and finally roots are harvested in autumn after two years of cultivation. Typical botanical characteristics of a three-year-old *P. notoginseng* include: growing to a height of 30–60 cm, determined by the height of floral stem; the stem is upright, unbranched, round with three to five dark green palmate compound leaves on the top and a cluster of white-green flowers or red fruit in the middle (Guo et al., 2010); the main root is conical or cylindrical in length from 1 to 6 cm and its diameter is from 1 to 4 cm, taupe or gray yellow surface with intermittent longitudinal wrinkles and branch root marks; stem marks are well pronounced on top of the root, surrounded by tuberous protuberances (Chinese Pharmacopoeia Commission, 2020).

P. notoginseng is a rare shade plant growing under the forest canopy, which might have originated from the wet evergreen broad-leaf forests of Southwest China (Yang, 2015). Although botanists have failed to find *P. notoginseng* growing in the wild in recent decades, it has been domesticated by local habitants in Yunnan and Guangxi Provinces of China. *P. notoginseng* is sensitive to sunshine, heat, and humidity, and suitable plantation areas are restricted to around N 23.5°, E 104° and at altitudes ranging from 1200 to 2000 m. It has been traditionally and widely cultivated in Wenshan county, Yunnan Province, where now 90% of the *P. notoginseng* raw materials in the whole country are produced with the best quality, and thus this region has been nominated as the homeland of *P. notoginseng* (Guo et al., 2010).

To date, serious continuous cropping obstacles have hindered the sustainable development of the *P. notoginseng* planting industry. The main manifestation of continuous cropping obstacles is due to severe root rot caused by fungi, often resulting in the death of the plant and thus economic losses. Extensive studies on the internal and external barrier mechanisms caused by continuous *P. notoginseng* cropping in recent decades are ongoing. Meanwhile, some composite approaches for continuous planting of *P. notoginseng* have been explored, including the breeding of new cultivars, soil improvement techniques, cultivation without pollution, the use of a quality traceability system, and treatment with microbial inoculum. Thus, in addition, an environmentally friendly cultivation technique and credible quality monitoring system might also improve any soil microbiological environment and promote the sustainable development of the *P. notoginseng* industry (Chen et al., 2017).

6.3 Ethnopharmacology

The roots and rhizomes of *P. notoginseng*, which have played an important role in Chinese healthcare for a long time, are recorded as herbal medicines in the Pharmacopeias of China, the USA, Britain, and Europe and were also included as a dietary supplement by the US Dietary Supplement Health and Education Act in 1994 (103rd United States Congress).

The initial curative description of the roots of *P. notoginseng* can be traced back to the *Compendium of Materia Medica*, written by Shizhen Li (Ming Dynasty). According to this record, *P. notoginseng* was used to treat all kinds of hematologic diseases, such as bleeding caused by swords or axes, blood stasis, hemoptysis, hematemesis, epistaxis, and blood dysentery. In a subsequent book after the *Compendium of Materia Medica*, descriptions of the function of *Sanqi* similar to that of *Panax ginseng* C. A. Meyer were recorded (*Supplements to Compendium of Materia Medica*, Qing Dynasty). The difference is that *P. ginseng* is good for tonifying *qi*, while *P. notoginseng* is especially helpful for nourishing the blood. In the *New Compilation of Materia Medica* (Qing Dynasty) records showed the tonic function of *P. notoginseng* as a folk medicine. In the latest 2020 version of the Chinese Pharmacopoeia, the properties of *Sanqi* are described as "warm in nature, sweet and slightly bitter in taste," attributed to the liver and spleen meridians, with the efficacy to dissipate blood stasis, stop bleeding, promote blood circulation, and alleviate pain, so it can be applied to the treatment of hemoptysis, hematemesis, metrorrhagia and metrostaxis, bleeding and pain caused by traumatic injury, rhinorrhagia, and hemafecia (Chinese Pharmacopoeia Commission, 2020).

Historically, *Sanqi* roots have been used in both the raw and processed forms due to their different therapeutic functions. The raw form has been used to arrest various internal or external hemorrhages, eliminate bruises, improve blood circulation, and treat swelling and pain, while the processed root (usually steamed with water or cooked with wine, and the juice of *Polygonatum cyrtonema*) was recorded to generate a tonic to nourish blood and increase the production of various blood cells in anemic conditions (Editorial Committee of National Administration of Traditional Chinese Medicine, 1998). Since processing might change the inherent chemical constituents with different pharmacological activities in Chinese Materia Medica, *Sanqi* has been thought to possess the characteristic of "the raw materials which eliminate diseases while the processed ones are for tonic" and different forms should be used for different therapeutic purposes (Ye and Zhang, 1997).

Besides the roots, its stems, leaves, and flowers, which can be processed into traditional Chinese medicinal drinks, foods, or cosmetics, also have certain medicinal value and healthcare functions. It is recorded in the *Compendium of Materia Medica* that the stems and leaves of *P. notoginseng* can stop bleeding caused by falling and broken bone injury, with the function of dispersing blood stasis and hemostasis, reducing swelling, and alleviating pain. *P. notoginseng* flowers are green, dried, unopened flower buds. Chinese herbal medicines in Yunnan record that the *P. notoginseng* flower is "sweet in flavor, cold in nature, with the efficacy to clear away heat, calm the liver and lower the blood pressure" and so could be used to treat hypertension and acute laryngopharyngitis. Local minority ethnic groups in Yunnan have the traditional habit of drinking the tea made from *Sanqi* stems, leaves, and flowers, and cooking them together with meat – a tradition since ancient times. In Southeast Asia, the stems, leaves, and flowers of *P. notoginseng* are mainly used in tea-bag form and in wine (Yang et al., 2017). The Yunnan Provincial Health and Family Planning Commission has officially approved the management of flowers, stems, and leaves of *P. notoginseng* as local characteristic raw food materials and initiated the establishment of local standards for food safety of the by-products. Some commercial enterprises have started to make full use of *P. notoginseng* stems, leaves, and flowers as raw materials to produce healthcare products, cosmetics, beverages, and food, such as candy and cookies, which are increasingly preferred by people with health awareness (Chen, 2016, Wang et al., 2019, Yang et al., 2017).

6.4 Phytochemistry

Since the 1980s, with the development of plant separation and structure identification techniques, phytochemical studies have been performed on all parts of *P. notoginseng*. Nearly 200 compounds have been isolated and identified from this plant, and the structures of the main chemical components have been elucidated, including saponins, amino acids, flavonoids, polysaccharides, polyacetylenes, sterols, and cyclodipeptides.

6.4.1 Saponins

The saponins dominate the chemical composition of *P. notoginseng*, and account for about 8% of its weight, which is twice as high as that found in *P. ginseng* and *P. quinquefolium*. Another difference in chemical constituents between *P. notoginseng* and other *Panax* plants is that the ginsenosides in *P. notoginseng* are the dammarane-type tetracyclic triterpenoid saponins, but no oleanane-type compounds. Ginsenosides Rb1, Re, Rd, Rg1, and notoginsenoside R1 (Figure 6.2) are the five major saponins in *P. notoginseng*, among which ginsenoside Rb1, Rg1, and notoginsenoside R1 are stipulated in the Chinese Pharmacopoeia as the standard compounds to evaluate the quality of *P. notoginseng*, and their total weight should not be less than 5%. Chemically, dammarane-type triterpenoid saponins are classified into two categories: 20(*S*)-protopanaxadiol (PPD) and 20(*S*)-protopanaxatriol (PPT), also referred to Rb-group and Rg-group. Compared with PPD-type saponins including ginsenoside Rb1, Rb2, Rb3, Rc, and Rd, there is one more hydroxyl group at the C-6 position, which could be substituted by glycosyl groups and a variety of aglycones in PPT-type saponins, such as ginsenoside Rg1, Rg2, Re, Rf, and notoginsenoside R1, R2 (Qiao et al., 2018) (Figure 6.2).

Unlike polar triterpenoid saponins, less polar ginsenosides (Figure 6.3), also known as rare saponins (such as Rg3, Rg5, Rg6, Rk1, and Rh2), are indeed rare in nature. The less polar ginsenosides have stronger bioactivities than their polar counterparts as the former are more easily absorbed and bind to cell membrane systems (Dou et al., 1999). 20(*S*)-PPD, reported with stronger cytotoxicity on leukemia THP-1 cells (Popovich and Kitts, 2002), has been developed as an anti-cancer drug candidate (Saklani and Kutty, 2008). Rb1 and compound K (CK), the major active metabolites of ginsenoside, exert multiple pharmaceutical activities, including anti-tumor, anti-diabetes, anti-inflammatory, anti-arthritis, skin protection, and neuroprotection activities (Oh and Kim, 2016, Lee et al., 2017, Hu et al., 2012, Chen et al., 2016, Chen et al., 2019, Kang et al., 2016).

Several studies on the newly transformed ginsenosides have successfully revealed both *in vitro* and *in vivo* pharmacological effects on a variety of diseases. For example, to expand the molecular diversity of ginsenosides and search for new promising therapeutic agents, several transformation processes were conducted with chemical (Teng et al., 2004, Gu et al., 2015a, Gu et al., 2015b), physical (Kim et al., 2000, Kim et al., 2013, Ki et al., 2013), or biological (Luo et al., 2013, Han et al., 2007, Zhang et al., 2009) methods to produce rare ginsenosides.

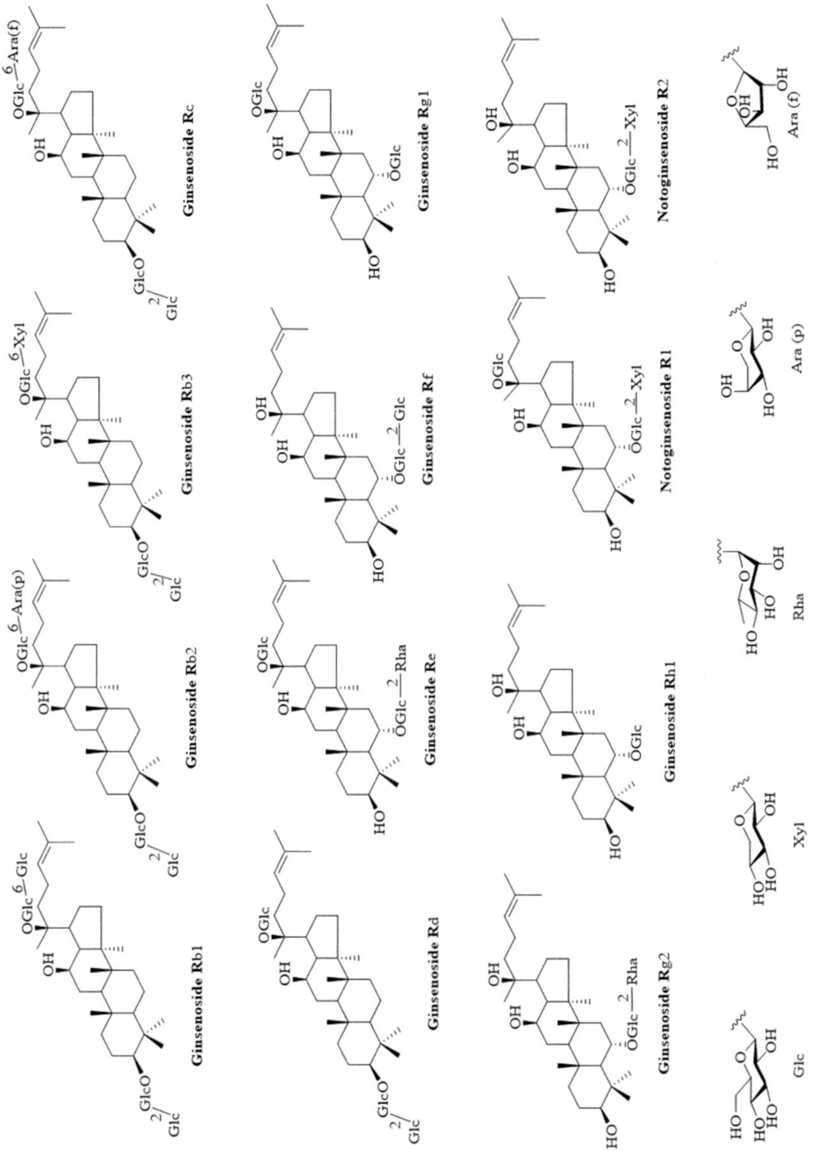

FIGURE 6.2 PPD and PPT type triterpenoid saponins from *Panax notoginseng*.

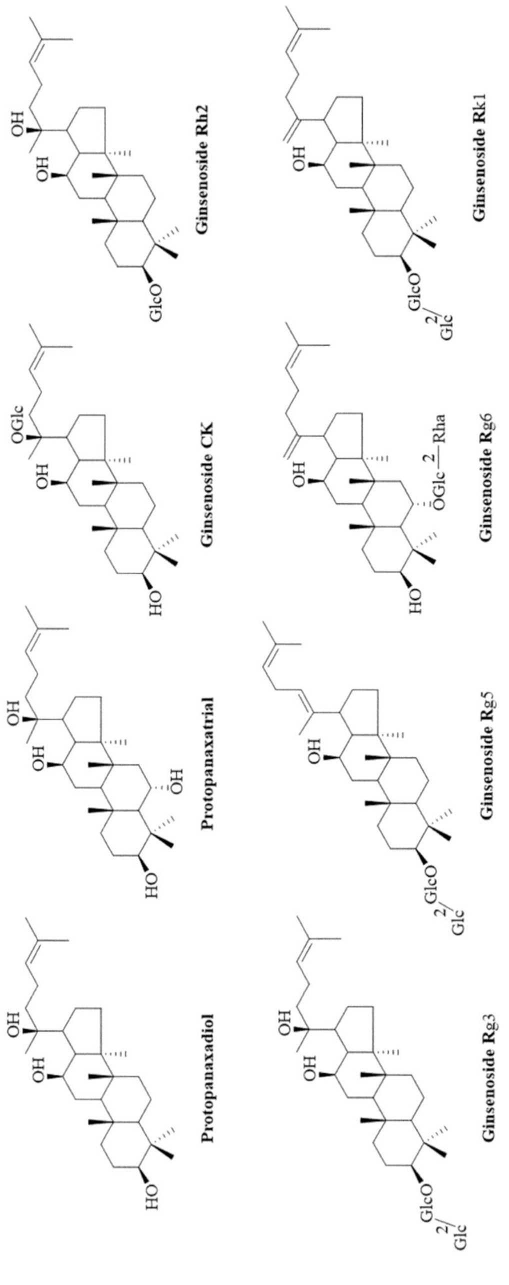

FIGURE 6.3 Less polar triterpenoid saponins from *P. notoginseng*.

Traditional methods for the extraction and separation of PNS, including water decoction, impregnation, and percolation methods, are time consuming and may result in degradation or rearrangement of the component. With the increasing demand for PNS and other bioactive compounds in *P. notoginseng*, new techniques have been developed. Among them, ethanol refluxing extraction is the most frequently used method due to its simple, cheap, environmentally friendly features achieving high extraction rates. The currently used purification methods include alcohol precipitation, membrane separation, aqueous phase extraction, ion exchange fiber method, and macroporous resin adsorption. In particular, membrane separation technology can be used for injection production, because the ultrafiltration membrane avoids the use of heat to remove protein, endotoxin, and macromolecular molecules. In addition to the above techniques, chromatography over silica gel can be used (Ma et al., 2016).

6.4.2 Amino Acids

Dencichine (Figure 6.4), a non-protein amino acid also called *Tianqi* amino acid, is the main hemostatic constituent in *P. notoginseng*. In 1980, Kosuge et al. isolated an amino acid with hemostatic activity from the water-soluble components of *P. notoginseng* for the first time (Kosuge et al., 1981). Its structure was then determined as β-*N*-oxalo-*L*-α-β-diaminopropionic acid by chemical degradation and spectral analysis. Lu et al. isolated dencichine from the root of *P. notoginseng* and determined its content in *P. notoginseng* grown in different years as an average yield of 0.87% (Lu and Li, 1988). The hemostatic mechanism of dencichine was explored and it was shown that it could increase histamine levels to constrict blood vessels; improve the level of platelets and promote blood coagulation; activate endogenous coagulation factor to increase fibrinogen content and inhibit fibrinolysis; release thromboxane A2 (TXA2) after binding to AMPA receptors on platelets and enhance platelets activation (Wang et al., 2014, Huang et al., 2014). Together with dencichine, at least 19 kinds of amino acids in *P. notoginseng* have been detected: seven are essential amino acids for humans, mainly arginine, aspartate, and glutamate (Chen et al., 2003).

The crude dencichine is usually extracted with methanol and water. After removing saponins by ethanol precipitation and *n*-butanol extraction, the extract is separated and purified by cation exchange resin column chromatography, then recrystallized to obtain dencichine (Xie et al., 2007). The earliest determination method for dencichine was carried out on an amino acid analyzer. In recent years, HPLC combined with simple preparative and various chromatographic columns have gradually replaced the original analyzer (Song and Zhang, 2010, Li et al., 2015).

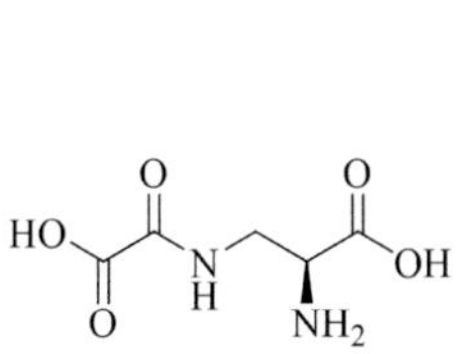

FIGURE 6.4 The structure of dencichine.

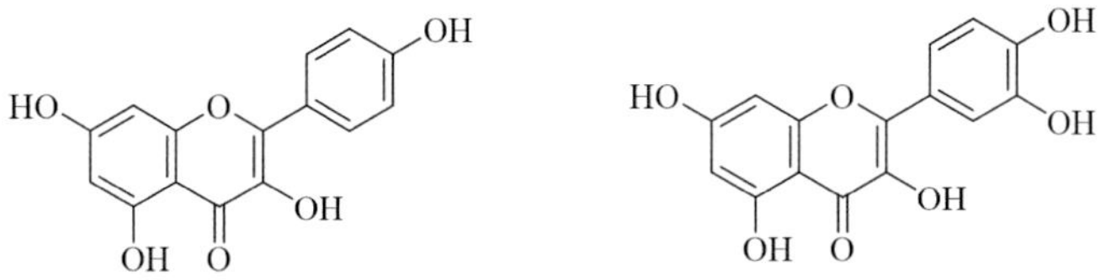

FIGURE 6.5 The structure of kaempferol (left) and quercetin (right).

6.4.3 Flavonoids

To date, 11 flavonoids have been isolated and identified from *P. notoginseng*, including quercetin, kaempferol (Figure 6.5), and their derivatives as the main flavonoids (Choi et al., 2010, Zheng et al., 2006, Huang et al., 2009, Xia et al., 2014). The contents of flavonoids in *Sanqi* vary throughout the stems, leaves, and in the flowers, which contain more flavonoids than the fibrous and main roots (Liu et al., 2015b). Studies on pharmacological activity indicated that total flavonoids of *P. notoginseng* exert anti-oxidant, anti-bacterial, anti-viral, anti-cancer, and liver-protecting activities (Huang et al., 2020).

6.4.4 Polysaccharides

P. notoginseng polysaccharides generally consist of mixtures of glucose, rhamnose, galactose, arabinose, mannose, and xylose. The total polysaccharide is 9.45% on average, with the highest content being in the root of *P. notoginseng* (Xiong et al., 2011). Normally, ion exchange resins and gel chromatography are used to separate and purify polysaccharides, and their content is determined by sulfuric acid-phenol, anthrone-sulfuric acid, and 3,5-dinitrosalicylic acid (DNS) detection methods (Liu et al., 2019). In 1987, sanchinan A was obtained from the root of *P. notoginseng* for the first time (Ohtani et al., 1987). Afterwards, scholars isolated and purified various polysaccharides from the polar extracts, including PBGA11, PBGA12, PF3111, PF3112, PNPS, PNPSIIa, and PNPSIIb (Gao et al., 1996, Sheng et al., 2007). In 2019, two novel polysaccharides MRP5 and MRP5A were discovered, with an average molecular weight of 11.38 × 104 Da (Feng et al., 2019). *P. notoginseng* polysaccharides possess significant immunomodulatory activity, as well as regulating immunity, anti-inflammatory, hypoglycemic, hypolipidemic, and anti-oxidant activities in *in vivo* and *in vitro* tests (Liu et al., 2019). Further exploitation of polysaccharides could contribute to more efficient utilization of this herb and reduce the waste of plant resources.

6.4.5 Polyacetylenes

Polyacetylenes are highly unsaturated long-chain molecules, which contain several hydroxyl groups or an epoxide, e.g. panaxytriol (Zhao et al., 1993), panaxynol, and panaxydol (Yoshikawa et al., 2001) (Figure 6.6). Because of the structural specialization, polyacetylenes exhibit a wide range of biological properties such as cytotoxicity, anti-tumor, anti-platelet, anti-microbial, and enzyme-inhibitory activities (Hisashi et al., 1989, Hisashi, 1995, Teng et al., 1989, Kobaisy et al., 1997). Reports on the natural polyacetylenes are rare, and their absolute configurations are difficult to confirm.

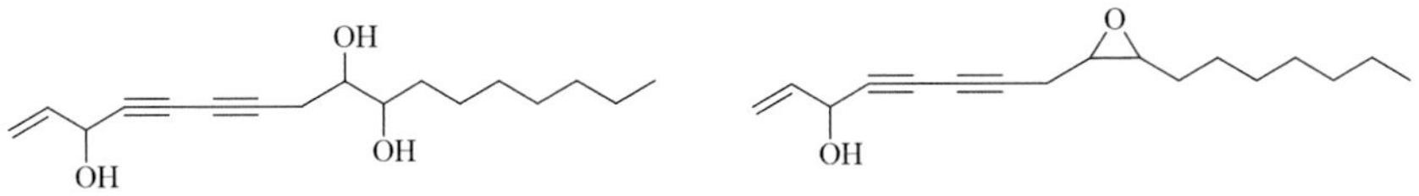

FIGURE 6.6 The structure of panaxytriol (left) and panaxynol (right).

Chemical synthesis strategies are now efficiently used to determine the structures and prepare their isomers, which possess promising activities and are valuable for further exploration (Mao et al., 2016a, Mao et al., 2016b, Mitsuru et al., 2002).

6.4.6 Other Chemical Compounds

Some special trace components in *P. notoginseng* have also been investigated, but there is little literature focusing on their biological activities. Volatile constituents including terpenes, alcohols, aldehydes, olefins, and alkanes were generally analyzed and identified by gas chromatography coupled with mass spectrometry (GC-MS) (Lv et al., 2005). Fourteen cyclodipeptides (Tan et al., 2003), a few sterols, phenolic glycosides, inorganic elements, and mineral salts have been isolated from *P. notoginseng* (Wang et al., 2016). Research in this area deserves more attention and may provide new directions in further studies.

6.5 Biological Activities

Modern pharmacological studies on *P. notoginseng* have indicated that this plant has a wide range of pharmacological activities such as hemostasis, blood-nourishing, anti-thrombotic, detumescence activities, and alleviating pain through the functions of scavenging free radicals, inhibiting platelet agglutination, lowering fat, dilating blood vessels, prolonging coagulation time, and improving microcirculation. *P. notoginseng* is mainly used in a clinical setting on the cardiovascular system, cerebrovascular system, immune system, nervous system, and so on. Most of the promising studies reporting the pharmacological effects of *P. notoginseng* on human diseases are *in vitro* cell-based or animal studies. Herein, we put together the recent studies on medicinal uses and adverse reactions of *P. notoginseng* in clinical trials.

6.5.1 Effects on the Cardiovascular System

Cardiovascular diseases (CVD) are very common with extremely high incidence (Jiang, 2018). Compared with only conventional drugs, the treatments of CVD patients in adjuvant therapy with *P. notoginseng* powder, granule, capsule, or PNS formulations have a more significant therapeutic effect on alleviating adverse symptoms, reducing blood pressure, eliminating swelling, promoting blood circulation, and blood stasis (Jiang, 2018, Jin and Wu, 2018, Wang, 2019, Bartle, 2020). Hao (2017) reported that after taking Sanqi powder orally in warm boiled water, the CVD symptoms of patients such as syncope, dizziness, palpitation, fatigue, and shortness of breath were relieved.

Coronary heart disease (CHD) refers to a serious heart disease caused by myocardial ischemia, hypoxia, and necrosis due to insufficient blood supply of coronary artery, stenosis, and obstruction of the vascular cavity, and angina pectoris (AP) is the most common clinical manifestation of CHD. Clinical studies have shown that the total effective rate of the treatment on CHD patients was significantly increased by adding *Xueshuantong* injection (total saponin of rhizome of *P. notoginseng*), with the AP being relieved and electrocardiogram improved (Cao et al., 2011, Wang, 2013, Luo and Wang, 2018, Wang and Ma, 2018). *Xueshuantong* injection was reported to effectively improve the clinical effect on unstable AP, while *Sanqi Guanxinning* has shown a safe and remarkable curative effect on stable AP (Bao, 2011).

Comparison of the clinical efficacy between a conventional blood-activating drug and Sanqi powder on patients with CHD, hypertension, or other CVDs indicated that oral administration of Sanqi powder can improve blood circulation and body metabolism, as well as promote postoperative recovery more effectively (Zheng and Yu, 2017). Sanqi powder also alleviates hypertension and hyperlipidemia by decreasing systolic and diastolic blood pressure, fasting blood glucose, total cholesterol, and triglyceride (Liu et al., 2017). The results of another clinical experiment showed that Sanqi powder combined with valsartan had a better effect in the treatment of hypertension with a lower adverse reaction rate and higher patient acceptance. Gao and Geng (2018) found that for 42 patients with hyperlipidemia, after taking Sanqi powder orally for 6 weeks, the levels of serum total cholesterol (TC), triglyceride (TG), and the low-density lipoprotein concentrations (LDL-C) were decreased, while the high-density lipoprotein concentrations (HDL-C) were significantly increased after treatment. The aforementioned studies (Hao, 2017, Zheng and Yu, 2017) also noted that Sanqi powder alone has a positive effect in the treatment of hypertension and lipid metabolism. *Xueshuantong* injection combined with benazepril were more active compared with the control group (with benazepril alone) in treating patients with hypertension by decreasing blood pressure and improving the hemodynamic indexes such as whole blood viscosity and erythrocyte specific volume (Li, 2010).

Except for the diseases mentioned above, *P. notoginseng* also exhibits definite effects on other human CVDs in clinical application. For example, *P. notoginseng* has anti-thrombotic, anti-hypertensive, and anti-atherosclerosis effects in clinical studies (Wang, 2019, Wang et al., 2008). *Xueshuantong* injection in combination with western medicine can effectively alleviate clinical symptoms of heart failure (Li, 2013), ischemic heart disease (Chen, 2006), and coronary microvascular disease (CMVD) (Luo, 2011).

6.5.2 Effects on the Cerebrovascular System

According to the literature, PNS and its preparations are the most frequently used forms of *P. notoginseng* to cure cerebrovascular diseases. Aspirin has been commonly used for the treatment of cerebral infarction (CI), but it has adverse effects on the human digestive system (Yang, 2016a). Whereas PNS in combination with aspirin could improve the efficacy rate and decrease gastrointestinal bleeding and recurrence rate (Yan and Cui, 2015, Chen et al., 2018). PNS and *Xueshuantong* injection were reported to alleviate the inflammatory response of ischemic brain tissue and protect the tissue in patients with acute cerebral infarction (ACI) (Gao, 2011, Yang, 2016b).

Sanqi Tongshu capsule combined with ganglioside injection exhibited clinical curative effects in the treatment of ACI by improving the condition of neurological deficit, and regulating the levels of MDA, NSE, TNF-α, IL-6, BDNF, and GSH-px (Tian et al., 2019). Additionally, PNS assists ischemic stroke (IS) and hypertensive cerebellar hemorrhage (HCH) patients in improving cerebral circulation and accelerating the absorption of hematoma (Zhou, 2016, Zhang, 2018).

6.5.3 Effects on the Immune System

Pharmacological studies have indicated that saponins and polysaccharides in *P. notoginseng* possess multiple immunomodulatory activities (Liu et al., 2015a, Zhong et al., 2016). PNS was found to ameliorate joint tenderness swelling and inflammation index in clinical trial when tested for the adjunctive therapy in treating rheumatoid arthritis (RA) patients (Zhang et al., 2007). On the basis of routine treatment, chronic nephritis hematuria patients taking *Yunnan Baiyao capsule* combined with Sanqi powder were found to have better kidney functions than the control group (Sheng et al., 2019). Compound preparation of processed Sanqi was reportedly effective in the treatment of systemic lupus erythematosus (SLE) complicated with dysmenorrhea, with fewer side effects (Le, 2017).

6.5.4 Effects on the Nervous System

PNS adjuvant conventional drug therapy for Alzheimer's disease (AD) can improve the cognitive function and daily behavior ability of patients (Li et al., 2019). *Sanqi Tongshu capsule* combined with anti-epileptic drugs revealed better therapeutic effects on refractory epilepsy (Feng, 2017). *Qiye Shenan tablet* showed favorable efficacy in prolonging sleeping time and alleviating symptoms of headache, dizziness, and palpitations of neurasthenia and anxiety disorder (Yang, 2016b).

6.5.5 Effects on Other Diseases

P. notoginseng is also used clinically to relieve pain, reduce postoperative bleeding, and prevent deep venous thrombosis after surgery (Wang, 2017, Lu and Hong, 2019, Zhang et al., 2018). Furthermore, combined administration of *P. notoginseng* and conventional medicine was shown to be more effective on digestive system diseases such as peptic ulcer, gastritis, gastrointestinal bleeding, ulcerative colitis, liver disease (Rong, 2013), and hemorrhoids (Lu, 2016). Moreover, the combination of *P. notoginseng* and western medicine significantly improves the curative effect and reduces the adverse reaction in patients, indicating that its clinical benefit is worthy of affirmation and practice.

6.6 Related Products

Some classical prescriptions of traditional Chinese medicine containing *P. notoginseng*, created by ancient doctors according to its clinical efficacy, have been processed into various pharmaceutical dosage forms for clinical use in modern society. Nearly

60 kinds of Chinese patent medicines are recorded in the Chinese Pharmacopoeia, and over 900 commercial products containing *P. notoginseng* have been officially approved by China Food and Drug Administration (CFDA), including 612 drugs, such as *Xueshuantong* capsule and injection, and 290 healthcare products, such as medicinal wine and scented tea of *P. notoginseng*. However, the well-known series of commercial products of *Pien Tze Huang* and *Yunnan Baiyao*, in which *P. notoginseng* is one of the main ingredients, has not been collected in this database. *P. notoginseng*-related daily chemical products and food, such as toothpaste, mask powder, facial cleanser, toner, lotion, and cream, as well as flavoring, tea, and wine, are also popular in Yunnan and surrounding provinces.

6.7 Conclusion

P. notoginseng has been used for many centuries, and dozens of its products are well-known and recognized. However, there is still plenty of room for further scientific and clinical exploration. Although many significant investigations have been made into its chemical constituents and pharmacological activities in the past decade, Sanqi powder, total extracts, or PNS rather than individual compounds are usually applied for the pharmacological activities and clinical research. A few of these compounds were reported with superior activities but have not been further studied due to insufficient amounts. Activity-oriented separation and efficient purification techniques might need to be developed for the separation stage. Together with its pharmacological activity, another important aspect of the medicinal potential of *P. notoginseng* is new clinical applications. Further research is required to ensure more comprehensive utilization of this plant can be achieved.

In addition, the continuous cropping obstacle radically reduces the quantity and quality of this herb and becomes a difficult problem in the cultivation process of *P. notoginseng*. Researchers have been developing individualized approaches, such as chemical, biological, and new cultivation methods to overcome this issue, but the problem is yet to be solved. Hence more attention should be paid to address this issue for the sustainable utilization of this resource.

REFERENCES

103rd United States Congress. Dietary Supplement Health and Education Act of 1994. The United States of America. https://www.govinfo.gov/content/pkg/STATUTE-108/pdf/STATUTE-108-Pg4325.pdf.

Bao LX, 2011. Clinical research of sanqi *Panax notoginseng* bolted treatment a stable angina. *China J Chin Med*, 26: 1371–1372.

Bartle SH, 2020. Clinical effect of *Panax notoginseng* in the treatment of patients with cardiovascular diseases. *J Clin Med Lit (Electronic Ed)*, 7: 161–162.

Cao H, Lu CH, Nie GL, 2011. Clinical effect on xueshuantong injection in the treatment of coronary heart disease and angina. *Chin Med Herald*, 8: 14–16.

Chen HH, 2016. Development of sanqi crisp flavor biscuit. *J Wenshan Univ*, 29: 35–38.

Chen JY, Si M, Wang Y, et al., 2019. Ginsenoside metabolite compound K exerts anti-inflammatory and analgesic effects via downregulating COX2. *Inflammopharmacology*, 27: 157–166.

Chen L, Chen SX, Yang M, et al., 2018. Clinical analysis of 120 cases of cerebral infarction treated by Panax notoginseng saponins combined with aspirin. *China J Modern Distance Education Tradit Chin Med*, 16: 54–56.

Chen WJ, Wang JL, Luo Y, et al., 2016. Ginsenoside Rb1 and compound K improve insulin signaling and inhibit ER stress-associated NLRP3 inflammasome activation in adipose tissue. *J Ginseng Res*, 40: 351–358.

Chen YY, 2006. Clinical observation on xueshuantong injection in the treatment of 40 cases of ischemic heart disease. *Hainan Med J*, 17: 46–47.

Chen ZJ, Sun YQ, Dong TX, et al., 2003. Comparison of amino acid contents in *Panax notoginseng* from different habitats. *Zhong Yao Cai*, 26: 86–88.

Chen ZJ, Wei FG, Dong LL, 2017. Pivotal technologies of *Panax notoginseng* continuous cropping. *World Science and Technology/Modernization of Traditional Chinese Medicine and Materia Medica*,19: 1629–1634.

Chinese Pharmacopoeia Commission, 2020. *Chinese Pharmacopoeia.* Notoginseng Radix et Rhizoma. The Medicine Science and Technology Press of China, Beijing.

Choi RCY, Zhu JTT, Leung KW, et al., 2010. A flavonol glycoside, isolated from roots of *Panax notoginseng*, reduces amyloid-beta-induced neurotoxicity in cultured neurons: signaling transduction and drug development for alzheimer's disease. *J Alzheimers Dis*, 19: 795–811.

Dou DQ, Jin LC, Chen YJ, 1999. Advances and prospects of the study on chemical constituents and pharmacological activities of *Panax ginseng. J Shenyang Pharm Univ*, 16: 151–156.

Editorial Committee of National Administration of Traditional Chinese Medicine, 1988. *Zhong Hua Ben Cao Jing Xuan Ben*. Shanghai Science and Technology Press, 1: 34–38.

Feng SL, Cheng HR, Xu Z, et al., 2019. Antioxidant and anti-aging activities and structural elucidation of polysaccharides from the root of *Panax notoginseng. Process Biochem*, 78: 189–199.

Feng SN, 2017. Clinical study on the treatment of 128 cases of intractable epilepsy with sanqi tongshu capsule and antiepileptic drugs. *J Dis Surveil Contr*, 11: 845–846.

Gao H, Wang FZ, Lien EJ, et al., 1996. Immunostimulating polysaccharides from *Panax notoginseng. Pharm Res*, 13: 1196–1200.

Gao JH, 2011. Clinical observation of *Panax notoginseng* saponins injection in the treatment of acute cerebral infarction. *Chin J Pract Neurol*, 14: 81–82.

Gao QR, Geng ZQ, 2018. Clinical observation of pseudo-ginseng in the treatment of hyperlipidemia by Chinese medicine. *Shanghai Med Pharm J*, 39: 23–25.

Gu CZ, Lv JJ, Zhang XX, et al., 2015a. Triterpenoids with promoting effects on the differentiation of PC12 cells from the steamed roots of *Panax notoginseng. J Nat Prod* 78: 1829–1840.

Gu CZ, Lv JJ, Zhang XX, et al., 2015b. Minor dehydrogenated and cleavaged dammarane-type saponins from the steamed roots of *Panax notoginseng. Fitoterapia*, 103: 97–105.

Guo HB, Cui XM, An N, et al., 2010. Sanchi ginseng (*Panax notoginseng* (Burkill) F. H. Chen) in China: distribution, cultivation and variations. *Genet Resour Crop Evol*, 57: 453–460.

Han Y, Jiang BH, Hu XM, et al., 2007. Biotransformatin of minor anticancer compound in *Panax notoginseng* stalks and leaves by *Fusarium sacchari. Chin Tradit Herb Drugs*, 38: 830–832.

Hao YL, 2017. Clinical effect of *Panax notoginseng* in the treatment of patients with cardiovascular diseases. *Guide China Med*, 15: 165–166.

Hisashi M, 1995. A possible mechanism for the cytotoxicity of a polyacetylenic alcohol, panaxytriol: inhibition of mitochondrial respiration. *Cancer Chemother Pharmacol*, 4: 291–296.

Hisashi M, Mitsuo K, Hiroshi Y, et al., 1989. Studies on the panaxytriol of *Panax* ginseng C. A. Meyer. isolation, determination and antitumor activity. *Chem Pharm Bull*, 37: 1279–1281.

Hu C, Song G, Zhang B, et al., 2012. Intestinal metabolite compound K of panaxoside inhibits the growth of gastric carcinoma by augmenting apoptosis via Bid-mediated mitochondrial pathway. *J Cell Mol Med*, 16: 96–106.

Huang F, Xiang FJ, Wu JX, 2009. Research progress in the extraction and separation of saponins and monomers from the leaves of *Panax notoginseng*. *Mater J Chin Med*, 32: 999–1005.

Huang LF, Shi HL, Gao B, 2014. Decichine enhances hemostasis of activated platelets via AMPA receptors. *Thromb Res*, 133: 848–854.

Huang ZY, Liu WJ, Chen Y, 2020. Research progress of *Panax notoginseng* flavonoids. *J Liaoning Univ Tradit Chin Med* 22: 81–84.

Jiang BJ, 2018. Clinical effect of *Panax notoginseng* in the treatment of patients with cardiovascular diseases. *Cardiovasc Dis J Integr Tradit Chin West Med*, 6: 163–166.

Jin P, Wu J, 2018. Research progress of *Panax notoginseng* total saponins in cardiovascular diseases. *Chin J Cardiovasc Med*, 23: 527–530.

Kang S, Siddiqi MH, Yoon SJ, et al., 2016. Therapeutic potential of compound K as an IKK inhibitor with implications for osteoarthritis prevention: an *in silico* and *in vitro* study. *In Vitro Cell Deve-An* 52: 895–905.

Ki SK, Ham J, Kim YJ, et al., 2013. Heat-processed *Panax ginseng* and diabetic renal damage: active components and action mechanism. *J Ginseng Res*, 37: 379–388.

Kim EJ, Oh HA, Choi HJ, et al., 2013. Heat-processed ginseng saponin ameliorates the adenine-induced renal failure in rats. *J Ginseng Res*, 37: 87–93.

Kim WY, Kim JM, Han SB, et al., 2000. Steaming of ginseng at high temperature enhances biological activity. *J Nat Prod*, 63: 1702–1704.

Kobaisy M, Abramowski Z, Lermer L, et al., 1997. Antimycobacterial polyynes of devil's club (*Oplopanax horridus*), a North American native medicinal plant. *J Nat Prod*, 60: 1210–1213.

Kosuge T, Yokota M, Ochiai A, 1981. Studies on antihemorrhagic principles in the crude drugs for hemostatics. II. on antihemorrhagic principle in sanchi Ginseng Radix (author's transl). *J Pharm Soc Jpn*, 101: 629–632.

Le YM, 2017. Clinical observation on the treatment of systemic lupus erythematosus complicated with dysmenorrhea using compound formula with processed sanqi. *J Rheum Arthritis*, 6: 30–32.

Lee SH, Kwon MC, Jang JP, et al., 2017. The ginsenoside metabolite compound K inhibits growth, migration and stemness of glioblastoma cells. *Int J Oncol*, 51: 414–424.

Li AD, Chen JL, Xia J, 2019. A multicenter randomized controlled clinical study of notoginsenoside in the treatment of Alzheimer's disease. *Geriatr Health Care*, 25: 501–505.

Li GS, 2010. Influence of xueshuantong injection to blood patients pressure with hypertensive disease and blood rheology. *National Conference on Integrative Chinese and Western Medicine for Prevention and Treatment of Cardiovascular and Cerebrovascular Diseases*. 19–21 November 2010, Beijing, China.

Li L, Wang CX, Qu Y, et al., 2015. Determination of dencichine in *Panax notoginseng* by reversed phase ion pair chromatography. *China J Chin Mater Med*, 40: 4026–4030.

Li XP, 2013. Clinical observation of xueshuantong combined with western medicine in elderly patients with coronary heart disease and heart failure. *Chin Foreign Med Res*, 11: 10–11.

Liu CM, Wang L, Ning YY, 2015a. The effect of sanqi powder on the immunoregulation of mice. *Asia Pac Tradit Med* 11: 7–8.

Liu H, Wang WJ, Lv JF, 2017. Discussion and evaluation of the therapeutic effect of *Panax notoginseng* on patients with three high diseases (hypertension, hyperglycemia and hyperlipidemia). *China Stand Health Manag*, 8: 104–107.

Liu WJ, Chen Y, Pang DQ, et al., 2019. Research progress of *Panax notoginseng* polysaccharides. *J Liaoning Univ TCM*, 21: 137–140.

Liu Y, Qu Y, Wang CX, et al., 2015b. Determination of total flavonoids in *Panax notoginseng* from different places. *J Anhui Agric Sci*, 43: 54–58.

Lu LF, 2016. Effect and safety analysis of sanqi huazhiwan in the treatment of hemorrhoids. *World Latest Med Inf*, 16: 145.

Lu Q, Li XG, 1988. Isolation, identification and determination of hemostatic components in *Panax notoginseng*. *Chin Tradit Pat Med*, 9: 34–35.

Lu ZQ, Hong ZH, 2019. Clinical effect of sanqi decoction combined with omeprazole in the treatment of gastric hemorrhage and its effect on gastric function. *J North Pharm*, 16: 99–100.

Luo JH, 2011. Study on treatment of cardiac syndrome x with xueshuantong injection. *J Clin Exp Med*, 10: 977–978.

Luo JH, Wang C, 2018. Effect of xueshuantong on serum inflammatory factors and hemorheology in elderly patients with coronary heart disease. *Chin J Gerontol*, 14: 3358–3360.

Luo SL, Dang LZ, Li JF, et al., 2013. Biotransformation of saponins by endophytes isolated from *Panax notoginseng*. *Chem Biodivers*, 10: 2021–2031.

Lv Q, Qing J, Zhang P, et al., 2005. Simultaneous distillation and solvent extraction and GC/MS analysis of volatile oil from flowers of *Panax notoginseng* (Burk.) F. H. Chen. *Chin J Pharm Anal*, 25: 284–287.

Ma ZQ, Pu JX, Qu YL, et al., 2016. Research progress in extraction, separation and purification of total saponins of *Panax notoginseng*. *China Mod Med*, 23: 19–22.

Mao JY, Li SN, Zhong JC, et al., 2016a. Total synthesis of panaxydol and its stereoisomers as potential anticancer agents. *Tetrahedron: Asymmetry*, 27: 69–77.

Mao JY, Zhong JC, Wang B, et al., 2016b. Synthesis of panaxytriol and its stereoisomers as potential antitumor drugs. *Tetrahedron: Asymmetry*, 27: 330–337.

Mitsuru S, Masayoshi I, Mitsuo W, et al., 2002. Absolute structure of panaxytriol. *Chem Pharm Bull*, 50: 126–128.

Oh J, Kim JS, 2016. Compound K derived from ginseng: neuroprotection and cognitive improvement. *Food Funct*, 7: 4506–4515.

Ohtani K, Mizutani K, Hatono S, et al., 1987. Sanchinan-A, a reticuloendothelial system activating arabinogalactan from sanchi-ginseng (roots of *Panax notoginseng*). *Planta Med*, 53: 166–169.

Popovich DG, Kitts DD, 2002. Structure-function relationship exists for ginsenosides in reducing cell proliferation and inducing apoptosis in the human leukemia (THP-1) cell line. *Arch Biochem Biophys*, 406: 1–8.

Qiao YJ, Shang JH, Wang D, et al., 2018. Research of *Panax* spp. in Kunming Institute of Botany, CAS. *Nat Prod Bioprospect*, 8: 245–263.

Rong XX, 2013. Recent research on the treatment of digestive diseases with *Panax notoginseng*. *J Med Pharm Chin Minor*, 19: 65–66.

Saklani A, Kutty Sk, 2008. Plant-derived compounds in clinical trials. *Drug Discov Today*, 13: 3–4.

Shang CB, Lowry IP, 2007. *Araliaceae. Flora of China*. Beijing: Science Press, 13: 435–491.

Sheng HP, Xu MF, Xu JL, 2019. Clinical observation on yunnanbaiyao capsule combined with sanqi powder in treating chronic nephritis hematuria. *Guangming J Chin Med*, 34: 2321–2324.

Sheng XH, Wang J, Guo JJ, et al., 2007. Study on the purification and physicochemical properties of *Panax notoginseng* polysaccharide. *Chin Tradit Herb Drugs*, 987–989.

Song LL, Zhang YP, 2010. Determination of dencichine in *Panax notoginseng* by HPLC-ELSD. *Beijing J Tradit Chin Med*, 29: 216–217.

Tan NH, Wang SM, Yang YB, et al., 2003. Cyclodipeptides of *Panax notoginseng* and lactams of *Panax ginseng*. *Acta Botanica Yunnanica*,25: 366–368.

Teng CM, Kuo SC, Ko FN, et al., 1989. Antiplatelet actions of panaxynol and ginsenosides isolated from *Panax ginseng* C. A. Meyer. *Biochimi Biophys Acta-Gen Subj*, 990: 315–320.

Teng RW, Li HZ, Wang DZ, et al., 2004. Hydrolytic reaction of plant extracts to generate molecular diversity: new dammarane glycosides from the mild acid hydrolysate of roots of *Panax notoginseng*. *Helv Chim Acta*, 87: 1270–1278.

Tian YQ, Huo JL, Zhao XF, et al., 2019. Clinical study of sanqitongshu capsule combined with ganglioside in the treatment of acute cerebral infarction. *Drugs Clinic*, 34: 3232–3235.

Wang AL, 2019. Clinical effect of sanqi capsule in the treatment of cardiocerebro vascular disease. *Clinic Res Pract*, 4: 129–130.

Wang D, 2013. Curative effect analysis of thrombus on unatable angina pectoris of coronary disease. *China J Pharm Econ*, 8: 205–207.

Wang HY, Guo DM, Yu WP, et al., 2019. Study on the candy of pressing the stem and leaf of *Panax notoginseng*. *J Xihua Univ (Nat Sci Ed)* 38: 69–72.

Wang N, Wan JB, Li MY, et al., 2008. Advances in studies on *Panax notoginseng* against atherosclerosis. *Chin Tradit Herb Drugs* 39: 787–790.

Wang T, Guo RX, Zhou GH, et al., 2016. Traditional uses, botany, phytochemistry, pharmacology and toxicology of *Panax notoginseng* (Burk.) F.H. Chen: a review. *J Ethnopharmacol*, 188: 234–258.

Wang TR, Ma W, 2018. Influence of xueshuantong injection on serum IL-6 and MMP-2 in patients with unstable angina pectoris. *Chin Herald Med*, 15: 127–130.

Wang YF, 2017. Analysis of the effect of sanqi powder combined with nursing intervention on postoperative pain of hip fracture. *Nei Mongol J Tradit Chin Med*, 36: 153–154.

Wang Z, Yang JY, Song SJ, 2014. Effect of *Panax notoginseng* on coagulation and hemostasis mechanism. *Chin J New Drugs* 23: 356–359.

Xia PG, Zhang SC, Liang ZS, et al., 2014. Research history and overview of chemical constituents of *Panax notoginseng*. *Chin Tradit Herb Drugs*, 45: 2564–2570.

Xie GX, Qiu MF, Zhao AH, et al., 2007. Separation, purification and structural analysis of dencichine from *Panax notoginseng*. *Nat Prod Res Dev*, 6: 1059–1061.

Xiong YH, Li J, Huang S, et al., 2011. Determination of total polysaccharides in *Panax notoginseng* by DNS. *Asia Pac Tradit Med*, 7: 7–9.

Yan XY, Cui DZ, 2015. Clinical observation of notoginsenoside combined with aspirin in the treatment of cerebral infarction. *Clinic Res Tradit Chin Med*, 7: 3–5.

Yang CR, 2015. The history and origin of *Panax notoginseng*. *J Mod Chin Medi Res Pract*, 29: 83–86.
Yang G, Cui XM, Chen M, et al., 2017. Study on new food raw materials of *Panax notoginseng* stem, leaf and flower. *J Chin Pharm Sci*, 52: 543–547.
Yang HY, 2016a. Adverse reactions of gastrointestinal bleeding in patients with coronary heart disease caused by enteric aspirin. *Chin J Mod Drug Appl*, 10: 154–155.
Yang WJ, 2016b. Clinical application of notoginsenoside in the treatment of nervous system diseases. *J North Pharm*, 13: 160–161.
Ye DJ, Zhang SC, 1997. *The processing of Chinese Materia Medica*. Beijing: People's Medical Publishing House.
Yoshikawa M, Morikawa T, Yashiro K, et al., 2001. Bioactive saponins and glycosides. XIX. Notoginseng (3): immunological adjuvant activity of notoginsenosides and related saponins: structures of notoginsenosides-L, -M, and -N from the roots of *Panax notoginseng* (Burk.) F. H. Chen. *Chem Pharm Bull*, 49: 1452–1456.
Zhang CX, 2018. Clinical observation on the treatment of cerebral hemorrhage after taking aspirin in adjuvant treatment of sanqi powder for a long time. *J Clin Med Lit (Electronic Ed)*, 5: 156.
Zhang DD, Cao JQ, Zhao YQ, 2009. Isolation and identification of a new biotransformation compound by *Fusarium sacchari* from saponin of *Panax notoginseng* leaves. *Chin Tradit Herb Drugs*, 40: 1863–1865.
Zhang JH, Wang JP, Wang HJ, 2007. Clinical study on effect of total saponins of *Panax notoginseng* on immunue related inner environment imbalance on rheumatoid arthritis patients. *Chin J Integr Med* 27: 589–592.
Zhang JL, Zhou LH, Chen YH, et al., 2018. Research progress of clinical application of sanqi powder. *Yunnan J Tradit Chin Med Mater Med*, 39: 80–82.
Zhao P, Liu YQ, Yang CR, 1993. Minor constituents from the root of *Panax notoginseng* (1). *Acta Botanica Yunnanica*, 15: 409–412.
Zheng S, Yu LM, 2017. Clinical analysis on the treatment of cardiovascular disease with *Panax notoginseng*. *Clinic J Chin Med*, 9: 71–72.
Zheng Y, Li XW, Gui MY, et al., 2006. Studies on flavonoids from stems and leaves of *Panax notoginseng*. *Chin Pharm J*, 41: 176–178.
Zhong YY, Yang XH, Zhang YW, et al., 2016. Effect of *Panax notoginseng* polysaccharide on immune function of mice. *West China J Pharm Sci*, 31: 573–576.
Zhou CL, 2016. Clinical research of total saponin of *Panax notoginseng* in the treatment of early cerebral hemorrhage. *Clinic Res Pract*, 1: 73–74.

7

Medicinal Orchids: *Dendrobium* Species and *Gastrodia elata* Blume

Jiang-Miao Hu*
State Key Laboratory of Phytochemistry and Plant Resources in West China, Kunming Institute of Botany, Chinese Academy of Sciences, Kunming, Yunnan, People's Republic of China

University of Chinese Academy of Sciences, Beijing, People's Republic of China.

Jie Yang
State Key Laboratory of Phytochemistry and Plant Resources in West China, Kunming Institute of Botany, Chinese Academy of Sciences, Kunming, Yunnan, People's Republic of China

University of Chinese Academy of Sciences, Beijing, People's Republic of China.

Liu Yang
State Key Laboratory of Phytochemistry and Plant Resources in West China, Kunming Institute of Botany, Chinese Academy of Sciences, Kunming, Yunnan, People's Republic of China

Meng-Ting Kuang
State Key Laboratory of Phytochemistry and Plant Resources in West China, Kunming Institute of Botany, Chinese Academy of Sciences, Kunming, Yunnan, People's Republic of China

University of Chinese Academy of Sciences, Beijing, People's Republic of China.

CONTENTS

* Corresponding author.

7.1 Introduction

"Shi-Hu" (*Dendrobium* species) was first recorded in *Shen Nong's Herbal Classic* (*Shennong Bencao Jing*) as a top-grade herbal medicine recognized for nourishing the *yin*, clearing heat, benefiting the stomach, and promoting fluid production. There are in fact more than 60 *Dendrobium* species distributed in the southern part of Yunnan, and the commonly cultivated varieties are *D. officinale*, *D. devonianum*, *D. chrysotoxum*, *D. nobile*, etc. They contribute to more than half of China's production. Chemical and pharmacological research has been performed on about 52 species. The chemical components mainly include polysaccharides, aromatic compounds (stilbenes, fluorenones, flavones, and phenylpropanoids), sesquiterpenes, and alkaloids. "Shi-Hu" possesses biological activity related to immunomodulation, antitumor, hypoglycemic, anti-oxidant, and anti-aging, etc.

"Tian-Ma", from the underground tuber of *Gastrodia elata* Blume, is also named "Chijian," "Limu," "Dingfengcao," etc. The name "Chijian" was first recorded as the top-grade herb in the *Shen Nong's Herbal Classic*, and then the name of "Tianma" was first recorded by "Kai Bao Materia Medica" in the Song Dynasty. As one of the famous herbs in China, "Tianma" has such function as an antispasmodic, improves circulation, and relieves headaches. Thus, "Tianma" was mainly used for limb numbness, dizziness, convulsions, stroke, hypertension, rheumatism, epilepsy, and so on. Prof. Zhou and his group began their pharmaceutical research with *G. elata* at the end of 1970s and described the main composition of the species. With the development of chromatography and spectroscopy, the chemical constituents of "Tianma" were further described, and to date, more than 97 compounds, including phenols, organic acid compounds, heterocycles, and sterols, have been isolated and elucidated from the species. For the pharmacological aspects, research on the sedative-hypnotic, anti-convulsion, pain-easing, and ischemic cerebral injury protection properties of *G. elata* has been conducted.

7.2 *Dendrobium*

"Shi-Hu," from the stems of *Dendrobium* species, is a perennial evergreen herbaceous also named "Huang-Cao." *Dendrobium* stems are erect, fleshy and thick, slightly flattened, cylindrical, 10–60 cm long, and 1.3 cm thick (Editorial Committee of Chinese Flora of Chinese Academy of Sciences, 1999b). The surface is yellow-green, smooth or with longitudinal stripes; the nodes are obvious, the color is dark, and there are membranous leaf sheaths on the nodes (Commission, 2020).

7.2.1 Resource Availability

Dendrobium is a large and diverse genus of the Orchidaceae family with approximately 1500 species distributed throughout the world and mainly distributed in tropical Asian and Pacific islands. Similar to other epiphytic orchids, *Dendrobium* has strict requirements on the ecological environment. Except for a few terrestrial plants (which are not produced in China), the rest are epiphytic. Trees or rocks in forests, especially tropical rain forests with high temperature and humidity, are places where they grow and reproduce.

Dendrobium is used differently in different places. About 30 species of *Dendrobium* are collected for medicinal use nationwide. There are several species (*D. officinale* Wall. ex Lindl., *D. nobile* Lindl., *D. huoshanense* C. Z. Tang et S. J. Cheng, *D. chrysotoxum* Lindl., and *D. fimbriatum* Hook.) (Figure 7.1) that have been included in the Chinese Pharmacopoeia over the years (Commission, 2020). In addition, fresh or dried products of similar species from the same genus can be used instead.

7.2.2 Plant Parts for Medicinal Uses

Traditionally, the stem of *Dendrobium* has been used for its traditional properties of nourishing the kidney, moisturizing the lung, benefiting the stomach, promoting the production of body fluids, and clearing heat. Recently, *Dendrobium* has been used for the treatment of cancer, chronic atrophic gastritis, skin aging, fever, and cardiovascular disease (Xu et al., 2019).

FIGURE 7.1 Photos of *Dendrobium* species. (A) *D. officinale*; (B) *D. nobile*; (C) *D. huoshanense*; (D) *D. chrysotoxum*; (E) *D. fimbriatum* (photos by Hu Jiang-Miao).

7.2.3 Phytochemistry

Dendrobium plants contain various types of chemical components and structures, such as alkaloids, sesquiterpenes, bibenzyls, phenanthrenes, fluorenones, flavonoids, phenylpropanoids, sterols, and polysaccharides, etc. The chemical composition varies widely among different species of *Dendrobium*. Herein, we briefly introduce the typical active compounds of *Dendrobium*.

7.2.3.1 Typical Compounds

7.2.3.1.1 Alkaloids

Suzuki Hideki first isolated alkaloids named dendrobine from *D. nobile* (Ye et al., 2002). As the main active compound, dendrobine was then listed as the quality control standard for *D. nobile* in the Chinese Pharmacopoeia. A total of 52 constituents of alkaloids were found in plants of *Dendrobium*. They could be divided into the following five types (Li et al., 2019).

(a) Sesquiterpene alkaloids

The basic skeleton is a 15-carbon picrotoxane-type sesquiterpene which is a tight four-ring system consisting of a nitrogen-containing pyrrole ring and a sesquiterpene moiety (Figure 7.2 and Table 7.1). The sesquiterpene alkaloids are mainly found in *D. nobile* and *D. findleyanum* and could be divided into dendrobine type, dendroxine type, nobiline type, and other types.

(b) Octahydroindolizine alkaloids

It is a kind of alkaloid with a five- and six-membered ring system (Figure 7.3A). These compounds exert various biological activities such as antiviral, anti-tumor, anti-inflammatory, anti-fungal, anti-oxidation, and immunomodulation activities (Li et al., 2019).

(c) Amide alkaloids

Amide alkaloids are a class of alkaloids whose nitrogen atoms are not bound in the ring (Figure 7.3B). Usually, the *Dendrobium* species with high polysaccharide content such as *D. officinale*, *D. moniliforme*, and *D. devonianum* contain amide alkaloids (Li et al., 2019).

(d) Pyrrolidine alkaloids

These alkaloids have been found in five *Dendrobium* species: *D. chrysanthum*, *D. loddigesii*, *D. aphyllum*, *D. lohohense*, and *D. primulinum* (Li et al., 2019) (Figure 7.3C).

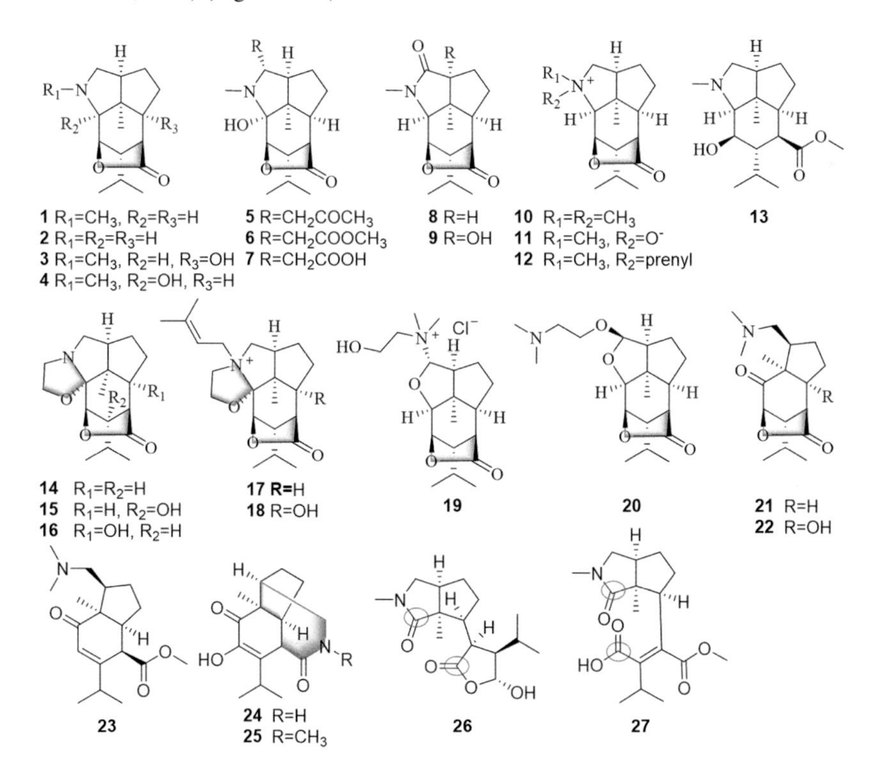

FIGURE 7.2 Reported picrotoxane type alkaloids in *Dendrobium* species.

TABLE 7.1

Reported Picrotoxane-Type Alkaloids in *Dendrobium*

No.	Chemical Names	References
1	Dendrobine	Ye et al., 2002
2	Mubironine B	Morita et al., 2000
3	Dendramine	Inubushi and Nakano, 1965
4	10-hydroxy-dendrobine	Inubushi et al., 1964
5	Dendronobiline A	Liu and Zhao, 2003
6	Dendrine	Inubushi and Nakano, 1965
7	(-)-(1R, 2S, 3R, 4S, 5R, 6S, 9S, 11R)-11-carboxymethyldendrobine	Meng et al., 2017
8	Mubironine A	Morita et al., 2000
9	3-hydroxy-2-oxodendrobine	Wang and Zhao, 1986
10	N-methyl-dendrobine	Inubushi et al., 1964
11	Dendrobine-N-oxide	Hedman and Leander, 1972
12	N-isopentenyl-dendrobine	Hedman and Leander, 1972
13	Mubironine C	Morita et al., 2000
14	Dendroxine	Okamoto et al., 1966
15	4-hydroxy-dendroxine	Okamoto et al., 1972
16	6-hydroxy-dendroxine	Okamoto et al., 1972
17	N-isopentenyl-dendroxine	Hedman and Leander, 1972
18	N-isopentenyl-6-hydroxy-dendroxine	Hedman and Leander, 1972
19	Dendrowardine	Glomqvist et al., 1973
20	Wardianumine A	Zhang et al., 2017
21	Nobiline	Liu et al., 2007
22	6-hydroxy-nobiline	Liu et al., 2007
23	Moniline	Liu et al., 2007
24	Findlayines A	Yang et al., 2018
25	Findlayines B	Yang et al., 2018
26	Findlayines C	Yang et al., 2018
27	Findlayines D	Yang et al., 2018

(e) Imidazole alkaloids

Imidazole is an aromatic five-membered heterocyclic compound with two meta nitrogen atoms in the first and third positions in the molecular structure (Figure 7.3D). Leader et al. isolated dendroparine from *D. anosmum*, which is the only imidazole found in *Dendrobium* alkaloids (Li et al., 2019).

7.2.3.1.2 Sesquiterpenes

The terpenoids reported in *Dendrobium* are mainly sesquiterpenes, but not all *Dendrobium* species contain sesquiterpenes. Sesquiterpenes in this genus can be divided into picrotoxane-type, copacamphane-type, cyclocopacamphane-type, cadinane-type, aromadendrane-type, emmotin-type, and others (Fan, 2013) (Figure 7.4).

Picrotoxane-type sesquiterpenes could be transformed into dendrobine-type alkaloids in plants. The natural picrotoxane skeleton sesquiterpene could form a lactone

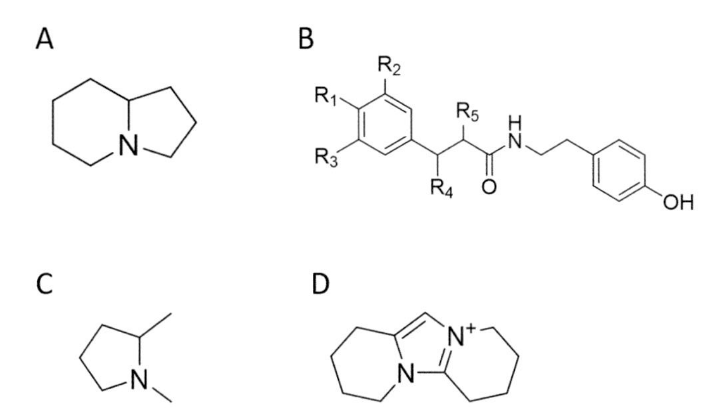

FIGURE 7.3 (A) Structural skeleton of octahydroindolizine alkaloids from *Dendrobium*; (B) Structural skeleton of amide alkaloids from *Dendrobium*; (C) Structural skeleton of pyrrolidine alkaloids from *Dendrobium*; (D) Structure of dendroparine.

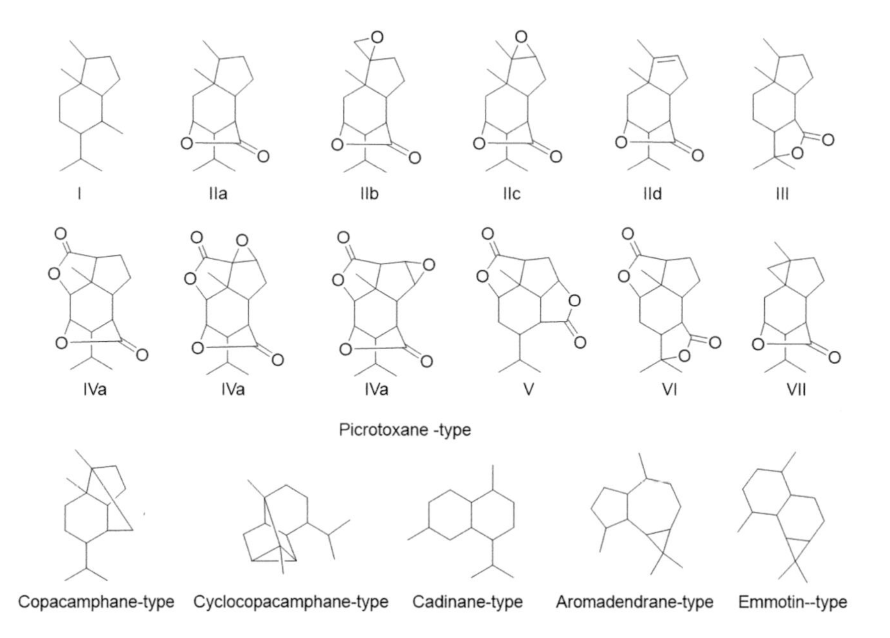

FIGURE 7.4 Structure skeleton of sesquiterpenes in *Dendrobium*.

ring or ternary epoxy ether, and its structure can be divided into I–VII. Most picrotoxane sesquiterpenes contain multiple chiral carbons. Besides, hydroxy substitution at different positions and numbers enriches the structure categories of picrotoxane-type sesquiterpenes. Picrotoxane-type sesquiterpenes from *Dendrobium* exhibit an angiogenesis effect against sunitinib-induced damage on intersegmental blood vessels (Meng et al., 2017), and inhibitory activity on nerve growth factor–mediated neurite outgrowth (Leon et al., 2019).

7.2.3.1.3 Aromatic Compounds

There are many aromatic compounds reported in *Dendrobium* species, with high contents of bibenzyls, phenanthrenes, and phenylpropanoids (Fan, 2013).

(a) Bibenzyls

Bibenzyls are active ingredients with anti-tumor effects and antiplatelet aggregation in *Dendrobium* plants and with wide existence (Figure 7.5A) (Ye et al., 2002, Chen and Guo, 2001).

(b) Phenanthrenes

The phenanthrenes are important active ingredients with anticancer effects obtained from *Dendrobium* plants (Ye et al., 2002, Chen and Guo, 2001) (Figure 7.5B). There are more than 40 phenanthrene compounds reported in *Dendrobium* plants (Hu et al., 2012).

(c) Fluorenone

The content of fluorenone in *Dendrobium* is not high. Generally, there are different degrees of substitution of hydroxyl or methoxy groups on the benzene ring of the basic skeleton (Ye et al., 2002, Fan, 2013) (Figure 7.5C).

(d) Flavonoids

Since the amount of flavonoids is relatively small and it is not easy to purify, not many of them have been reported (Ye et al., 2002) (Figure 7.5D).

(e) Phenylpropanoids

Phenylpropanoids from *Dendrobium* mainly include the types of coumarin, lignan, p-hydroxycinnamate, etc. (Ye et al., 2002) (Figure 7.5E).

(f) Anthraquinones

There are some reports about anthraquinones in *D. chrysanthum* and *D. polyanthum* (Yang et al., 2004, Hu et al., 2009) (Figure 7.5F).

7.2.3.1.4 Polysaccharides

Polysaccharides are the main components of some dried *Dendrobium* species. Structures of polysaccharides from *Dendrobium* species always vary from each other, due to their different sources, culture conditions, growing periods, plant parts, extraction methods, and treatment processes. The monosaccharide compositions mainly include glucose and mannose, and also contain arabinose, galacturonic acid, xylose, rhamnose, galactose, etc. The differences are mainly reflected in the ratio of the constituent monosaccharides, the composition of the branched glycosyl groups, and the position of substitution of O-acetyl groups. Recently, Kuang et al. (2020) isolated two polysaccharides, named DOP-1 and DOP-2, with molecular weights of 6.8 kDa and 14.3 kDa, respectively, from the stems of *D. officinale*. Subsequently, they identified that DOP-1 and DOP-2 may have a backbone consisting of →4)-β-D-Glcp-(1→, →4)-β-D-Manp-(1→,→4)-2–O -acetyl-β-D-Manp-(1→ and →4)-3-O-acetyl-β-D-Manp-(1→ (Kuang et al., 2020) (Figure 7.6).

7.2.3.1.5 Other Compounds

Apart from those phytochemicals mentioned above, there are also some other compounds such as fatty acids, esters, and sterols in *Dendrobium* plants (Ye et al., 2002).

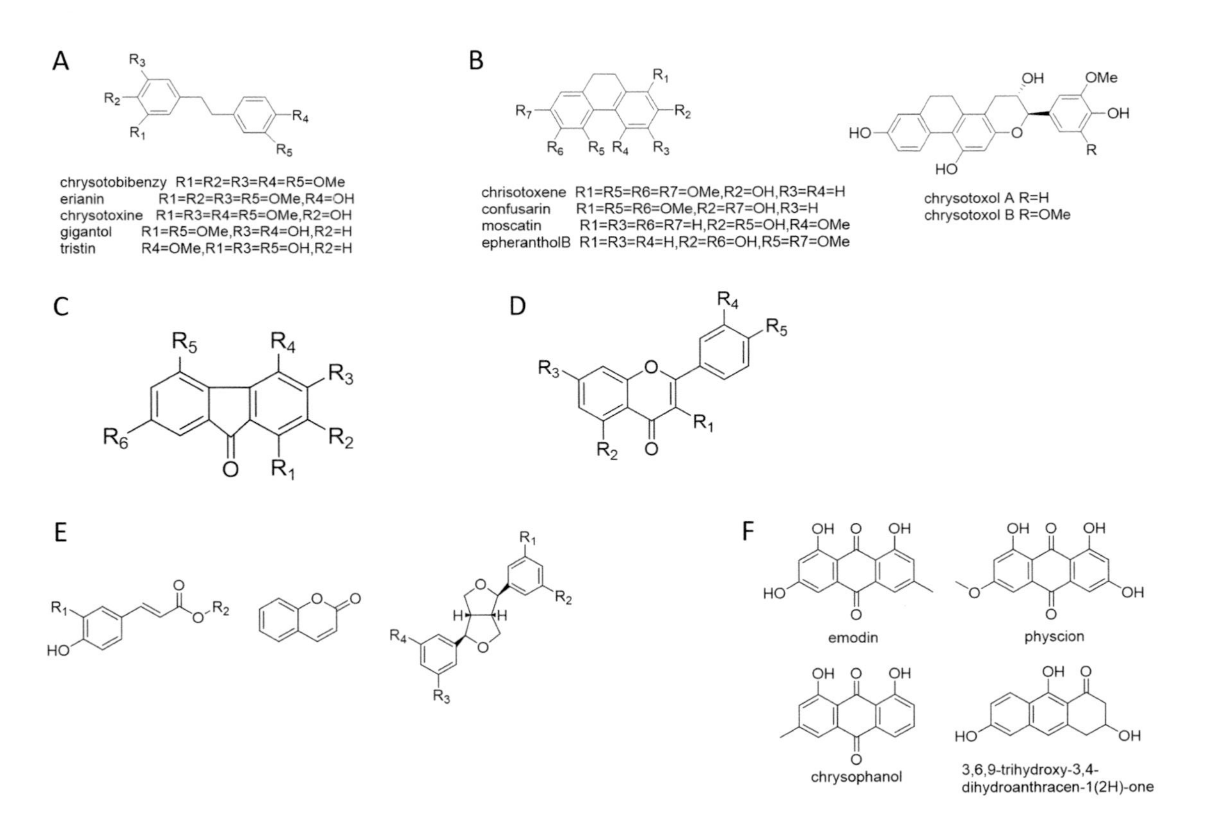

FIGURE 7.5 (A) Some typical bibenzyls from *Dendrobium*; (B) Some typical phenanthrenes from *Dendrobium*; (C) Structural skeleton of fluorenone from *Dendrobium*; (D) Structural skeleton of flavonoids from *Dendrobium*; (E) Structural skeleton of phenylpropanoids from *Dendrobium*; (F) The reported anthraquinones from *Dendrobium*.

OAc ↓ 3 (on second residue); OAc ↓ 3 (on fifth residue)

→4)-β-D-Manp-(1→4)-β-D-Manp-(1→4)-β-D-Glcp-(1→4)-β-D-Manp-(1→4)-β-D-Manp-(1→4)-β-D-Manp-(1→

4)-β-D-Manp-(1→4)-β-D-Manp-(1→4)-β-D-Manp-(1→4)-β-D-Glcp-(1→4)-β-D-Manp-(1→4)-β-D-Manp-(1→

2 ↑ OAc (on first residue); 2 ↑ OAc (on fifth residue)

FIGURE 7.6 Possible repeated unit of DOP-1 and DOP-2.

7.2.4 Biological Activities Related to Medicinal Uses

7.2.4.1 *Immune Activities*

D. officinale can be used as an immune enhancer, which has a significant protective effect on immunodeficiency caused by immune suppressants. Polysaccharides from *D. officinale* (DOP) could enhance the secretion of T helper type 1 (Th1) cytokines and the proportion of CD_4 T lymphocytes, and promote the subpopulation of T lymphocytes by up-regulating T-bet and GATA-3 mRNA expressions. In addition, DOP could increase the secretion of immunoglobulins, and stimulate the differentiation of plasmocytes by up-regulating IRF4, Blimp-1, and XBP-1 mRNA expressions (Li et al., 2020c). DOP could also stimulate cytokines production (TNF-α, IL-1β) in cells via up-regulating NF-κB and ERK1/2 (He et al., 2016). Li et al. also found that DOP could increase the diversity of gut microbiota and improve the immunity response of mice (Li et al., 2020b).

7.2.4.2 *Anti-Tumor Activities*

Polysaccharides, phenanthrenes, and benzyls are the main anti-tumor active substances of *Dendrobium*. There are several main aspects for the anti-tumor activity of *Dendrobium* including improving body immunity, inducing apoptosis, and inhibiting tumor cell proliferation (Xu et al., 2017a). Zhu's study showed that erianin present in *Dendrobium* extract could induce cell cycle G2/M-phase arrest and apoptosis via the JNK signaling pathway in bladder cancer (Zhu et al., 2019). Li's study also showed that ethyl acetate extract from metabolites of endophytic fungi from *D. officinale* induced apoptosis via the increase in Bax and the decrease in Bcl-2 (Li et al., 2018).

7.2.4.3 *Cardiovascular Regulatory Activities*

D. officinale exerts cardiovascular protection effects through inhibiting vascular intimal hyperplasia and reducing vascular damage (Wang et al., 2015). Zhong et al. found that *Dendrobium* polysaccharides could inhibit the injury by high glucose of endothelium-dependent relaxation of thoracic aorta rings and the mechanisms may be related to the inhibition of endoplasmic reticulum stress and NF-κB signaling pathway (Zhong et al., 2017). The aqueous extract of *Dendrobium* flowers reduces the blood pressure of spontaneously hypertensive rats in a dose-dependent manner. The mechanisms include reducing the plasma angiotensin II level and down-regulating the expression level of lung ACE mRNA (He et al., 2017). The effects on the cardiovascular system also reflect good anticoagulant and antithrombotic effects (Hu et al., 2008).

7.2.4.4 Anti-Oxidant and Anti-Aging Activities

In vitro experiments showed that five *Dendrobium* species (*D. moniliforme*, *D. nobile*, *D. officinale*, *D. devoninum*, and *D. aphyllum*) had the effect of scavenging active oxygen, which may be related to the phenanthrenequinones, bibenzyls, and fluorenone (Li et al., 2004). Liang et al. found that the polysaccharides from *D. officinale* could scavenge the DPPH free radicals in a dose-dependent manner (Liang et al., 2019). Li's study showed that *D. officinalis* flower could alleviate brain aging and improve the spatial learning abilities in senescent rats, with the possible underlying mechanisms of attenuating oxidative stress and thus reducing hippocampal damage and balancing the release of neurotransmitters (Li et al., 2020a).

7.2.4.5 Hypoglycemic Activities

According to Chinese medicine theory, *Dendrobium* has been used for consumptive thirst syndrome. Clinical studies have shown that the *Dendrobium* mixture is effective in preventing and treating type 2 diabetes and its complications. *Dendrobium* polysaccharides play a key role in decreasing blood sugar. Liu et al. found that DOP could promote hepatic glycogen synthesis and inhibit hepatic glycogen degradation and hepatic gluconeogenesis via affecting the glucagon-mediated signaling pathways, cAMP-PKA and Akt/FoxO1, thus ameliorating diabetic hepatic glucose metabolism (Liu et al., 2020). Kuang et al. also found that DOP may decrease fasting blood sugar levels by stimulating GLP-1 secretion and that intracellular DOP-induced GLP-1 secretion involved the Ca^{2+}/calmodulin/CaMKII and p38-mitogen-activated protein kinases pathways (Kuang et al., 2020). In addition, Xu et al. found that *D. nobile* alkaloids could increase glucose metabolism gene Glut2 and FoxO1 expressions and play integrated roles in regulating metabolic disorders (Xu et al., 2017b).

7.2.4.6 Other Activities

Wu et al. found that gigantol isolated from *D. chrysotoxum* could be used for the treatment of diabetic cataract through inhibiting aldose reductase activity and gene expression (Wu et al., 2017). An et al. found that *Dendrobium* polysaccharides could reduce liver enzymes and improve liver fibrosis in liver fibrosis model induced by CCl_4 combined with ethanol. The anti-fibrosis mechanism may include increasing the activity of anti-oxidant SOD, inhibiting the lipids in liver tissue, and liver damage caused by peroxidation (An et al., 2016). Besides, the alkaloids of *D. nobile* can improve the hepatic abnormal lipid profile of high-fat diet-fed mice via enhancing the taurine-conjugated bile acids which are highly hydrophilic and contribute to the excretion of cholesterol absorption (Huang et al., 2019).

7.2.5 Commercial Products

Due to significant pharmacological activities as described above, *Dendrobium* has been used as folk medicine and health food, such as Fengdou, for a long time. In addition, *Dendrobium* also has ornamental value due to its unique appearance.

FIGURE 7.7 *Gastrodia elata* (photo by Hu Jiang-Miao).

7.3 *Gastrodia elata*

"Tianma," from the tuber of *Gastrodia elata* Blume (family Orchidaceae) (Figure 7.7), is a perennial parasitic herb also named "Chijian," "Limu," "Dingfengcao," etc. Interestingly, during the growing process of *G. elata*, it is necessary to establish a symbiotic relationship with at least two types of fungi: *Mycena* for seed germination and *Armillaria mellea* for plant growth (Yuan et al., 2018).

7.3.1 Resource Availability

"Tianma" grows in the forest at 400–3200 m above sea level. Twenty-three species of *Gastrodia* and five variations of *G. elata* are recorded in China, and they are mainly distributed in the southwest, northwest and northeast of China (Editorial Committee of Chinese Flora of Chinese Academy of Sciences, 1999a). However, only *G. elata* is used as the medicinal herb in the clinical practice of Chinese medicine and registered in the Chinese Pharmacopoeia.

7.3.2 Plant Parts for Medicinal Uses

According to the 2020 edition of the Chinese Pharmacopoeia, the tuber of *G. elata* is recorded as the useful part, which suppresses a hyperactive liver for calming endogenous wind, dredging the meridians, and relieving pain. It can be used for infantile convulsion and headache, dizziness, numbness, and other symptoms (Commission, 2020).

7.3.3 Phytochemistry

To date, gastrodins, parishins, phenolics, organic acid compounds, sterols, and others have been isolated from *G. elata*. Among them, gastrodin and its aglycone gastrodigenin (4-hydroxybenzyl alcohol) are considered as the characteristic and main active constituents of *G. elata*.

7.3.3.1 The Main Active Components

Zhou Jun and his group at the Kunming Institute of Botany began the pharmaceutical research work on *G. elata* in the late 1970s. The active compound of *G. elata*, gastrodin, was discovered by them (Zhou et al., 1979) and the chemical structure was elucidated with IR, NMR, and MS. They were also the first to develop the method of preparation of gastrodin derivatives.

Gastrodin, the main active component of *G. elata*, possesses the function of sedative, anticonvulsant, antiepileptic, and analgesic (Dang et al., 2017). It can increase the cerebral blood flow, and improve the blood supply of the vertebral basilar artery, anterior inferior cerebellar artery, posterior inferior cerebellar artery, labyrinthine artery, and inner ear blood artery (Cheng et al., 2019, Wang et al., 2020). It can protect nerve cells, promote the function of the myocardial cells' energy metabolism, and improve the myocardial cell hypoxia. Clinically, it is widely used in vertigo (Meniere's disease, drug toxicity dizziness, vestibular neuronitis, vertebral basilar artery blood supply deficiency, etc.), headaches (nervous exhaustion and neurasthenia syndrome, vascular headaches, tension headaches, brain injury syndrome and migraine, etc.), and to aid the treatment of epilepsy (Zhan et al., 2016) (Figure 7.8A).

Parishins are another set of main active ingredients of *G. elata*. The basic structure is an ester formed by the condensation of a molecule of citric acid with different molecules of gastrodin (or gastrodin aglycone). In the processing of medicinal materials, parishins can be hydrolyzed into gastrodin, which is more easily metabolized in the body. A list of parishins from *G. elata* have been reported (Taguchi et al., 1981, Lin et al., 1996, Yang et al., 2007, Wang et al., 2012, Li et al., 2015).

Although gastrodin and parishins are the main active components of *G. elata*, there are also other bioactive substances present in *G. elata*, such as phenolic compounds, toluene derivatives, organic acid compounds, sterols, and polysaccharides, etc. (Figure 7.8B).

7.3.3.2 Other Types of Compounds

7.3.3.2.1 Phenolics and Its Glycosides

Abundant phenolics in *G. elata* exert significant activities such as neuroprotective, anti-inflammatory, and anti-oxidant activities, and many others. These compounds usually have a unique structure with variants of 4-hydroxybenzyl alcohol (Zhan et al., 2016). Phenolics are listed in Figure 7.9 and Table 7.2.

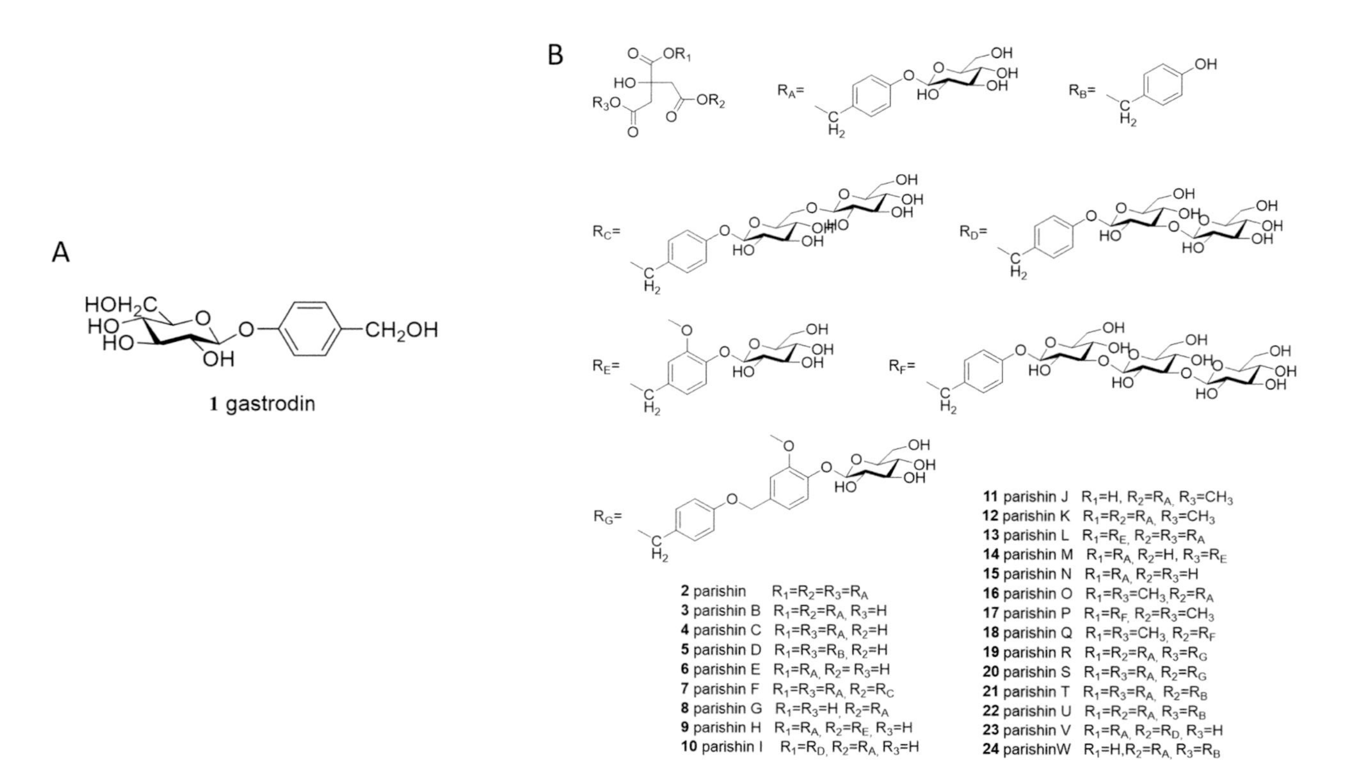

FIGURE 7.8 (A) The chemical structure of gastrodin; (B) The reported parishins from *G. elata*.

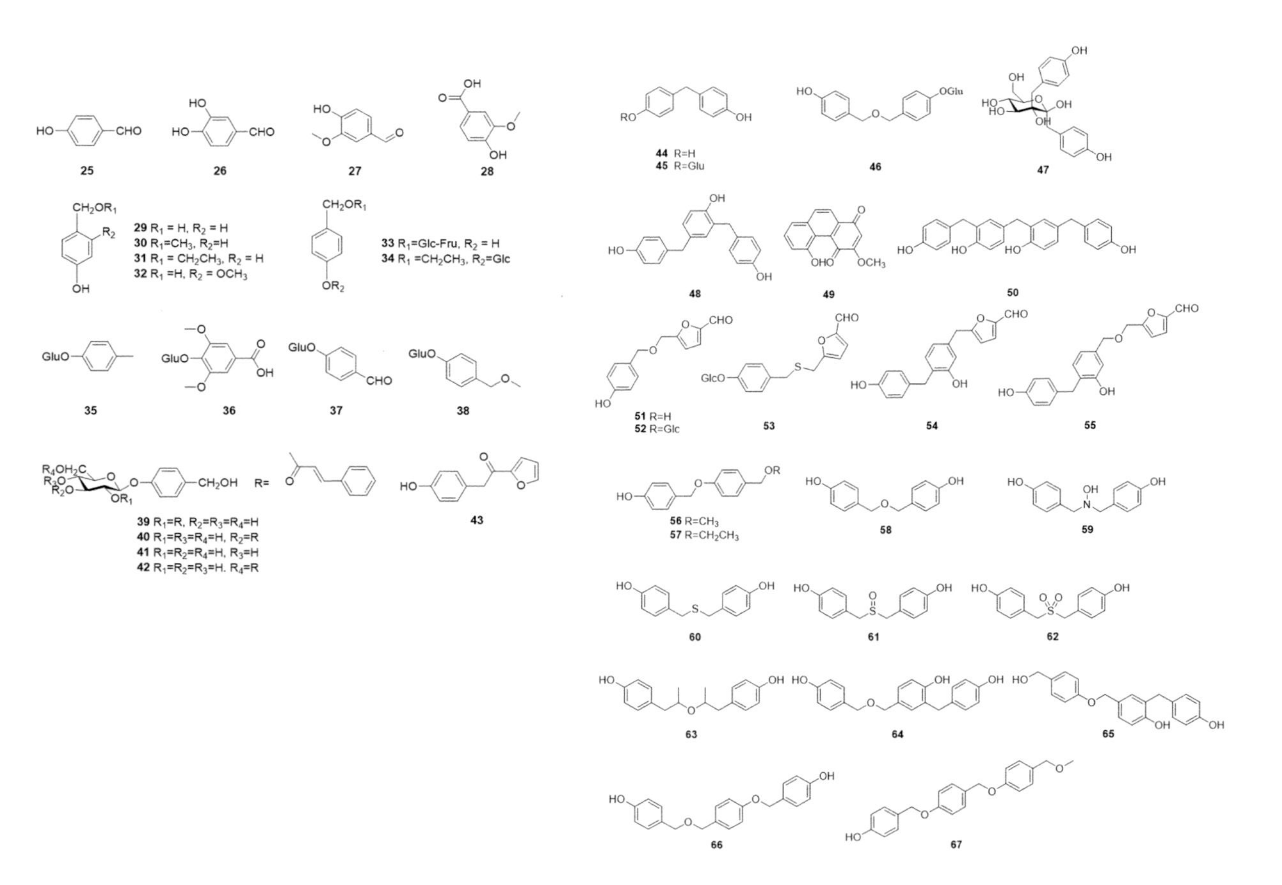

FIGURE 7.9 The phenolics from *G. elata.*

TABLE 7.2

The Phenolics from *G. elata*

No.	Chemical Names	References
25	4-hydroxybenzaldehyde	Zhou et al., 1980
26	3,4-dihydroxybenzaldehyde	Zhou et al., 1980
27	4-hydroxy-3-methoxybenzaldehyde	Lee et al., 2006
28	4-hydroxy-3-methoxybenzoic acid	Lee et al., 2006
29	4-hydroxybenzyl alcohol	Zhou et al., 1980
30	4-(methoxymethyl)phenol	Taguchi et al., 1981
31	4-ethoxymethyl phenol	Zhou et al., 1979
32	4-(hydroxymethyl)-3-methoxyphenol	Lee et al., 2006
33	Gastrodin A	Li et al., 2007
34	p-Ethoxymethyl phenyl-O-β-D-glucoside	Huang et al., 2006
35	p-methylphenyl-1-O-β-D-glucopyranoside	Huang et al., 2005
36	3,5-dimethoxybenzoic acid-4-O-β-D-glucopyranoside	Huang et al., 2005
37	4-O-glucopyranosyl-benzaldehyde	Ma et al., 2015
38	4-(methoxymethyl) phenyl-1-O-beta-D-glucopyranoside	Wang et al., 2012
39	1-O-(4-hydroxymethylphenoxy)-2-O-trans-cinnamoyl-beta-D-glucoside	Wang et al., 2019
40	1-O-(4-hydroxymethylphenoxy)-3-O-trans-cinnamoyl-beta-D-glucoside	Wang et al., 2019
41	1-O-(4-hydroxymethylphenoxy)-4-O-trans-cinnamoyl-beta-D-glucoside	Wang et al., 2019
42	1-O-(4-hydroxymethylphenoxy)-6-O-trans-cinnamoyl-beta-D-glucoside	Wang et al., 2019
43	1-furan-2-yl-2-(4-hydroxyphenyl)-ethanone	Lee et al., 2007
44	4,4'-dihydroxydiphenyl-methane	Lee et al., 2006
45	Gastrodin B	Zhang et al., 2013
46	Bis(4-hydroxybenzyl) mono-*β*-D-glucopyranoside	Taguchi et al., 1981
47	Gastrodioside(bis(4-hydroxybenzyl) mono-*β*-D-glucopyranoside	Taguchi et al., 1981
48	2,4-bis(4-hydroxybenzyl)phenol	Noda et al., 1995
49	Cymbinodin A	Xiao et al., 2002
50	2-[4-hydroxy-3-(4-hydroxybenzyl)benzyl]-4-(4-hydroxybenzyl)phenol	Huang et al., 2016
51	5-(4-hydroxybenzyloxymethyl)-furan-2-carbaldehyde	Lee et al., 2007
52	5-((4-O-β-D-glucopyranosylbenzyloxy)methyl)-furan-2-carbaldehyde	Li et al., 2016
53	5-((4-O-β-D-glucopyranosylbenzylsulfide)methyl)-furan-2-carbaldehyde	Li et al., 2016
54	5-[4'-(4''-hydroxybenzyl)-3'-hydroxybenzyl]-furan-2-carbaldehyde	Huang et al., 2015
55	5-[4'-(4''-hydroxybenzyl)-3'-hydroxybenzyloxymethyl]furan-2-carbaldehyde	Huang et al., 2015
56	4-((4-(methoxymethyl)phenoxy)methyl)phenol	Lee et al., 2006
57	4-((4-(ethoxymethyl)phenoxy)methyl)phenol	Taguchi et al., 1981
58	4,4'-dihydroxydibenzyl ether	Zhou et al., 1980
59	Di-(*p*-hydroxybenzyl) hydroxylamine	Hao et al., 2000
60	4,4'-thiobis(methylene)bisphenol	Xiao et al., 2002
61	4,4'-sulfinylbis(methylene)diphenol	Yun-Choi et al., 1997
62	4,4′-Dihydroxybenzyl sulfone	Pyo et al., 2004
63	p-hydroxybenzylethyl ether	Ruan et al., 1988
64	Gastrol	Hayashi et al., 2002
65	Gastrol B	Zhang et al., 2013
66	Gastrol A	Li et al., 2007
67	4-[4′-(4″-hydroxybenzyloxy)benzyloxy]benzyl methyl ether	Yun-Choi et al., 1998

7.3.3.2.2 *Toluene Derivatives*

There are also some benzyl alcohol, benzaldehyde, and benzoic acid compounds without phenolic hydroxyl group (Lee et al., 2006, Liu et al., 2002, Wang et al., 2003, Zhan et al., 2016) in *G. elata* (Figure 7.10A).

7.3.3.2.3 *Organic Acid Compounds*

Citric acid ester compounds including parishin are typical compounds in *G. elata* (Hao et al., 2000, Choi and Lee, 2006, Feng et al., 1979) (Figure 7.10B). There are also some chain compounds such as succinic acid, palmitic acid, and so on (Liu et al., 2002, Xiao et al., 2002, Wang et al., 2006, Pyo et al., 2006, Choi and Lee, 2006) in *G. elata* (Figure 7.10C).

7.3.3.2.4 *Sterols*

There are relatively few reports on sterol constituents in *G. elata* (Liu et al., 2002, Wang et al., 2006, Yun-Choi et al., 1998) (Figure 7.11).

7.3.3.2.5 *Polysaccharides*

As the main components of *G. elata*, polysaccharides have received wide attention especially on the monosaccharide composition, ratio, and glycosidic linkages and biological activities. *G. elata* polysaccharides are water-soluble polysaccharides, which contain more than three kinds of monosaccharides including glucose, rhamnose, fucose, galactose, and mannose, etc. The main structure is 1→6 keyed branched chain of α-(1→4)-D-glucan. *G. elata* polysaccharides have been evidenced to possess immunomodulatory, anti-oxidant, and neuroprotective activities (Zhu et al., 2019). Recently, many researchers have focused on the isolation of *G. elata* polysaccharides (GEP). Usually, crude GEP was obtained by boiling *G. elata* in hot water, precipitating with ethanol, and removing proteins using the Sevage method. Then the crude polysaccharides were applied to a macroporous resin D101 and further purified through a DEAE-52 cellulose column and a Sephadex G-100 column, followed by dialysis and lyophilization.

7.3.3.2.6 *Other Compounds*

Other compounds include furan derivatives (Pyo et al., 2004, Yun-Choi et al., 1997), adenosine derivatives (Huang et al., 2006, Huang et al., 2005), acylamide derivatives (Andersson et al., 1995, Hao et al., 2000), and so on (Huang et al., 2005, Hao et al., 2000, Pyo et al., 2006, Choi and Lee, 2006) in the tuber of *G. elata* (Figure 7.12). In particular, researchers identified some new compounds such as gastrodinol and indole derivatives from the flower branch of *G. elata* (Yang et al., 2020, Lv et al., 2020).

7.3.4 Biological Activities Related to Medicinal Uses

7.3.4.1 Sedative and Hypnotic Activities

Clinically, gastrodin preparations have significant sedative and hypnotic effects. Usually, *G. elata* can play a synergistic role when co-administered with pentobarbital

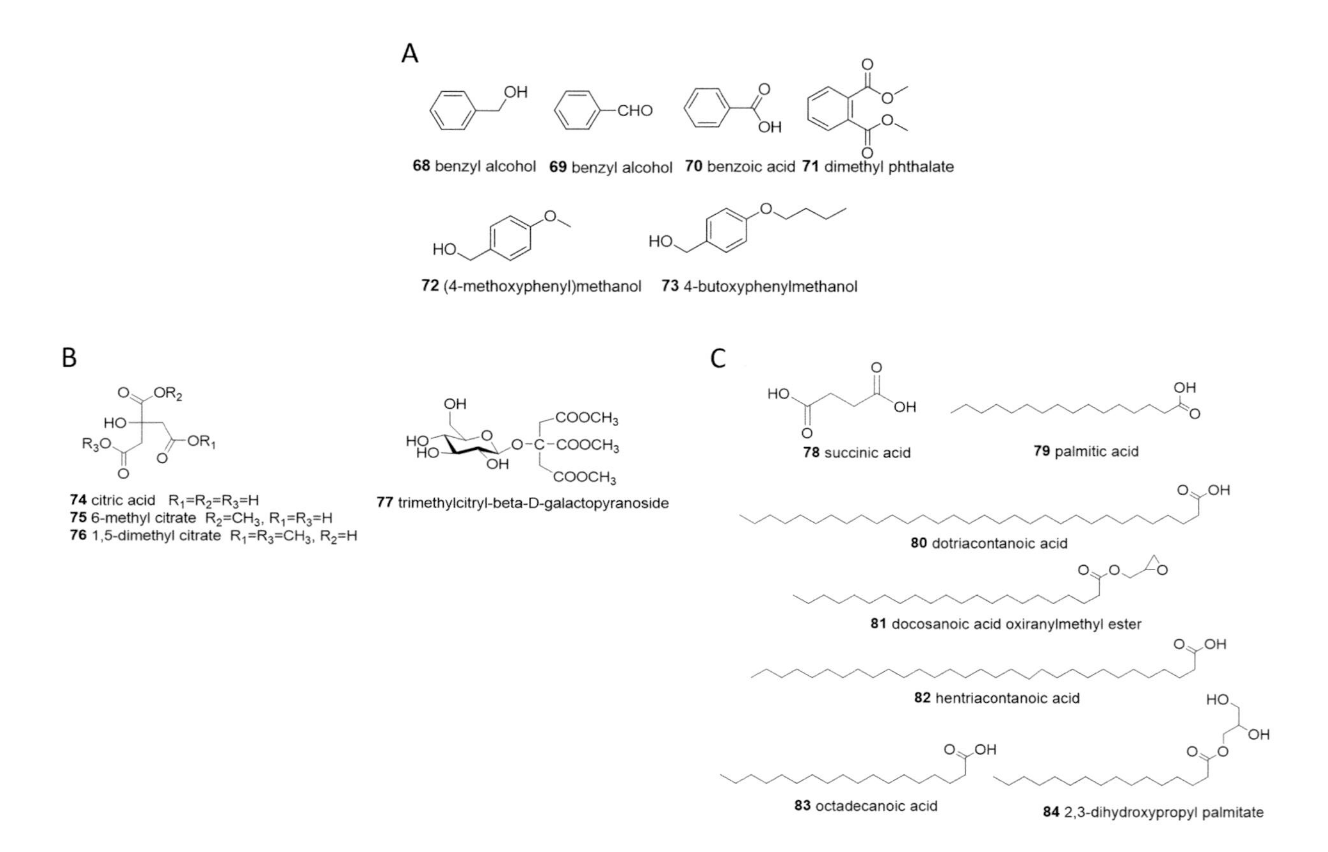

FIGURE 7.10 (A) The toluene derivatives from *G. elata*; (B) The citric acid ester from *G. elata*; (C) The chain compounds from *G. elata*.

85 beta-sitosterol R = H
86 daucosterol R = Glc
87 3-O-(4'-hydroxybenzyl)-beta-sitosterol
R= H_2C–C6H4–OH
RO

FIGURE 7.11 Sterols from *G. elata*.

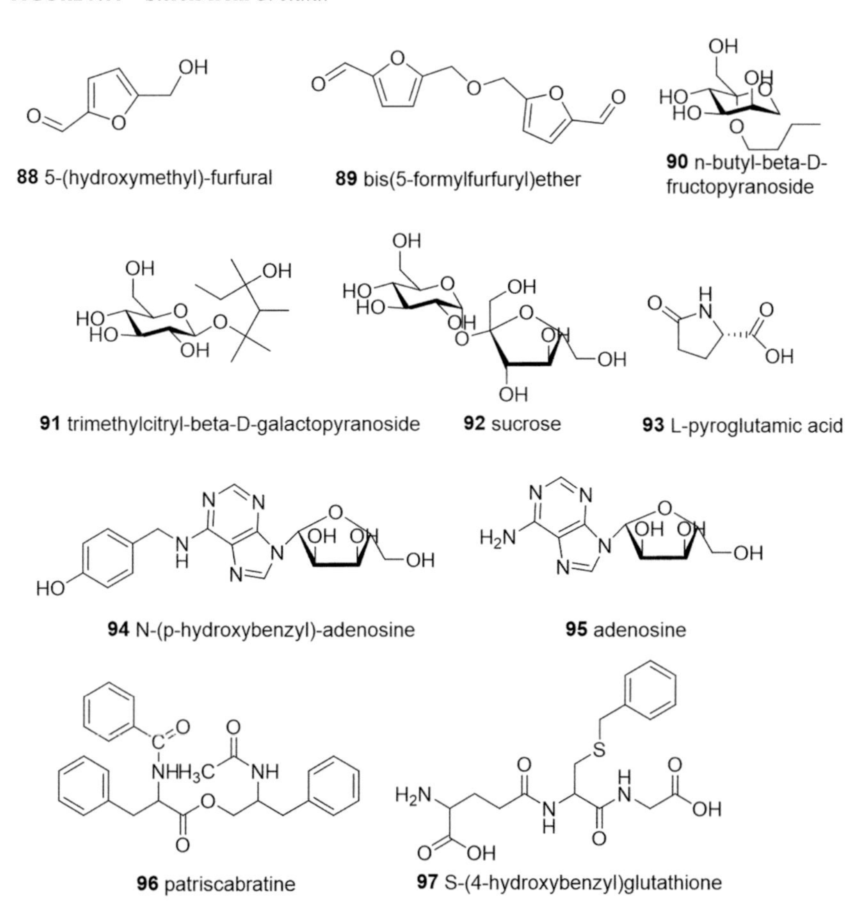

FIGURE 7.12 Other compounds from *G. elata*.

sodium, chloral hydrate, and thiopentone, and its sedative hypnotic effect is related to the increase of central neurotransmitter dopamine and the expression of dopamine receptors of various subtypes (Hu et al., 2019). N-(p-hydroxybenzyl)-adenosine isolated from *G. elata* exerts significant sedative and hypnotic effects which might be mediated by the activation of adenosine A1/A2A receptors and stimulation of the sleep center ventrolateral preoptic area (Zhang et al., 2012).

7.3.4.2 Antiepileptic and Anticonvulsive Activities

The *Gastrodia* capsule can increase the antiepileptic effect of carbamazepine, which is related to the reduction of hippocampal neuronal damage and the high expression of multidrug resistance associated protein gene (mrpl) (Dang et al., 2017). Furthermore, Ha et al. found that 4-hydroxybenzaldehyde isolated from *G. elata* exerted antiepileptic and anticonvulsive activity by anti-oxidation and positive modulation of GABAergic neuromodulation (Ha et al., 2000).

7.3.4.3 Anti-Anxiety and Antidepressant Activities

G. elata is a valuable Chinese medicine for treating depression. Studies showed that *G. elata* aqueous ethanol extract exerted antidepressant-like effects comparable to those of fluoxetine in experimental animal models (Zhou et al., 2006). Chen et al. also found that the water extracts of *G. elata* exhibited antidepressant effects in rats by decreasing monoamine metabolism and modulating cytoskeleton remodeling-related protein expression in the Slit-Robo pathway (Chen et al., 2016).

7.3.4.4 Antipsychotic Activities

Clinical research showed that gastrodin injection could improve the cognitive function of patients with schizophrenia. 5-HT_{1A} receptors play an important role in the pathophysiology of schizophrenia. Shin et al. found that *G. elata* attenuated the phencyclidine-induced abnormal behaviors via the activation of 5-HT_{1A} in mice (Shin et al., 2011).

7.3.4.5 Anti-Vertigo Activities

Clinically, gastrodin is a basic medicine for dizziness. Yu's study showed that polysaccharides of *G. elata* and *Armillaria mellea* could clearly shorten the "escape time" of electrical shock and increase the food intake of vertiginous mice induced by machinery rotation (Yu et al., 2006).

7.3.4.6 Neuroprotective Activities

There are three main underlying mechanisms of *G. elata*'s neuroprotective effect, including the protection of neuronal cells, the protection of neuro-synaptic plasticity, and anti-oxidative activities (Zhan et al., 2016). Zhou et al. found that the polysaccharides from *G. elata* exerted a neuroprotective effect against corticosterone-induced apoptosis in PC12 cells. This effect may be related to the inhibition of the endoplasmic reticulum stress-mediated pathway (Zhou et al., 2018). Arulmani et al. hypothesized that *G. elata* could promote neuro-regenerative processes by inhibiting stress-related proteins and mobilizing neuroprotective genes such as Nxn, Dbnl, Mobkl3, Clic4, Mki67, and Bax with various regenerative modalities and capacities related to neuro-synaptic plasticity (Manavalan et al., 2012). Furthermore, Han et al. found that the methanol extract from *G. elata* exerted neuroprotective effect against oxidative glutamate toxicity via the inhibition of glutamate-induced apoptotic death and the suppression of the production of ROS (Han et al., 2014).

7.3.4.7 Memory-Improving and Anti-Aging Activities

Choi et al. found that *G. elata*-derived vanillyl alcohol (GEVA) could inhibit acetylcholinesterase activity and amyloid-peptide (A(25-35))-induced caspase (CASP)-3/7 activities in SH-SY5Y cells. These results indicated that GEVA may be effective in improving memory (Choi et al., 2019). Morris water maze tests also showed that *G. elata* treatment significantly improved the spatial memory of Alzheimer's disease rats. Further study showed *G. elata* could reduce the number of amyloid deposits in the hippocampus. In addition, *G. elata* could increase choline acetyltransferase expression in the medial septum and hippocampus and decrease the activity of acetylcholinesterase in all three regions (Huang et al., 2013).

7.3.4.8 Cardiovascular Regulatory Activities

The effects of *Gastrodia* on the cardiovascular system mainly include effects on the heart, blood vessels, blood pressure, and microcirculation. Gastrodin plays a protective role against myocardial ischemia-reperfusion injury by eliminating oxygen-free radicals, reducing inflammation, and inhibiting calcium overload (Cheng et al., 2019). Wang et al. found that parishins isolated from *G. elata* exhibited myocardial protection through down-regulation of the level of cleaved-caspase-3 and cytochrome C in the cytoplasm and Bax, and up-regulation of cytochrome C in the mitochondria and Bcl-2 (Wang et al., 2020). Wang's study also showed that Tianma Gouteng Decoction could reduce the homocysteine levels of H-type hypertension with left ventricular hypertrophy and improve left ventricular diastolic and systolic functions (Wang and Liu, 2016). Song's study showed that Tianma Xifeng Decoction could significantly improve the local microcirculation disorder (Song et al., 2011).

7.3.4.9 Anti-Inflammatory and Analgesic Activities

Phenolic compounds isolated from *G. elata* exerted anti-inflammatory and analgesic activities via inhibiting COX activity and ROS generation in a dose-dependent manner (Lee et al., 2006). He et al. assessed the therapeutic effect of the n-butanol extract of *G. elata* (BGE) on complete Freund's adjuvant-induced arthritis rats. Results showed that BGE protected cartilage from destruction and reduced inflammatory cell infiltration and synovial proliferation. A further study confirmed that BGE decreased the production of nitric oxide and inflammatory cytokines in LPS-stimulated RAW264.7 cells (He et al., 2020). Furthermore, Sun et al. suggested that an ethanol extract of *G. elata* rhizome suppressed the TNF-α-induced vascular inflammatory process via the inhibition of oxidative stress and NF-κB activation in HUVEC (Hwang et al., 2009).

7.3.4.10 Anti-Tumor Activities

Kim et al. found that the extrusion of *G. elata* could decrease cell viability and induce the expression of Caspase-3 and Bax in colon HT29 cells (Kim et al., 2017). Shu's study showed that gastrodin could enhance cytotoxic activities of natural killer and $CD8^+$ T cells against H22 hepatic cancer via ameliorating tumor cell transplantation-induced activation of endogenous pro-apoptotic pathway in $CD4^+$ T cells

(Shu et al., 2013). Vanillin and p-hydroxybenzaldehyde isolated from *G. elata* could significantly inhibit both intracellular Ca^{2+} rise and glutamate-induced apoptosis in human neuronal cells (Lee et al., 1999).

7.3.4.11 Other Activities

Polysaccharides and phenolic compounds of *G. elata* were reported for anti-oxidant activities (Ahmad et al., 2019). Jian et al. also reported that the purified *G. elata* polysaccharide exhibited potential lipid-lowering effects in hyperlipidemia rats (Ming et al., 2012). Furthermore, Chen et al. reported that *G. elata* polysaccharides showed broad-spectrum bacteriostatic activity, especially for Gram-negative bacteria (Chen et al., 2018).

7.3.5 Commercial Products

According to the National Medical Products Administration of the People's Republic of China, *G. elata*-related medicines include *Gastrodia* pill, *Gastrodia* tablet, and gastrodin injection, etc. Due to the remarkable effects on the nervous system, cardiovascular system, and so on, many health products including foods related to *G. elata* have been developed.

7.4 Conclusion

Orchid plants have a long history of medicinal use in China. This chapter briefly summarizes the chemical compounds and pharmacological activities of *Dendrobium* and *G. elata*. For future studies researchers should pay more attention to the sustainable utilization of orchid resources through interdisciplinary methods. With the development of new technology and the progress of further research, it is anticipated that more active components and their mechanisms of action will be discovered, which will lead to wider clinical applications of "Shi-Hu" and "Tianma."

REFERENCES

Ahmad O, Wang B, Ma KJ, et al., 2019. Lipid modulating anti-oxidant stress activity of gastrodin on nonalcoholic fatty liver disease larval zebrafish model. *International Journal of Molecular Sciences*, 20: 11.

An ZX, He YL, Chen YJ, 2016. Therapeutic effect and mechanism of *Dendrobium nobile* polysaccharides on liver damage in rats with liver fibrosis. *Lishizhen Medicine and Materia Medica Research*, 27: 1865–1867.

Andersson M, Bergendorff O, Nielsen M, et al., 1995. Inhibition of kainic acid binding to glutamate receptors by extracts of *Gastrodia*. *Phytochemistry*, 38: 835–836.

Chen C, Li XX, Fu YD, 2018. Study on antibacterial activity of polysaccharides from *Gastrodia elata* in Hanzhong area. *Jiangsu Agricultural Sciences*, 46: 156–159.

Chen WC, Lai YS, Lin SH, et al., 2016. Anti-depressant effects of *Gastrodia elata* Blume and its compounds gastrodin and 4-hydroxybenzyl alcohol, via the monoaminergic system and neuronal cytoskeletal remodeling. *Journal of Ethnopharmacology*, 182: 190–199.

Chen XM, Guo SX, 2001. Advances in the research of constituents and pharmacology of *Dendrobium*. *Natural Product Research and Development*, 13: 70–75.

Cheng QQ, Yang WM, Liu X, 2019. Research progress on pharmacological mechanism of Gastrodiae Rhizoma in cardiovascular and metabolic diseases. *Academic Journal of Shanghai University of Traditional Chinese Medicine*, 33: 96–100.

Choi JH, Lee DU, 2006. A new citryl glycoside from *Gastrodia elata* and its inhibitory activity on GABA transaminase. *Chemical and Pharmaceutical Bulletin*, 54: 1720–1721.

Choi YJ, Kim MH, Park IS, et al., 2019. *Gastrodia elata* Blume-derived vanillyl alcohol suppresses amyloid beta-peptide-induced caspase activation in SH-SY5Y cells. *Neurochemical Journal*, 13: 110–112.

Commission CP, 2020. *Chinese Pharmacopoeia*, China Medical Science and Technology Press. Beijing.

Dang XJ, Wang YX, Jiao HS, 2017. Effects of *Gastrodia* capsule and gastrodin combined with carbamazepine on the hippocampal neurons and the expression level of mrp1 gene of PTZ-kindled mice. *Chinese Journal of New Drugs*, 26: 1556–1561.

Editorial Committee of Chinese Flora of Chinese Academy of Sciences, 1999a. *Flora of China*, Vol. 18. Science Press, Beijing.

Editorial Committee of Chinese Flora of Chinese Academy of Sciences. 1999b. *Flora of China*, Vol. 19. Science Press, Beijing.

Fan WW, 2013. Chemical constituents of four dendrobium plants and their bioactivities. PhD thesis, Kunming Institute of Botany, Chinese Academy of Sciences.

Feng XZ, Chen YW, Yang JS, 1979. Study on the chemical constituents of *Gastrodia*. *Chinese Pharmaceutical Journal*: 80.

Glomqvist L, Brandänge S, Gawell L, et al., 1973. Studies on orchidaceae alkaloids. XXXVII. Dendrowardine, a quaternary alkaloid from *Dendrobium wardianum* Wr. *Acta Chemica Scandinavica*, 27: 1439–1441.

Ha JH, Lee DU, Lee JT, et al., 2000. 4-Hydroxybenzaldehyde from *Gastrodia elata* Bl. is active in the antioxidation and GABAergic neuromodulation of the rat brain. *Journal of Ethnopharmacology*, 73: 329–33.

Han YJ, Je JH, Kim SH, et al., 2014. *Gastrodia elata* shows neuroprotective effects via activation of PI3K signaling against oxidative glutamate toxicity in HT22 cells. *American Journal of Chinese Medicine*, 42: 1007–1019.

Hao XY, Tan NH, Zhou J, 2000. Constituents of *Gastrodia elata* in Guizhou. *Plant Diversity*, 22: 81–84.

Hayashi J, Sekine T, Deguchi S, et al., 2002. Phenolic compounds from Gastrodia rhizome and relaxant effects of related compounds on isolated smooth muscle preparation. *Phytochemistry*, 59: 513–519.

He P, Hu Y, Huang C, et al., 2020. N-butanol extract of *Gastrodia elata* suppresses inflammatory responses in lipopolysaccharide-stimulated macrophages and complete freund's adjuvant- (CFA-) induced arthritis rats via inhibition of MAPK signaling pathway. *Evidence-Based Complementary and Alternative Medicine*, 2020: 1658618.

He TB, Huang YP, Yang L, et al., 2016. Structural characterization and immunomodulating activity of polysaccharide from *Dendrobium officinale*. *International Journal of Biological Macromolecules*, 83: 34–41.

He XY, Wu RZ, Long HQ, 2017. Study on antihypertensive effect and mechanism of *Dendrobium* flowers in spontaneously hypertensive rats. *China Journal of Traditional Chinese Medicine and Pharmacy*, 32: 1836–1840.

Hedman K, Leander K, 1972. Studies on orchidaceae alkaloids. XXVII. Quaternary salts of the dendrobine type from *Dendrobium nobile* Lindl. *Acta Chemica Scandinavica*, 26: 3177–3180.

Hu JM, Chen JJ, Yu H, et al., 2008. Two novel bibenzyls from Dendrobium trigonopus. *Journal of Asian Natural Products Research*, 10: 647–651.

Hu JM, Fan W, Dong F, et al., 2012. Chemical components of *Dendrobium chrysotoxum*. *Chinese Journal of Chemistry*, 30: 1327–1330.

Hu JM, Zhao YX, Miao ZH, et al., 2009. Chemical components of *Dendrobium polyanthum*. *Bulletin of the Korean Chemical Society*, 30: 2098–2100.

Hu PC, Wang J, Liu MX, 2019. Effect of *Gastrodia elata* on improving sleep in mice and its mechanism. *Chinese Traditional and Herbal Drugs*, 50: 3140–3146.

Huang GB, Zhao T, Muna SS, et al., 2013. Therapeutic potential of *Gastrodia elata* Blume for the treatment of Alzheimer's disease. *Neural Regeneration Research*, 8: 1061–1070.

Huang JY, Yuan YH, Yan JQ, et al., 2016. 20C, a bibenzyl compound isolated from *Gastrodia elata*, protects PC12 cells against rotenone-induced apoptosis via activation of the Nrf2/ARE/HO-1 signaling pathway. *Acta Pharmacologica Sinica*, 37: 731–740.

Huang LQ, Li ZF, Wang Q, et al., 2015. Two new furaldehyde compounds from the rhizomes of *Gastrodia elata*. *Journal of Asian Natural Products Research*, 17: 352–356.

Huang S, Wu Q, Liu H, et al., 2019. Alkaloids of *Dendrobium nobile* Lindl. altered hepatic lipid homeostasis regulation of bile acids. *Journal of Ethnopharmacology*, 241: 111976.

Huang ZB, Song DM, Chen FK, 2005. The chemical constituents isolated from *Gastrodia elata* Bl. *Chinese Journal of Medicinal Chemistry*, 4: 227–229.

Huang ZB, Wu Z, Chen FK, et al., 2006. The protective effects of phenolic constituents from *Gastrodia elata* on the cytotoxicity induced by KCl and glutamate. *Archives of Pharmacal Research*, 29: 963.

Hwang SM, Lee YJ, Kang DG, et al., 2009. Anti-inflammatory effect of *Gastrodia elata* rhizome in human umbilical vein endothelial cells. *American Journal of Chinese Medicine*, 37: 395–406.

Inubushi Y, Ishii H, Yasui B, et al., 1964. Isolation and characterization of alkaloids of the Chinese drug "Chin-Shih-Hu". *Chemical and Pharmaceutical Bulletin*, 12: 1175–1180.

Inubushi Y, Nakano J, 1965. Structure of dendrine. *Tetrahedron Letters*, 6: 2723–2728.

Kim NH, Xin MJ, Cha JY, et al., 2017. Antitumor and immunomodulatory effect of *Gastrodia elata* on colon cancer *in vitro* and *in vivo*. *American Journal of Chinese Medicine*, 45: 319–335.

Kuang MT, Li JY, Yang XB, et al., 2020. Structural characterization and hypoglycemic effect via stimulating glucagon-like peptide-1 secretion of two polysaccharides from *Dendrobium officinale*. *Carbohydrate Polymers*, 241: 116326.

Lee JY, Jang YW, Kang HS, et al., 2006. Anti-inflammatory action of phenolic compounds from *Gastrodia elata* root. *Archives of Pharmacal Research*, 29: 849–58.

Lee YK, Woo MH, Kim CH, et al., 2007. Two new benzofurans from *Gastrodia elata* and their DNA topoisomerases I and II inhibitory activities. *Planta Medica*, 73: 1287–1291.

Lee YS, Ha JH, Yong CS, et al., 1999. Inhibitory effects of constituents of *Gastrodia elata* Bl. on glutamate-induced apoptosis in IMR-32 human neuroblastoma cells. *Archives of Pharmacal Research*, 22: 404–9.

Leon RM, Ravi D, An JS, et al., 2019. Synthesis of C14-desmethylene corialactone D and discovery of inhibitors of nerve growth factor mediated neurite outgrowth. *Organic Letters*, 21: 3193–3197.

Li LZ, Lei SS, Li B, et al., 2020a. *Dendrobium officinalis* flower improves learning and reduces memory impairment by mediating antioxidant effect and balancing the release of neurotransmitters in senescent rats. *Combinatorial Chemistry & High Throughput Screening*, 23: 402–410.

Li M, Yue H, Wang Y, et al., 2020b. Intestinal microbes derived butyrate is related to the immunomodulatory activities of *Dendrobium officinale* polysaccharide. *International Journal of Biological Macromolecules*, 149: 717–723.

Li MZ, Huang XJ, Hu JL, et al., 2020c. The protective effects against cyclophosphamide (CTX)-induced immunosuppression of three glucomannans. *Food Hydrocolloids*, 100: 10.

Li N, Wang KJ, Chen JJ, et al., 2007. Phenolic compounds from the rhizomes of Gastrodia elata. *Journal of Asian Natural Products Research*, 9: 373–377.

Li RQ, Li JF, Zhou ZY, et al., 2018. Antibacterial and antitumor activity of secondary metabolites of endophytic fungi Ty5 from *Dendrobium officinale. Journal of Biobased Materials and Bioenergy*, 12: 184–193.

Li Y, Zhao YP, Chen BY, 2004. Scavenging effect of 5 kinds of *Dendrobium* water extracts on active oxygen. *Chinese Traditional and Herbal* Drugs, 11: 44–46.

Li Z, Wang Q, Ouyang H, et al., 2016. New compounds with neuroprotective activities from *Gastrodia elata. Phytochemistry Letters*, 15: 94–97.

Li ZF, Wang YW, Ouyang H, et al., 2015. A novel dereplication strategy for the identification of two new trace compounds in the extract of *Gastrodia elata* using UHPLC/Q-TOF-MS/MS. *Journal of Chromatography. Part B, Analytical Technologies in the Biomedical and Life Sciences*, 988: 45–52.

Li ZJ, Wang YC, Han B, 2019. Research progress on constituents of alkaloids in plants from *Dendrobium* Sw. *Chinese Traditional and Herbal Drugs*, 50: 3246–3254.

Liang J, Zeng Y, Wang H, et al., 2019. Extraction, purification and antioxidant activity of novel polysaccharides from *Dendrobium officinale* by deep eutectic solvents. *Natural Product Research*, 33: 3248–3253.

Lin JH, Liu YC, Hau JP, et al., 1996. Parishins B and C from rhizomes of *Gastrodia elata. Phytochemistry*, 42: 549–551.

Liu QF, Zhao WM, 2003. A new dendrobine-type alkaloid from *Dendrobium nobile. Chinese Chemical Letters*, 14: 278–279.

Liu WH, Hua YF, Zhan ZJ, 2007. Moniline, a new alkaloid from *Dendrobium moniliforme. Journal of Chemical Research*, 2007: 317–318.

Liu XQ, Baek WS, Kyun AD, et al., 2002. The constituents of the aerial part of *Gastrodia elata* blume. *Natural Product Sciences*, 8: 137–140.

Liu Y, Yang L, Zhang Y, et al., 2020. *Dendrobium officinale* polysaccharide ameliorates diabetic hepatic glucose metabolism via glucagon-mediated signaling pathways and modifying liver-glycogen structure. *Journal of Ethnopharmacology*, 248: 112308.

Lv YF, Ren FC, Kuang MT, et al., 2020. Total synthesis of gastrodinol via photocatalytic 6π electrocyclization. *Organic Letters*, 22: 6822–6826.

Ma Q, Wan Q, Huang S, et al., 2015. Phenolic constituents with inhibitory activities on acetylcholinesterase from the rhizomes of *Gastrodia elata. Chemistry of Natural Compounds*, 51: 158–160.

Manavalan A, Ramachandran U, Sundaramurthi H, et al., 2012. *Gastrodia elata* Blume (tianma) mobilizes neuro-protective capacities. *International Journal of Biochemistry and Molecular Biology*, 3: 219–241.

Meng CW, He YL, Peng C, et al., 2017. Picrotoxane sesquiterpenoids from the stems of *Dendrobium nobile* and their absolute configurations and angiogenesis effect. *Fitoterapia*, 121: 206–211.

Ming J, Liu J, Wu S, et al., 2012. Structural characterization and hypolipidemic activity of a polysaccharide PGEB-3H from the fruiting bodies of *Gastrodia elata* Blume. *Procedia Engineering*, 37: 169–173.

Morita H, Fujiwara M, Yoshida N, et al., 2000. New picrotoxinin-type and dendrobine-type sesquiterpenoids from *Dendrobium snowflake* 'Red Star'. *Tetrahedron*, 56: 5801–5805.

Noda N, Kobayashi Y, Miyahara K, et al., 1995. 2,4-Bis(4-hydroxybenzyl) phenol from *Gastrodia elata*. *Phytochemistry*, 39: 1247–1248.

Okamoto T, Natsume M, Onaka T, et al., 1966. The structure of dendroxine. The third alkaloid from Dendrobium nobile. *Chemical and Pharmaceutical Bulletin*, 14: 672–675.

Okamoto T, Natsume M, Onaka T, et al., 1972. Further studies on the alkaloidal constituents of *Dendrobium nobile* (Orchidaceae)-structure determination of 4-hydroxy-dendroxine and nobilomethylene. *Chemical and Pharmaceutical Bulletin*, 20: 418–421.

Pyo MK, Jin JL, Koo YK, et al., 2004. Phenolic and furan type compounds isolated from *Gastrodia elata* and their anti-platelet effects. *Archives of Pharmacal Research*, 27: 381.

Pyo MK, Yun-Choi HS, et al., 2006. Isolation of n-butyl-beta-D-fructopyranoside from *Gastrodia elata* blume. *Natural Product Sciences*, 12: 101–103.

Ruan DC, Yang CR, Pu XY, 1988. HPLC quantitative analysis of phenolic components in *Gastrodia* and its relatives. *Plant* Diversity, 2: 231–237.

Shin EJ, Kim JM, Nguyen XKT, et al., 2011. Effects of *Gastrodia elata* Bl on phencyclidine-induced schizophrenia-like psychosis in mice. *Current Neuropharmacology*, 9: 247–250.

Shu GW, Yang TM, Wang CY, et al., 2013. Gastrodin stimulates anticancer immune response and represses transplanted H22 hepatic ascitic tumor cell growth: Involvement of NF-kappa B signaling activation in CD4+T cells. *Toxicology and Applied Pharmacology*, 269: 270–279.

Song ZY, Liu SW, Hao LJ, 2011. Research of Tianma Xifeng oral liquid on microcirculation of mice. *Hebei Journal of Traditional Chinese Medicine*, 33: 1712–1714.

Taguchi H, Yosioka I, Yamasaki K, et al., 1981. Studies on the constituents of *Gastrodia elata* Blume. *Chemical and Pharmaceutical Bulletin*, 29: 55–62.

Wang HQ, Zhao YF, Li TZ, 2015. Effects of *Dendrobium officinale* Kimura et Migo on cardiac function and changes of blood vessel in rabbits with coronary heart disease. *Modernization of Traditional Chinese Medicine and Materia Medica-World Science and Technology*, 17: 856–860.

Wang L, Wang YP, Xiao HB, 2006. Study on the chemical components of gastrodia(II). *Chinese Traditional and Herbal* Drugs, 11: 1635–1637.

Wang L, Xiao H, Liang X, 2003. Studies on chemical constituents of *Gastrodia elata*. *Chinese Traditional and Herbal Drugs*, 7: 11–12.

Wang L, Xiao HB, Yang L, et al., 2012. Two new phenolic glycosides from the rhizome of *Gastrodia elata*. *Journal of Asian Natural Products Research*, 14: 457–62.

Wang Q, Li Z, Wang D, et al., 2020. Myocardial protection properties of parishins from the roots of *Gastrodia elata* Bl. *Biomedicine & Pharmacotherapy*, 121: 109645.

Wang XK, Zhao TF, 1986. The chemical constituents of *Dendrobium* plants and the traditional Chinese medicine Dendrobium. *Chinese Pharmaceutical Journal*, 11: 666–669.

Wang YH, Liu YJ, 2016. Effect of Tianmagouteng decoction on heart function and cardiovascular events in hypertensive patients with left ventricular hypertrophy. *Chinese Journal of Evidence-Based Cardiovascular Medicine*, 8: 997–998.

Wang ZW, Li Y, Liu DH, et al., 2019. Four new phenolic constituents from the rhizomes of *Gastrodia elata* Blume. *Natural Product Research*, 33: 1140–1146.

Wu J, Li X, Wan WC, et al., 2017. Gigantol from *Dendrobium chrysotoxum* Lindl. binds and inhibits aldose reductase gene to exert its anti-cataract activity: An *in vitro* mechanistic study. *Journal of Ethnopharmacology*, 198: 255–261.

Xiao YQ, Li L, You XL, 2002. Studies on chemical constituents of effective part of *Gastrodia elata*. *China Journal of Chinese Materia Medica*, 27: 35–6.

Xu WQ, Wang YB, Sun ZR, 2017a. Antitumor research status analysis of *Dendrobium*. *Chinese Journal of Modern Applied Pharmacy*, 34: 130–134.

Xu Y, Liu HC, Li X, 2019. Research progress in chemical composition, fingerprint and pharmacological activity of *Dendrobii caulis*. *Chinese Journal of Information on Traditional Chinese Medicine*, 26: 129–132.

Xu YY, Xu YS, Wang Y, et al., 2017b. *Dendrobium nobile* Lindl. alkaloids regulate metabolism gene expression in livers of mice. *Journal of Pharmacy and Pharmacology*, 69: 1409–1417.

Yang D, Cheng ZQ, Yang L, et al., 2018. Seco-dendrobine-type alkaloids and bioactive phenolics from *Dendrobium findlayanum*. *Journal of Natural Products*, 81: 227–235.

Yang L, Jiang R, Li HH, et al., 2020. Three new compounds from the flower branch of *Gastrodia elata* Blume and anti-microbial activity. *Rsc Advances*, 10: 14644–14649.

Yang L, Wang Y, Bi ZM, 2004. Studies on chemical constituents of *Dendrobium chrysanthum*. *Chinese Journal of Natural Medicines*, 5: 27–29.

Yang XD, Zhu J, Yang R, et al., 2007. Phenolic constituents from the rhizomes of *Gastrodia elata*. *Natural Product Research*, 21: 180–186.

Ye QH, Zhao WM, Qin GW, 2002. Progress on chemical and biological studies of *Dendrobium* plants. *Progress in Medicinal Chemistry*, 3: 113–143.

Yu L, Shen YS, Miao HC, 2006. Study on the anti-vertigo function of polysaccharides of *Gastrodia elata* and polysaccharides of *Armillaria mellea*. *Chinese Journal of Information on Traditional Chinese Medicine*. 8: 29–36.

Yuan Y, Jin XH, Liu J, et al., 2018. The *Gastrodia elata* genome provides insights into plant adaptation to heterotrophy. *Nature Communications*, 9: 11.

Yun-Choi HS, Pyo MK, Park KM, 1997. Cirsiumaldehyde from *Gastrodia elata*. *Natural Products Sciences*, 3: 104–105.

Yun-Choi HS, Pyo MK, Park KM, 1998. Isolation of 3-O(4′-hydroxybenzyl)-β-sitosterol and 4-[4′-(4″-hydroxybenzyloxy)benzyloxy]benzyl methyl ether from fresh tubers of *Gastrodia elata*. *Archives of Pharmacal Research*, 21: 357–360.

Zhan HD, Zhou HY, Sui YP, et al., 2016. The rhizome of *Gastrodia elata* Blume - An ethnopharmacological review. *Journal of Ethnopharmacology*, 189: 361–85.

Zhang C, Liu SJ, Yang L, et al., 2017. Sesquiterpene amino ether and cytotoxic phenols from *Dendrobium wardianum* Warner. *Fitoterapia*, 122: 76–79.

Zhang Y, Li M, Kang RX, et al., 2012. NHBA isolated from *Gastrodia elata* exerts sedative and hypnotic effects in sodium pentobarbital-treated mice. *Pharmacology, Biochemistry, and Behavior*, 102: 450–457.

Zhang ZC, Su G, Li J, et al., 2013. Two new neuroprotective phenolic compounds from *Gastrodia elata*. *Journal of Asian Natural Products Research*, 15: 619–623.

Zhong HJ, Chen L, Zhou J, 2017. Effect of *Dendrobium* polysaccharides on injury of endothelium-dependent relaxation induced by high glucose. *Chinese Journal of New Drugs*, 26: 1443–1449.

Zhou B, Tan J, Zhang C, et al., 2018. Neuroprotective effect of polysaccharides from *Gastrodia elata* Blume against corticosteroneinduced apoptosis in PC12 cells via inhibition of the endoplasmic reticulum stress mediated pathway. *Molecular Medicine Reports*, 17: 1182–1190.

Zhou BH, Li XJ, Liu M, et al., 2006. Antidepressant-like activity of the *Gastrodia elata* ethanol extract in mice. *Fitoterapia*, 77: 592–4.

Zhou J, Yang YB, Yang CR, 1979. New phenolic glycoside in gastrodia--Gastrodin. *Chinese Science Bulletin*, 7: 335–336.

Zhou J, Yang YB, Yang CR, 1980. Study on the chemistry of gastrodia II--synthesis of gastrodin and its analogs. *Acta Chimica Sinica*, 38: 162–166.

Zhu H, Liu C, Hou J, et al., 2019. *Gastrodia elata* Blume polysaccharides: A review of their acquisition, analysis, modification, and pharmacological activities. *Molecules*, 24: 2436.

Zhu Q, Sheng Y, Li W, et al., 2019. Erianin, a novel dibenzyl compound in *Dendrobium* extract, inhibits bladder cancer cell growth via the mitochondrial apoptosis and JNK pathways. *Toxicology and Applied Pharmacology*, 371: 41–54.

8

Garcinia Plants

Fengke Lin
College of Life and Environmental Sciences, Minzu University of China, Beijing, People's Republic of China

Key Laboratory of Ethnomedicine (Minzu University of China), Ministry of Education, Beijing, People's Republic of China

Ping Li
Key Laboratory of Agro-Environment in the Tropics, Ministry of Agriculture, South China Agricultural University, Guangzhou, Guangdong, People's Republic of China

Grace Gar-Lee Yue
Institute of Chinese Medicine and State Key Laboratory of Research on Bioactivities and Clinical Applications of Medicinal Plants, The Chinese University of Hong Kong, Shatin, New Territories, Hong Kong SAR, China

Clara Bik-San Lau
Institute of Chinese Medicine and State Key Laboratory of Research on Bioactivities and Clinical Applications of Medicinal Plants, The Chinese University of Hong Kong, Shatin, New Territories, Hong Kong SAR, China

Edward Kennelly
Department of Biological Sciences, Lehman College and The Graduate Center, City University of New York, New York City, New York, USA

Chun-Lin Long*
College of Life and Environmental Sciences, Minzu University of China, Beijing, People's Republic of China

Key Laboratory of Ethnomedicine (Minzu University of China), Ministry of Education, Beijing, People's Republic of China

Kunming Institute of Botany, Chinese Academy of Sciences, Kunming, Yunnan, People's Republic of China

* Corresponding author.

CONTENTS

8.1 Introduction

Garcinia L. is a large genus of the family Clusiaceae (also called Guttiferae) with approximately 450 species in the world. They are evergreen fruit trees or shrubs, naturally occurring in tropical and South Africa, Madagascar, northeastern Australia, tropical Asia, western Polynesia, and tropical America (Li et al., 2007). *Garcinia* plants are commonly known as saptrees, garcinias, or "monkey fruit" (Patil and Appaiah, 2015). In China, 20 species of this genus have been recorded, 13 of which are endemic. *Garcinia* species are most abundant in Yunnan Province of southwestern China, with 14 taxa, followed by Guangxi and Guangdong provinces in southern China (Li et al., 2007).

Garcinia plants are well-known ethnobotanically as edible fruits and traditional medicines (Yapwattanaphun et al., 2002). Many species bear edible juicy fruits, which are appreciated by local people. In addition, various plant parts including roots, stems, leaves, stem bark, twigs, and fruits have long been used as ethnomedicines to treat many human diseases such as cancer, diabetes, and inflammation (Hemshekhar et al., 2011). The most famous example is *G. mangostana* commonly known as mangosteen. The fruit of this species is regarded as the *Queen of the fruits* and is a popular edible worldwide (Ovalle-Magallanes et al., 2017). Besides its edible pulps, the seeds and pericarps of the fruits have been used in traditional medicinal practices for hundreds of years for gastrointestinal and urinary tract infections. Other parts of the plant such as the bark and the roots also have been used ethnomedicinally in southeast Asia to treat a wide variety of human disorders such as chronic ulcers and dysentery (Obolskiy et al., 2009, Ovalle-Magallanes et al., 2017).

Due to the ethnopharmacological importance of *Garcinia* species, the genus is of great interest for researchers worldwide. In the past decades, extensive phytochemical and pharmacological research has been conducted demonstrating that *Garcinia* species in Yunnan Province are rich sources of bioactive compounds such as xanthones, benzophenones, and flavonoids possessing various bioactivities, e.g. anti-cancer and antimicrobial activities (Hu et al., 2014a, Trisuwan et al., 2014, Fang et al., 2018, Hassan et al., 2018).

The information for this review chapter was obtained from classic medicinal books and various scientific databases up until August 2020, including Google Scholar, Science Direct, Web of Science (WoS), PubMed, and the Chinese databases such as Wanfang, China National Knowledge Infrastructure (CNKI), and China Science and Technology Journal Database (WP). To obtain botanical and distribution information, records from the *Flora of China* (http://flora.huh.harvard.edu/china), the Chinese Virtual Herbarium (CVH, http://www.cvh.ac.cn/), and the Herbarium at Kunming Institute of Botany (KUN) were acquired. In addition, the results from our previous ethnobotanical surveys conducted in Yunnan Province during 2019 are also summarized and documented to complement the ethnomedicinal uses. In this chapter, we particularly focus on the therapeutically significant constituents and the relevance of traditional practices and modern biological evidence, aiming to highlight the medicinal importance of *Garcinia* species in Yunnan Province as well as to stimulate the development and sustainability of these species.

8.2 Botanical Description and Distribution

Garcinia species in Yunnan have evolved diverse morphological characteristics. In general, they can be obviously characterized by monopodial form and yellow or white latex originated from cut plant parts such as boles and twigs (Nazre et al., 2018). *Garcinia* plants in Yunnan are usually evergreen trees, rarely shrubs, with leaves opposite, glabrous, and leathery to membranous. The plants are dioecious, sometimes monoecious, and apparently flowers are bisexual. Sepals are usually four or five, free and decussate or imbricate; petals usually four or five. Stamens in male flowers, many to few, are accompanied by filaments, free or completely united, and by anthers one-, two-, four-, or many-celled; pistillode can be present or absent. Staminode fascicles in female flowers are free or united at the base of the ovary; stigmas are absent or present (short, peltate two–five-lobed, or entire). The fruits, smooth or sulcate, have exocarp leathery to thin and large seeds usually less than four per fruit, embedded in endocarpic pulps (Li et al., 2007). Some specific morphological characteristics of selected *Garcinia* species in Yunnan Province are shown in Figures 8.1 and 8.2.

Even though Yunnan Province represents the richest source of *Garcinia* species in China, with 14 taxa distributed, the geographical distribution of *Garcinia* species in Yunnan is not uniform (Ma et al., 2013). They are mainly distributed in the southeastern, southwestern, and southern parts of Yunnan Province (Table 8.1). In addition to one imported species, namely *G. mangostana*, all other species are native to Yunnan Province. Among the 13 native species, 6 taxa, i.e. *G. esculenta*, *G. lancilimba*, *G. erythrosepala*, *G. tetralata*, *G. xipshuanbannaensis*, and *G. yunnanensis*, are endemic species to the province. However, other than *G. yunnanensis* and *G. esculenta*, all the endemic plants have been currently documented as vulnerable species in the *Threatened Species List of China's Higher Plants*, based on International Union for Conservation of Nature (IUCN) red list categories and criteria (Qin et al., 2017).

FIGURE 8.1 Photos of (A) *G. xanthochymus* and (B) *G. yunnanensis* (photos by Fengke Lin).

Additionally, the native *Garcinia* species are still under wild or semi-wild status (Table 8.1), highlighting the importance of resource conservation and sustainability.

8.3 Ethnomedicinal Uses

Yunnan Province is well known as a hotspot of medicinal plants, and the local people who come from diverse ethnic groups, such as *Dai*, *Miao*, and *Yao* nationalities, have garnered considerable traditional knowledge and practices related to the medicinal plants (Long and Li, 2004). Other than the edible fruits of *Garcinia* species, people in Yunnan also employ several *Garcinia* species as ethnomedicines. Based on a literature review (Chinese Herbal Medicinal Company, 1994, Editorial Committee of National Compilation of Chinese Herbal Medicine, 1996, Editorial Committee of the Administration Bureau of Traditional Chinese Medicine, 1999, Editorial Committee of Chinese Ethnomedicines, 2005, Liu et al., 2016) and our field surveys, six species, including *G. cowa*, *G. esculenta*, *G. multiflora*, *G. paucinervis*, *G. xanthochymus*, and *G. yunnanensis*, are used as traditional medicines (Table 8.2).

Garcinia cowa and *G. xanthochymus* are used as traditional medicines by the *Dai* ethnic group. The stem sap and leaves of both *G. cowa* and *G. xanthochymus* can be used to remove leeches from the nose (Liu et al., 2016). In addition, the stems, and leaves of *G. cowa* are used for the treatment of eczema, stomatitis, periodontitis, and burns, while its fruits are used for analgesia and antiphlogosis (Chinese Herbal Medicinal Company, 1994). Based on our surveys, the bark of *G. xanthochymus* are useful for treating stomachache and dysentery by the ethnic *Miao* people, and the fruits of *G. esculenta* can be employed to treat rheumatism by the ethnic *Dulong* people (Table 8.2).

FIGURE 8.2 Morphological characteristics of different plant parts from selected *Garcinia* species (photos by Fengke Lin).

The medicinal uses of *G. multiflora* range from simple disorders such as cough and emesis to complex ailments including spleen deficiency. Different parts can be used as ethnomedicines including the bark, fruit, pericarps, and seeds (Table 8.2). For instance, the bark can be used to fight inflammation such as stomatitis, periodontitis enteritis, and gastric, lower limb and duodenal ulcers, and the pericarps are used to relieve coughs (Editorial Committee of the Administration Bureau of Traditional Chinese Medicine, 1999). Therefore, understanding the bioactive constituents of different plant parts is critical for effectively using them as traditional medicines.

Different parts of *G. paucinervis* are also used as traditional medicines. The roots are used to treat stomachaches, while the bark, twigs, or leaves have some therapeutic effects such as analgesia, antiphlogosis, and detumescence (Table 8.2). Notably, *G. paucinervis* has been classified as a Class II endangered species in

TABLE 8.1

Distribution of *Garcinia* Species in Yunnan Province, China

Species	Distribution Areas (County or County-Level Zone)	Status
G. bracteata	Mengla, Menghai, Jinghong; Cangyuan, Fengqing Gengma, Zhenkang, Linxiang; Malipo, Funing, Xichou; Jinping, Yuanjiang	Semi-wild
G. cowa	Mengla, Menghai, Jinghong; Gengma, Lancang, Cangyuan; Jinping, Lüchun, Hekou, Pingbian; Malipo; Ruili	Semi-wild
*G. esculenta**	Ruili, Yingjaing, Longchuan; Gongshan; Malipo	Wild
*G. lancilimba**	Mengla, Menghai, Jinghong	Semi-wild
G. mangostana	Longchuan; Mengla	Cultivated
G. multiflora	Mengla, Menghai, Jinghong; Shuangjiang; Jinping, Lüchun, Hekou, Pingbian, Yuanyang; Xichou, Malipo, Maguan; Xinping; Ruili	Semi-wild
G. nujiangensis	Yingjiang, Longchuan; Gongshan	Wild
G. paucinervis	Malipo; Yingjiang	Semi-wild
G. pedunculata	Ruili, Yingjaing; Gongshan	Semi-wild
*G. erythrosepala**	Yingjiang; Mengla	Semi-wild
*G. tetralata**	Mengla, Jinghong, Cangyuan, Gengma	Semi-wild
G. xanthochymus	Mengla, Menghai, Jinghong; Menglian, Jingdong, Lancang, Jinggu, Yunxian, Gengma, Yongde; Jinping, Hekou, Lüchun; Yingjiang, Lianghe; Xichou	Semi-wild
*G. xipshuanbannaensis**	Mengla; Yuanjiang; Yingjiang	Semi-wild
*G. yunnanensis**	Mengla, Menghai, Jinghong; Gengma, Zhenkang, Cangyuan, Lancang; Mangshi; Jinping, Yuanyang	Wild

*Endemic to Yunnan

China (Li, et al., 2016c). Thus, for the management of *G. paucinervis*, it is urgent to find a balance between resource utilization and conservation, and further research conducted with regards to population enlargement and management techniques is urgently needed.

8.4 Phytochemistry

The therapeutic efficacy of medicinal plants is directly associated with the phytochemical constituents. *Garcinia* species have aroused tremendous interest from researchers since a variety of pharmacologically active compounds has been discovered (Hemshekhar et al., 2011). Phytochemical investigations into Yunnan *Garcinia* species also led to the discovery of potential drug-lead metabolites, such as xanthones, benzophenones, and flavonoids.

8.4.1 Xanthones

Xanthones, derived from 9*H*-xanthen-9-one, are a major phytochemical class in the genus *Garcinia*. Natural xanthones, in general, are classified into six groups including

TABLE 8.2

Medicinal Uses of *Garcinia* Species in Yunnan Province, China

Species	Local Names (Ethnic Groups)	Medicinal Parts	Traditional Uses
G. cowa	Guo mu bang, ge ha hao (*Dai* people); ma na (*Jinuo* people)	Stem, leaf	Clearing heat and detoxication, eczema, stomatitis, periodontitis, burns and scalding, expelling leech out of the nose
		Fruit	Analgesia, antiphlogosis
G. esculenta	Mai ren (*Dulong* people); wang du xi man xi, di rang xi (*Jingpo* people); bu nan bao (*De'ang* people)	Fruit	Clearing heat and detoxication, rheumatism, periodontitis
G. multiflora	Mi kou, luo wang (*Zhuang* people); bu nang wa (*Dulong* people); si lan shen (*Lisu* people); dan luo dan (*Yao* people)	Bark	Clearing heat, anti-alcoholism, analgesia, hemostasis, lower limb ulcer, eczema, stomatitis, periodontitis, enteritis, gastric ulcer, duodenal ulcer, dyspepsia, stomachache, diarrhea, asthma, burns and scalding
		Fruit	Emesis, cough, anti-alcoholism, spleen deficiency
		Pericarp	Cough
		Seed	Eczema, stomatitis, periodontitis, burns and scalding, and analgesia, hematocele
G. paucinervis	Mai gui, miyou bo, mei lu dun (*Zhuang* people); suimian (*Yao* people)	Root	Stomachache
		Bark, twig, leaf	Clearing heat and detoxication, analgesia, antiphlogosis, detumescence, burns and scalding
G. xanthochymus	Guo man da, guo ma la, ge ma la, mai ma la (*Dai* people); wai pi guo guo (*Yi* people); ma la (*Jinuo* people)	Stem, leaf	Expelling worm, expelling leech from noses
		Bark	Stomachache, dysentery
G. yunnanensis	Ma gei an (*Wa* people); duo bu xie (*Hani* people)	Twig, leaf, fruit	Cough, bronchitis, kidney deficiency, rheumatism

simple oxygenated xanthones, glycosylated xanthones, prenylated xanthones, xanthone dimers, xanthonolignoids, and some miscellaneous compounds (Pinto et al., 2005). In addition to fungi and lichens, xanthones occur in higher plant families, mainly Clusiaceae and Gentianaceae (Negi et al., 2013). *Garcinia* species contain many different types of xanthones. Approximately 360 xanthones have been isolated from the genus *Garcinia* from 2012 to 2019 (Han et al., 2020, Klein-Júnior et al., 2020). Yunnan

Garcinia species are also rich sources of bioactive xanthones. To date, more than 300 xanthones have been isolated from *Garcinia* from Yunnan (Wang et al., 2016). Among them, α-mangostin (Figure 8.3a) is the most well-known metabolite for its versatile bioactivities.

α-Mangostin, a prenylated xanthone, was first isolated from *G. mangostana* in 1855 (Fei et al., 2014) and has been found in other species such as *G. cowa* (Panthong et al., 2006) and *G. pedunculata* (Vo et al., 2015). Current *in vitro* and *in vivo* evidence has demonstrated that α-mangostin possesses extensive bioactivities including anti-cancer, antimalarial, antimicrobial, antiobesity, anti-inflammatory, anti-oxidant,

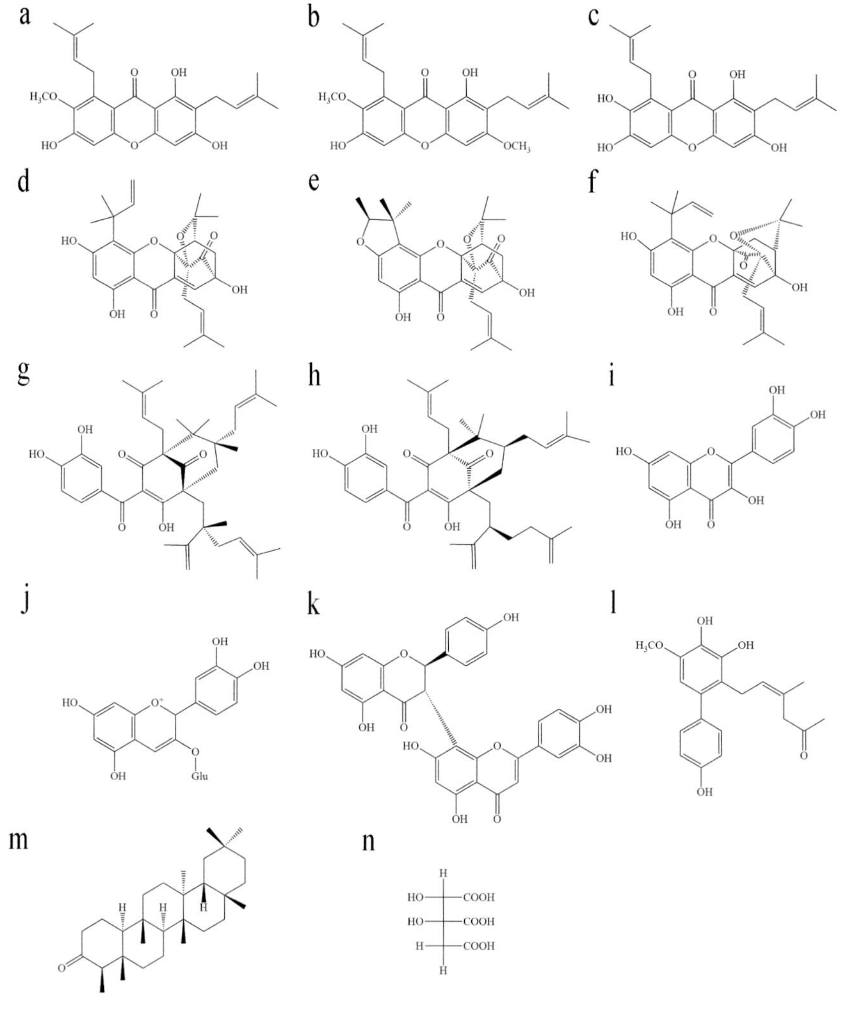

FIGURE 8.3 Chemical structures of pharmacologically active constituents isolated from *Garcinia* species in Yunnan Province of China. Keys: (a) α-mangostin; (b) β-mangostin; (c) γ-mangostin; (d) bractatin; (e) isobractatin; (f) neobractatin; (g) garcinol; (h) xanthochymol; (i) quercetin; (j) cyanidin-3-glucoside; (k) morelloflavone; (l) multibiphenyl A; (m) friedelan-3-one; (n) (–)-hydroxycitric acid.

and antiviral activities, as well as hepatoprotective and cardioprotective potentials, and neuroprotection in Alzheimer's disease (AD) (Wang et al., 2017b, Zhang, et al., 2017b, Chen et al., 2018). Due to the notable bioactivities of α-mangostin, its derivatives have also been studied. β-Mangtostin (Figure 8.3b), which has been found in *G. cowa* (Panthong et al., 2006) and *G. mangostana* (Obolskiy et al., 2009), possesses anti-cancer, anti-malarial, and anti-microbial bioactivities (Hemshekhar et al., 2011, Aizat et al., 2019). In addition, γ-mangostin (Figure 8.3c), which has been identified in several Yunnan *Garcinia* species including *G. cowa* (Na et al., 2013b), *G. mangostana* (Aizat et al., 2019), *G. esculenta* (Zhang et al., 2014), and *G. xipshuanbannaensis* (Han et al., 2008), possesses anti-HIV, anti-cancer, and anti-inflammatory effects as well as hepatoprotective, neuronal, and cardiovascular protection properties (Aizat et al., 2019, Hemshekhar et al., 2011).

Caged xanthones, a special group of prenylated xanthones with a unique 4-oxotricyclo [4.3.1.0^{3,7}] dec-8-en-2-one scaffold, have attracted increasing attention due to their promising pharmacological activities, in particular, anti-cancer activity (Anantachoke et al., 2012). They have been isolated and purified from *Garcinia* species by column chromatography over silica gel or Sephadex LH-20 (Jia et al., 2015). Even though the genus *Garcinia* contains over 400 species, caged xanthones are limited to less than ten *Garcinia* species such as *G. morella* and *G. hanburyi* (Anantachoke et al., 2012). The representative xanthone, gambogic acid, is considered as a potent anti-cancer agent (Wang and Chen, 2012). However, it has never been found in *Garcinia* species in Yunnan. From *Garcinia*, present in Yunnan, only *G. bracteata* has been reported to produce caged xanthones. Considerable *in vitro* assays showed that the caged xanthones from *G. bracteata* possessed potent anti-cancer capacity. For example, bractatin (Figure 8.3d), isobractatin (Figure 8.3e), and neobractatin (Figure 8.3f) showed strong cytotoxicity against several human cancer cell lines such as human lung cancer (A549) and human colon cancer (SW480) cell lines (Na et al., 2013a). As caged xanthones have not been discovered in other *Garcinia* species in Yunnan, these compounds have great potential to be used as chemotaxonomic markers to distinguish *G. bracteata* from other *Garcinia* species.

8.4.2 Benzophenones

Benzophenones, sharing a common structure with a phenol-carbonyl-phenol skeleton, are known for their biological activities including anti-cancer, anti-fungal, anti-HIV, anti-microbial, anti-oxidant, anti-malaria, and anti-viral activities (Acuna et al., 2009). According to our previous review (Wu et al., 2014), more than 300 natural benzophenones have been isolated and identified, most of which were derived from Clusiaceae plants. These benzophenones can be classified into two groups, namely basic benzophenones and polyprenylated benzophenones (Wu et al., 2014). *Garcinia* species in Yunnan Province produce more than 100 polyprenylated benzophenones (Li et al., 2016b,c). Garcinol (Figure 8.3g) and xanthochymol (Figure 8.3h) are the most cited polyprenylated benzophenones in the literature with various pharmacological activities.

Garcinol has been isolated from the fruits of *G. pedunculata* and *G. xanthochymus* (Sahu et al., 1989, Zhang et al., 2017a), and the twigs of *G. multiflora* and *G. esculenta* (Zhang et al., 2017a). It is a multi-therapeutic natural product possessing anti-cancer, anti-microbial, anti-inflammatory, and anti-oxidant bioactivities (Hemshekhar et al., 2011). In addition to classical isolation techniques using silica gel column chromatography and recrystallization, garcinol has been rapidly isolated from *Garcinia* extracts by high-speed counter-current chromatography (HSCCC) (Zhang et al., 2017a, Schobert and Biersack, 2019). Xanthochymol has been isolated from the fruits of *G. xanthochymus* (Baggett et al., 2005) and the twigs of *G. xipshuanbannaensis* (Han et al., 2008). This compound displays anti-cancer, anti-inflammatory, anti-microbial, and anti-protozoal effects (Hemshekhar et al., 2011). Although xanthochymol is difficult to separate from its π-bond isomer, guttiferone E, using a normal column chromatography system, these two isomers can be well separated using a combination with high-performance liquid chromatography (HPLC), HSCCC, and silver nitrate coordination reaction (Li et al., 2017b).

8.4.3 Flavonoids

Flavonoids are a group of polyphenols with low molecular weight, consisting of C_6–C_3–C_6 rings (Wang et al., 2018a). These naturally occurring compounds are subdivided into several types such as flavanols, anthocyanidins, chalcones, and biflavonoids (Hemshekhar et al., 2011). They have been well investigated for various pharmacological activities such as anti-oxidant, anti-inflammatory, and anti-cancer effects (Wang et al., 2018a, Owona et al., 2020). Flavonoids have been isolated and identified in different parts of *Garcinia* species in Yunnan such as *G. bracteata*, *G. mangostana*, and *G. xanthochymus* (Konoshima et al., 1970, Yu et al., 2007, Hu et al., 2014b). Quercetin (Figure 8.3i), cyanidin-3-glucoside (Figure 8.3j), and morelloflavone (Figure 8.3k) are three examples of therapeutically important flavonoids isolated from *Garcinia* species that grow in Yunnan Province of China.

Quercetin, a common flavanol, has been isolated from the stems of *G. bracteata* (Hu et al., 2014b) and *G. cowa* (Shen et al., 2007), the pericarps of *G. mangostana* (Azima et al., 2017), the bark of *G. xanthochymus* (Ji et al., 2012), and the twigs of *G. xipshuanbannaensis* (Na and Xu, 2009). It has been well-investigated for various pharmacological activities including anti-cancer, cardioprotective, and antidiabetic activities (Wang et al., 2018a). Cyanidin-3-glucoside is characterized as a major anthocyanin in the pericarp of *G. mangostana* (Azima et al., 2017). Numerous studies indicate that it possesses potent anti-oxidant capacity as well as anti-inflammatory, anti-cancer, neuroprotective, and antihyperglycemic effects (Hu et al., 2019, Yoon et al., 2019). Morelloflavone, a biflavonoid, has been discovered in *G. cowa* (Panthong et al., 2009), *G. multiflora* (Liu et al., 2010), *G. xanthochymus* (Hassan et al., 2018), and *G. yunnanensis* (Fang et al., 2018), exhibiting a wide range of biological activities including anti-HIV, anti-oxidant, and anti-inflammatory activities (Pereañez et al., 2014).

Our previous structure–activity relationship study of seven biflavonoids, amentoflavone, volkensiflavone, morelloflavone, fukugeside GB 1a, GB 1a glucoside, GB 2a, and GB 2a glucoside, isolated from *G. xanthochymus* and *G. paucinervis*, showed that amentoflavone and morelloflavone exerted anti-angiogenic effects in zebrafish (Li et al., 2017c). These results indicated that *Garcinia* species found in Yunnan can serve as natural sources of various types of bioactive flavonoids.

8.4.4 Other Compounds

Other types of pharmacologically active metabolites have also been isolated from *Garcinia* species in Yunnan Province such as biphenyls, triterpenes, depsidones, and organic acids. For instance, multibiphenyl A (Figure 8.3l), a novel biphenyl isolated from *G. multiflora*, showed potent rotavirus activity comparable to that of ribavirin (Gao et al., 2016). Friedelan-3-one (Figure 8.3m), a triterpene isolated from *G. multiflora* also exhibited anti-inflammatory activity in RAW 264.7 cells (Cheng et al., 2018). In addition, hydroxycitric acid (Figure 8.3n), an abundant organic acid in the leaves and fruits of *G. cowa* (Jena et al., 2002), has been well studied for its anti-obesity potential (Hemshekhar et al., 2011). Altogether, the phytochemical investigations indicated that *Garcinia* species in Yunnan Province generated a broad range of secondary metabolites with therapeutic significance.

8.5 Bioactivity

Local plants serve as important sources of traditional medicines for humans to combat various disorders and ailments. *Garcinia* plants have long been recognized as traditional medicines by the local people in Yunnan Province of China. Modern biological studies *in vitro* or *in vivo* demonstrated that the extracts or pure compounds from *Garcinia* species in Yunnan Province possess diverse therapeutic effects such as anti-cancer, anti-inflammatory, and anti-microbial activities (Figure 8.4), which could help us to gain insights into the validation of related ethnomedicinal practices with a scientific basis.

8.5.1 Anti-Cancer Activity

Garcinia species are a well-known, rich source of anti-cancer natural compounds, such as xanthones and benzophenones, which have exerted cytotoxicity, apoptosis,

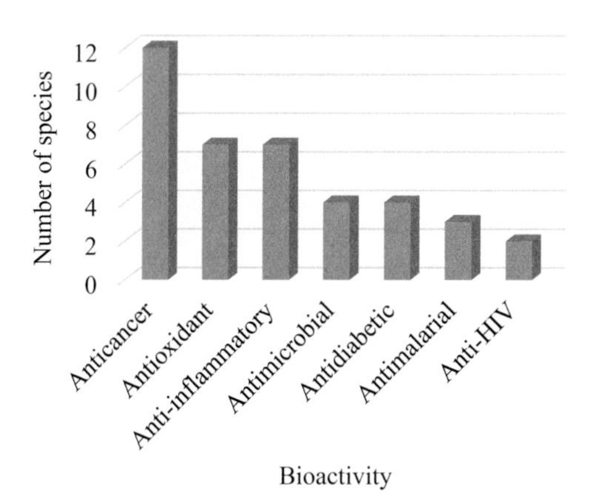

FIGURE 8.4 Selected biological activities of *Garcinia* species in Yunnan Province of China.

cell cycle arrest in cancer cells, and anti-tumor activities in animal studies over the past decade (Han and Xu, 2009, Wang and Chen, 2012, Xu et al., 2015, Brito et al., 2017, Aizat et al., 2019). The best studied caged xanthone, gambogic acid, is derived from the brownish gamboge resin of *G. hanburyi* and *G. morella* trees (Banik et al., 2018). Its growth-inhibitory activities in human hepatoma and lung cancer *in vitro* and *in vivo* have been reported since 2004 (Guo et al., 2004, Wu et al., 2004). Despite the large evidence of the anti-tumor potential and efficacy of gambogic acid, it has not yet been found in Yunnan *Garcinia* species. Nonetheless, another potent anti-cancer drug candidate, mangostin, the xanthone that can be isolated from Yunnan *Garcinia* species, such as *G. mangostana* (Fei et al., 2014), *G. cowa* (Panthong et al., 2006), and *G. pedunculata* (Vo et al., 2015), possesses various intracellular and extracellular actions. Many studies demonstrate that α-mangostin could regulate cell cycle, induce apoptosis, autophagy, interrupt angiogenesis and metastasis, reverse multidrug resistance, and inhibit tumor growth, which in turn contributes to the potential of being a suppressing agent in cancer (Wang et al., 2017a, Zhang et al., 2017b, Aizat et al., 2019, Klein-Júnior et al., 2020).

Furthermore, garcinol, a polyisoprenylated benzophenone derivative, has been isolated from the fruit of *G. indica* (Kaur et al., 2012), and from Yunnan *Garcinia* species *G. multiflora* (Liu et al., 2017) and *G. yunnanensis* (Zheng et al., 2017). The anti-tumor activities of garcinol involve the induction of apoptosis and cell cycle arrest, anti-angiogenesis, anti-metastasis, and modulation of gene expressions (Liu et al., 2015). Garcinol displays significant inhibitory activities in different tumor models, in which the inhibition of NF-kappa B signaling, and STAT-signaling pathways are involved (Schobert and Biersack, 2019).

Apart from the well-studied compounds mentioned above, some potent anti-tumor and/or anti-metastatic compounds have been recently isolated from the Yunnan *Garcinia* species. For example, new prenylated and scalemic caged xanthones from the leaf extract of *G. bracteata* exhibit strong cytotoxic activity against ten human cancer cell lines (Niu et al., 2018, Zhang et al., 2019a). A caged prenylxanthone isolated from *G. bracteata*, neobractatin, exerted an anti-proliferative effect on various cancer cells and reduced tumor burden in HeLa xenograft model (Zheng et al., 2019), as well as inhibiting metastasis through the modulation of RNA-binding-protein MBNL2 expression (Zhang et al., 2019b). Compounds isolated from *G. cowa*, such as chamuangone (Sae-Lim et al., 2020), garcicowins (Xia et al., 2015), and cowanin (Chowchaikong et al., 2018), inhibit cell proliferation and induce cell apoptosis in human cervical cancer and colon cancer cells, respectively. A dimeric xanthone isolated from *G. esculenta*, griffipavixanthone, inhibits tumor migration, invasion, and proliferation *in vitro* and *in vivo* in esophageal cancer model (Ding et al., 2016). In addition, nujiangexanthone A (NJXA) derived from *G. nujiangensis* induces cell apoptosis by activating the reactive oxygen species (ROS)-mediated JNK signaling pathway in Hela and SiHa cells. Its tumor inhibitory effect was confirmed also in tumor-bearing model (Zhang et al., 2016b), while new nujiangxanthones G-P showed potent cytotoxicity in breast and lung cancer cells with IC_{50} values <10 μM (Liu et al., 2020).

Polyprenylated benzoylphloroglucinol derivatives isolated from *G. multiflora* were shown to induce apoptosis in HeLa-C3 cells (Liu et al., 2010). Our previous study also showed that lavandulyl benzophenones garcimultiflorone L and garcimultiflorone

O exhibited cytotoxic activities *in vitro* against five cancer cell lines (Wang et al., 2018b). Garcimultiflorone K from the stems of *G. multiflora* exhibit anti-angiogenic activities in human endothelial progenitor cells (Yang et al., 2019). In the studies on *G. paucinervis*, researchers found that new xanthones displayed antiproliferative effect against HL-60 with IC_{50} values ranging from 0.80 to 30 μM (Li et al., 2016a; Jia et al., 2018).

For the natural compounds isolated from Yunnan commonly found species, *G. yunnanensis*, oblongifolin C (OC) and guttiferone K (GUTK) would be the representatives (Sui et al., 2020). In fact, the underlying mechanisms of action in human cancer cells of OC have been well illustrated in a series of studies (Xu et al., 2008, Feng et al., 2012, Kan et al., 2013, Lao et al., 2014, Shen et al., 2016, Zhang et al., 2016a, Li et al., 2017a).

8.5.2 Anti-Inflammatory Activity

Several medicinal *Garcinia* species in Yunnan such as *G. cowa*, *G. esculenta*, and *G. multiflora* have been used to treat different inflammatory disorders including stomatitis and rheumatism (Table 8.2), which suggests that the extracts or the chemical compounds from these species might possess anti-inflammatory activities.

The methanol extract of *G. cowa* stems was tested for its anti-inflammatory activity in activated murine RAW 264.7 macrophage cells by determining nitric oxide (NO) production, and the results revealed that the extract showed significant inhibition against NO production with an IC_{50} value of 25.0 μg/mL, without affecting the viability of RAW 264.7 macrophage cells (Jabit et al., 2009). Further investigations demonstrated that several xanthones from this species possessed anti-inflammatory properties (Panthong et al., 2009, Wahyuni et al., 2017). α-Mangostin isolated from the bark of *G. cowa* could reduce NO production by 83.4% at a concentration of 50 μM (Wahyuni et al., 2017). In addition, when compared with the positive control of phenylbutazone, six xanthones including cowaxanthones A–D, cowanin, and α-mangostin isolated from the fruits of *G. cowa* exhibited *in vivo* anti-inflammatory activity against rat ear edema induced by phenylpropiolate (EPP), with edema inhibition >50 %, after treatment for 30 min, 1 h, or 2 h (Panthong et al., 2009). These results, in part, support the traditional uses of the stems and fruits of *G. cowa* as anti-inflammatory agents (Table 8.2).

The ethyl acetate fraction of methanol extracts from *G. multiflora* fruits is considered as a potent inhibitor against formyl peptide receptors (FPRs), which could trigger inflammation by the mediation of neutrophil activation. *In vitro*, the fraction could significantly reduce the release of superoxide anion, and *in vivo*, it could effectively improve some inflammatory conditions such as neutrophil infiltration, pulmonary edema, and alveolar damage in mice with acute lung injury induced by lipopolysaccharide (LPS) (Tsai et al., 2018). Various benzophenones isolated from the fruits of *G. multiflora*, such as garcimultiflorone B and 13-hydroxygarcimultiflorone B, possess anti-inflammatory potential *in vitro* (Chen et al., 2009), indicating that benzophenones might be responsible for the antiphlogistic property of the fruit extracts from *G. multiflora*. In addition, two flavonoids namely volkensiflavone and fukugetin isolated from *G. multiflora* bark exhibited anti-inflammatory activity against NO production in LPS induced RAW 264.7 macrophage cells, with IC_{50} values of 25.48 and

44.17 μM, respectively (Xu et al., 2016). These results might be useful to justify that the stem bark and seeds derived from *G. multiflora* can be ethnomedicinally used to treat ailments related to inflammation such as asthma and eczema. However, the specific anti-inflammatory activities of separated seeds from *G. multiflora* fruits remain to be investigated for better understanding of the associated traditional uses.

Even though the fruits of *G. esculenta* are traditionally used to fight inflammatory diseases such as rheumatism, its anti-inflammatory information has not been available. Additionally, the antiphlogistic property of medicinal plant parts of *G. yunnanensis* and *G. paucinervis* are limited. Therefore, related studies need to be conducted for the validation of associated traditional practices.

On the other hand, the anti-inflammatory activities of some other Yunnan *Garcinia* such as *G. mangostana* and *G. xanthochymus* have been studied by researchers. Pongphasuk et al., 2003) investigated the anti-inflammatory effects of pericarp extracts from *G. mangostana* in albino mice. The results showed that the extracts could significantly reduce edema in a dose-dependent manner, with inhibition of 45.7% after treatment for 3 h at dose of 5.0 g/kg. The pericarp extracts also showed remarkable anti-inflammatory activity against NO production in RAW 264.7 macrophage cells with an IC_{50} value of 1.0 μg/mL (Tewtrakul et al., 2009). Ongoing research revealed that various xanthones, especially α- and γ-mangostin from mangosteen pericarps, exhibited inflammation-reducing activities (Tewtrakul et al., 2009, Ovalle-Magallanes et al., 2017). The extracts of the fruits, roots, bark, and leaves of *G. xanthochymus* also showed anti-inflammatory potential (Pal et al., 2005, Hamidon et al., 2016, Winata et al., 2018). The ethanolic fruit extract, and the methanolic and petroleum ether leaf extracts could significantly reduce paw edema induced by carrageenan in rats (Pal et al., 2005, Winata et al., 2018), while the dichloromethane and n-hexane extracts of roots and bark compromised NO generation in LPS-induced cells in a dose-dependent manner (Hamidon et al., 2016). These findings suggest that various parts of *G. xanthochymus* serve as good sources for the development of anti-inflammatory agents.

8.5.3 Antimicrobial Activity

Some *Garcinia* plants in Yunnan including *G. cowa*, *G. multiflora*, *G. paucinervis*, and *G. xanthochymus* are ethnomedicines used to relieve several human diseases associated with microbial infections such as diarrhea, stomachache, and dysentery (Table 8.2). Thus, the anti-microbial activities of these species facilitate a better understanding of their traditional uses.

Various parts of *G. cowa* have been tested for antimicrobial activities. The methanol extract of *G. cowa* stems was examined for antimicrobial activity and the extract showed moderate inhibition against *Salmonella typhi*, *Escherichia coli*, *Pseudomonas aureus*, *Bacillus subtilis*, and *Bacillus megaterium* with zones of inhibition ranging from 7.0 to 8.0 mm at a concentration of 400 μg/disc (Zahan, 2017). In addition, the methanolic extract of *G. cowa* leaves showed strong inhibitory activities against *Staphylococcus aureus*, *Bacillus cereus*, *Shigella flexneri*, and *Vibrio cholerae* with MIC values ranging from 0.63 to 1.25 mg/mL (Panda et al., 2016). These studies might be useful to explain why the stems and leaves of *G. cowa* have been traditionally used for treating stomatitis, periodontitis, and burns.

Both extracts and pure compounds from *G. xanthochymus* possess antimicrobial potential *in vitro*. The fruit extracts showed significant antimicrobial activities against bacteria and fungi, such as *Salmonella typhimurium* and *Streptococcus mutans*, using a disc diffusion method (Murmu et al., 2016). The seed oils also displayed antimicrobial activity against Gram-positive bacteria *Staphylococcus aureus* and *Streptococcus mutans* with MIC values of 15.0–18.0 μg/mL (Manohar et al., 2014). Additionally, several xanthones, benzophenones, and flavonoids isolated from *G. xanthochymus*, such as garcinexanthone A, xanthochymol, and morelloflavone, also possess significant anti-microbial activity (Hassan et al., 2018). *Garcinia xanthochymus* could be considered as a promising source of antimicrobial agent. However, the antimicrobial activity of *G. xanthochymus* bark, which is ethnomedicinally used to relieve stomachache and dysentery, requires further evaluation. Furthermore, the antimicrobial effects from medicinal plant parts of *G. multiflora* and *G. paucinervis* need further study to better understand their ethnopharmacological uses for treating ailments associated with microbial infections.

In addition to medicinal *Garcinia* plants in Yunnan, other species including *G. mangostanna* and *G. pedunculata* also inhibit certain types of microorganisms (Cunha et al., 2014, Sarma et al., 2016). The crude extracts from various parts of *G. mangostanna* including fruits, resins, and leaves, and the water extracts of *G. pedunculata* have strong antimicrobial activities, providing information for the further isolation and identification of antibiotic constituents (Cunha et al., 2014; Sarma et al., 2016, Aizat et al., 2019). However, the antimicrobial properties of other non-medicinal *Garcinia* species such as *G. bracteata*, *G. lancilimba*, and *G. tetralata* are yet to be explored.

8.5.4 Anti-Oxidant Activity

Oxidative stress contributes to diverse acute and chronic human diseases including inflammatory diseases, microbial infection disorders, cardiovascular diseases, and cancers (Dalle-Donne et al., 2006). Thus, the anti-oxidant activity of medicinal *Garcinia* plants in Yunnan, to some extent, helps to explain their ethnomedicinal uses.

Different extracts from *Garcinia* species in Yunnan including *G. cowa*, *G. multiflora*, *G. paucinervis*, *G. xanthochymus*, *G. yunnanensis*, *G. pedunculata*, and *G. mangostana* have been investigated for their anti-oxidant effects by various free radical scavenging assays (Wu et al., 2008, Obolskiy et al., 2009, Sharma et al., 2015, Jia et al., 2017, Fang et al., 2018, Hassan et al., 2018). The presence of phenolic compounds is mainly responsible for the anti-oxidant activities (Hassan et al., 2018). For example, the leaves, roots, and fruit extracts from *G. xanthochymus* have been tested for the free radical scavenging property *in vitro* by 2,2'-diphenyl-1-picrylhydrazyl (DPPH) and 2,2'-azino-bis (3-ethylbenzo-thiazoline-6-sulphonic acid) (ABTS) methods and the results indicated that the anti-oxidant activity of the extracts was closely associated with the phenolic contents. The ethyl acetate extract of leaves with the highest level of phenolic compounds exhibited the strongest anti-oxidant activities in DPPH ($IC_{50} = 6.10 \pm 0.01$ μg/mL) and ABTS ($IC_{50} = 6.74 \pm 0.09$ μg/mL) assays which were comparable to that of a standard compound, 2,6-di-tert-butyl-4-methyl phenol (BHT) (Fu et al., 2012). Another study indicated that the fruits of *G. pedunculata* and *G. xanthochymus* showed anti-oxidant effects in DPPH, ABTS, and reducing power assays, which were positively related to the total phenolic content (Sharma et al., 2015).

In addition to the crude extracts, the monomeric compounds especially xanthones and flavonoids from different *Garcinia* species in Yunnan possess potent anti-oxidant activities. For instance, four biflavones and one flavone from the pericarps of *G. yunnanensis* showed significant scavenging activity against DPPH with IC_{50} values in the range of 0.17–2.21 μM (Fang et al., 2018), and numerous xanthones from *G. xanthochymus* such as garcinenone, abigarcinenone B, and subeliptenone B possessed marked anti-oxidant activity using the same methods (Hassan et al., 2018). In consideration of potential anti-oxidant activities of crude extracts or pure compounds from *Garcinia* plants, these species which have not been investigated for radical scavenging effects such as *G. esculenta* and *G. nujiangensis* are good candidates for further study in order to fill the research gap and to discover more potential anti-oxidant substitutes for preventing related human diseases.

8.5.5 Other Bioactivities

Modern biological studies showed that *Garcinia* species in Yunnan also possessed other important bioactivities including anti-malarial, anti-diabetic, and antiviral effects, even though traditional practices by Yunnan people may be limited. Therefore, the current biological research could complement the depth and breadth of the ethnomedicinal uses of Yunnan *Garcinia* species.

Several studies demonstrated that some *Garcinia* species in Yunnan possessed antidiabetic activity, including *G. mangostana*, *G. pedculanta*, *G. xanthochymus*, and *G. cowa* (Obolskiy et al., 2009, Ali et al., 2017, Payamalle et al., 2017, Phukhatmuen et al., 2020). For instance, the leaf extracts of *G. cowa* showed anti-diabetic activity with the effects on α-glucosidase inhibition (IC_{50} = 21.4 μg/mL), glucose consumption (39.8 μg/mL), and glucose uptake (1.7-fold induction) in 3T3-L1 cells. Bioassay-guided isolation resulted in the discovery of prenylated xanthones and benzophenones with potent α-glucosidase inhibitory activity, including guttiferone I (13.1 μM), α-mangostin (15.0 μM), and cowanol (18.0 μM) (Phukhatmuen et al., 2020). Other research showed that the seed extracts of *G. xanthochymus* displayed strong hypoglycemic effects without acute oral toxicity in diabetic mice induced by alloxan with a dose-dependent response against blood glucose levels (Payamalle et al., 2017).

Several *Garcinia* species in Yunnan have anti-malarial activities. Five xanthones isolated from the barks of *G. cowa* including β-mangostin, cowanin, cowaxanthone, cowanol, and 7-*O*-methylgarcinone E possess antimalarial activity *in vitro* against *Plasmodium falciparum* with IC_{50} values in a range of 1.5–3.0 μg/mL, which was equivalent to that of pyrimethamine (IC_{50} = 2.8 μg/mL) (Likhitwitayawuid et al., 1998). The crude extracts of *G. mangostana* pericarps together with the pure compounds, i.e. α-mangostin, β-mangostin, gartanin, and 9-hydroxycarbaxathone, showed *in vitro* anti-plasmodial activity against *P. falciparum* 3D7 and K1 strains with IC_{50} values ranging from 0.5 to 47.3 μg/mL, which might due to the interruption of tricarboxylic acid cycle (TCA) metabolism in malaria parasites (Chaijaroenkul et al., 2014). Further *in vivo* studies in mice infected by *P. berghei* showed that the greatest therapeutic effects of α-mangostin acted through intraperitoneal administration with inhibition of about 80% after treatment with 100 mg/kg per day for a week (Upegui et al., 2015). Our previous research showed that the methanolic extracts of *G. mangostana* and *G. xanthochymus* fruit pulps displayed antiplasmodial activities against *P. falciparum* D6 and some benzophenones and xanthones such as guttiferone E and 3-isomangostin were the active compounds with IC_{50} values ranging from 4.71 to 11.40 μM (Lyles et al., 2014).

Two *Garcinia* species in Yunnan named *G. mangostana* and *G. multiflora* exhibited anti-HIV activity. The fruit extracts of *G. mangostana* showed significant inhibition against HIV-1 protease, which resulted in the isolation of bioactive prenylated xanthones such as mangostin (IC_{50} = 5.12 ± 0.41 μM) and γ-mangostin (4.81 ± 0.32 μM) (Chen et al., 1996). The methanol extracts of twigs and leaves from *G. multiflora* possess significant inhibitory activity against HIV-1 reverse transcriptase, which resulted in the isolation of active bioflavonoids such as morelloflavone (Lin et al., 1997).

In summary, the modern pharmacological investigations on *Garcinia* species in Yunnan Province of China indicate that *Garcinia* plants possess a wide spectrum of biological activities, some of which could support related ethnomedicinal uses. Therefore, they could be considered as promising sources for the discovery of drug-lead compounds to fight against human diseases. However, some species especially endemic plants such as *G. tetralata* and *G. yunnanensis* need to undergo further pharmacological investigation. In consideration of the versatile bioactivities of *Garcinia* species including antiviral potential, some researchers pointed out that *Garcinia* species such as *G. kola* might be useful against coronavirus disease 2019 (COVID-19), a global pandemic caused by severe acute respiratory syndrome coronavirus 2 (SARS-CoV-2) (Gbadamosi, 2020, Wu et al., 2020). Thus, it will be of great interest to extend the therapeutic potentials of Yunnan *Garcinia* species against SARS-CoV-2 for human health.

REFERENCES

Acuna UM, Jancovski N, Kennelly EJ, 2009. Polyisoprenylated benzophenones from Clusiaceae: potential drugs and lead compounds. *Curr Top Med Chem*, 9: 1560–1580.

Aizat WM, Jamil IN, Ahmad-Hashim FH, et al., 2019. Recent updates on metabolite composition and medicinal benefits of mangosteen plant. *PeerJ*, 7: e6324.

Ali M, Paul S, Tanvir E, et al., 2017. Antihyperglycemic, antidiabetic, and antioxidant effects of *Garcinia pedunculata* in rats. *Evid-Based Compl Alt*, 2017: 1–15.

Anantachoke N, Tuchinda P, Kuhakarn C, et al., 2012. Prenylated caged xanthones: chemistry and biology. *Pharm Biol*, 50: 78–91.

Azima AS, Noriham A, Manshoor N, 2017. Phenolics, antioxidants and color properties of aqueous pigmented plant extracts: *Ardisia colorata* var. *elliptica*, *Clitoria ternatea*, *Garcinia mangostana* and *Syzygium cumini*. *J Funct Foods*, 38: 232–241.

Baggett S, Protiva P, Mazzola EP, et al., 2005. Bioactive benzophenones from *Garcinia xanthochymus* fruits. *J Nat Prod*, 68: 354–360.

Banik K, Harsha C, Bordoloi D, et al., 2018. Therapeutic potential of gambogic acid, a caged xanthone, to target cancer. *Cancer Lett*, 416: 75–86.

Brito LC, Berenger ALR, Figueiredo MR, 2017. An overview of anticancer activity of *Garcinia* and *Hypericum*. *Food Chem Toxicol*, 109: 847–862.

Chaijaroenkul W, Mubaraki MA, Ward SA, et al., 2014. Metabolite footprinting of *Plasmodium falciparum* following exposure to *Garcinia mangostana* Linn. crude extract. *Exp Parasitol*, 145: 80–86.

Chen GQ, Li Y, Wang W, et al., 2018. Bioactivity and pharmacological properties of α-mangostin from the mangosteen fruit: a review. *Expert Opin Ther Pat*, 28: 415–427.

Chen JJ, Ting CW, Hwang TL, et al., 2009. Benzophenone derivatives from the fruits of *Garcinia multiflora* and their anti-inflammatory activity. *J Nat Prod*, 72: 253–258.

Chen SX, Wan M, Loh BN, 1996. Active constituents against HIV-1 protease from *Garcinia mangostana*. *Planta Med*, 62: 381–382.

Cheng LY, Tsai YC, Fu SL, et al., 2018. Acylphloroglucinol derivatives from *Garcinia multiflora* with anti-inflammatory effect in LPS-induced RAW264.7 macrophages. *Molecules*, 23: 2587.

Chinese Herbal Medicinal Company, 1994. Guttiferae. In: Zeng MY, Zeng JF (eds.) *Resources of Traditional Chinese Medicine*. Beijing: Science Press, pg. 419–423.

Chowchaikong N, Nilwarangkoon S, Laphookhieo S, et al., 2018. p38 inhibitor inhibits the apoptosis of cowanin-treated human colorectal adenocarcinoma cells. *Int J Oncol*, 52: 2031–2040.

Cunha BLA, França JPd, Moraes AadFS, et al., 2014. Evaluation of antimicrobial and antitumoral activity of *Garcinia mangostana* L. (mangosteen) grown in Southeast Brazil. *Acta Cir Bras*, 29 Supplement 2: 21–28.

Dalle-Donne I, Rossi R, Colombo R, et al., 2006. Biomarkers of oxidative damage in human disease. *Clin Chem*, 52: 601–623.

Ding Z, Lao Y, Zhang H, et al., 2016. Griffipavixanthone, a dimeric xanthone extracted from edible plants, inhibits tumor metastasis and proliferation via downregulation of the RAF pathway in esophageal cancer. *Oncotarget*, 7: 1826–1837.

Editorial Committee of the Administration Bureau of Traditional Chinese Medicine, 1999. Clusiaceae. In: Song LR (ed.) *Chinese Materia Medica (Zhonghua bencao)*. Shanghai: Shanghai Science and Technology Press, pg. 586–594.

Editorial Committee of Chinese Ethnomedicines, 2005. *Garcinia*. In: Jia MR, Li XW (eds.) *Chinese Ethnomedicines (Zhongguo Minzu Yaozhi Yao)*. Beijing: China Medical Science Press, pg. 286.

Editorial Committee of National Compilation of Chinese Herbal Medicine, 1996. *Garcinia*. In: Xie ZW (ed.) *National Compilation of Chinese Herbal Medicine*. Beijing: People's Medical Publishing House, pg. 101–102.

Fang X, Fu Z, Zhang H, et al., 2018. Chemical constituents of *Garcinia yunnanensis* and their scavenging activity against DPPH radicals. *Chem Nat Compd*, 54: 232–234.

Fei X, Jo M, Lee B, et al., 2014. Synthesis of xanthone derivatives based on α-mangostin and their biological evaluation for anti-cancer agents. *Bioorg Med Chem Lett*, 24: 2062–2065.

Feng C, Zhou LY, Yu T, et al., 2012. A new anticancer compound, oblongifolin C, inhibits tumor growth and promotes apoptosis in HeLa cells through Bax activation. *Int J Cancer*, 131: 1445–1454.

Fu M, Feng HJ, Chen Y, et al., 2012. Antioxidant activity of *Garcinia xanthochymus* leaf, root and fruit extracts *in vitro*. *Chin J Nat Med*, 10: 129–134.

Gao XM, Ji BK, Li YK, et al., 2016. New biphenyls from *Garcinia multiflora*. *J Brazil Chem Soc*, 27: 10–14.

Gbadamosi IT, 2020. Stay safe: helpful herbal remedies in COVID-19 infection. *Afr J Biomed Res*, 23: 131–133.

Guo QL, You QD, Wu ZQ, et al., 2004. General gambogic acids inhibited growth of human hepatoma SMMC-7721 cells *in vitro* and in nude mice. *Acta Pharmacol Sin*, 25: 769–774.

Hamidon H, Taher M, Jaffri JM, et al., 2016. Cytotoxic and anti-inflammatory activities of *Garcinia xanthochymus* extracts on cell lines. *Makara J Health Res*, 20: 3.

Han QB, Xu HX, 2009. Caged *Garcinia* xanthones: development since 1937. *Curr Med Chem*, 16: 3775–3796.

Han QB, Yang NY, Tian HL, et al., 2008. Xanthones with growth inhibition against HeLa cells from *Garcinia xipshuanbannaensis*. *Phytochemistry*, 69: 2187–2192.

Han YT, Li XY, Yuan CN, et al., 2020. Chemical constituents from the bark of *Garcinia oblongifolia*. *Nat Prod Commun*, 15: 1–4.

Hassan NKNC, Taher M, Susanti D, 2018. Phytochemical constituents and pharmacological properties of *Garcinia xanthochymus* - a review. *Biomed Pharmacother*, 106: 1378–1389.

Hemshekhar M, Sunitha K, Santhosh MS, et al., 2011. An overview on genus *Garcinia*: phytochemical and therapeutical aspects. *Phytochem Rev*, 10: 325–351.

Hu QF, Meng YL, Yao JH, et al., 2014a. Flavonoids from *Garcinia paucinervis* and their biological activities. *Chem Nat Compd*, 50: 994–997.

Hu QF, Niu DY, Wang SJ, et al., 2014b. New flavones from *Garcinia bracteata* and their biological activities. *Chem Nat Compd*, 50: 985–988.

Hu ZR, He YJ, He ZY, et al., 2019. Absorption, metabolism and physiological functions of cyanidin-3-glucoside. *Chin J Anim Nutr*, 31: 2052–2062.

Jabit ML, Wahyuni FS, Khalid R, et al., 2009. Cytotoxic and nitric oxide inhibitory activities of methanol extracts of *Garcinia* species. *Pharm Biol*, 47: 1019–1026.

Jena BS, Jayaprakasha GK, Sakariah KK, 2002. Organic acids from leaves, fruits, and rinds of *Garcinia cowa*. *J Agri Food Chem*, 50: 3431–3434.

Ji F, Li ZL, Niu SL, et al., 2012. Studies on the chemical constituents of the barks of *Garcinia xanthochymus*. *Chin J Med Chem*, 22: 507.

Jia BY, Li SS, Hu XR, et al., 2015. Recent research on bioactive xanthones from natural medicine: *Garcinia hanburyi*. *Aaps Pharmscitech*, 16: 742–758.

Jia CC, Han T, Xu J, et al., 2017. A new biflavonoid and a new triterpene from the leaves of *Garcinia paucinervis* and their biological activities. *J Nat Med*, 71: 642–649.

Jia CC, Xue JJ, Gong C, et al., 2018. Chiral resolution and anticancer effect of xanthones from *Garcinia paucinervis*. *Fitoterapia*, 127: 220–225.

Kan WL, Yin C, Xu HX, et al., 2013. Antitumor effects of novel compound, guttiferone K, on colon cancer by p21Waf1/Cip1-mediated G(0) /G(1) cell cycle arrest and apoptosis. *Int J Cancer*, 132: 707–716.

Kaur R, Chattopadhyay SK, Tandon S, et al., 2012. Large scale extraction of the fruits of *Garcinia indica* for the isolation of new and known polyisoprenylated benzophenone derivatives. *Ind Crops Prod*, 37: 420–426.

Klein-Júnior LC, Campos A, Niero R, et al., 2020. Xanthones and cancer: from natural sources to mechanisms of action. *Chem Biodivers*, 17: e1900499.

Konoshima M, Ikeshiro Y, Miyahara S, 1970. The constitution of biflavanoids from *Garcinia* plants. *Tetrahedron Lett*, 48: 4203–4206.

Lao Y, Wan G, Liu Z, et al., 2014. The natural compound oblongifolin C inhibits autophagic flux and enhances antitumor efficacy of nutrient deprivation. *Autophagy*, 10: 736–749.

Li DH, Li CX, Jia CC, et al., 2016a. Xanthones from *Garcinia paucinervis* with *in vitro* anti-proliferative activity against HL-60 cells. *Arch Pharm Res*, 39: 172–177.

Li H, Meng XX, Zhang L, et al., 2017a. Oblongifolin C and guttiferone K extracted from *Garcinia yunnanensis* fruit synergistically induce apoptosis in human colorectal cancer cells *in vitro*. *Acta Pharmacol Sin*, 38: 252–263.

Li HH, Zhang H, Fu WW, et al., 2016b. Progress on polycyclic polyprenylated acylphloroglucinols from *Garcinia* species in China. *World Chin Med*, 11: 1195–1201.

Li J, Gao RX, Zhao D, et al., 2017b. Separation and preparation of xanthochymol and guttiferone E by high performance liquid chromatography and high speed counter-current chromatography combined with silver nitrate coordination reaction. *J Chromatogr A*, 1511: 143–148.

Li P, Anandhi Senthilkumar H, Figueroa M, et al., 2016c. UPLC-QTOFMS^{E}-guided dereplication of the endangered Chinese species *Garcinia paucinervis* to identify additional benzophenone derivatives. *J Nat Prod*, 79: 1619–1627.

Li P, Yue GGL, Kwok HF, et al., 2017c. Using ultra-performance liquid chromatography quadrupole time of flight mass spectrometry-based chemometrics for the identification of anti-angiogenic biflavonoids from edible *Garcinia* species. *J Agric Food Chem*, 65: 8348–8355.

Li XW, Li J, Robson NK, et al., 2007. Clusiaceae. In: *Flora of China*. Beijing: Science Press, and St. Louis: Missouri Botanical Garden Press, pg. 40–47.

Likhitwitayawuid K, Phadungcharoen T, Krungkrai J, 1998. Antimalarial xanthones from *Garcinia cowa*. *Planta Med*, 64: 70–72.

Lin YM, Anderson H, Flavin MT, et al., 1997. *In vitro* anti-HIV activity of biflavonoids isolated from *Rhus succedanea* and *Garcinia multiflora*. *J Nat Prod*, 60: 884–888.

Liu B, Zhang XB, Bussmann RW, et al., 2016. *Garcinia* in Southern China: ethnobotany, management, and niche modeling. *Econ Bot*, 70: 416–430.

Liu C, Ho PC, Wong FC, et al., 2015. Garcinol: current status of its anti-oxidative, anti-inflammatory and anti-cancer effects. *Cancer Lett*, 362: 8–14.

Liu H, Gan F, Jin S, et al., 2017. Acylphloroglucinol and tocotrienol derivatives from the fruits of *Garcinia multiflora*. *RSC Adv*, 7: 29295–29301.

Liu X, Yu T, Gao XM, et al., 2010. Apoptotic effects of polyprenylated benzoylphloroglucinol derivatives from the twigs of *Garcinia multiflora*. *J Nat Prod*, 73: 1355–1359.

Liu XJ, Hu X, Peng XH, et al., 2020. Polyprenylated xanthones from the twigs and leaves of *Garcinia nujiangensis* and their cytotoxic evaluation. *Bioorg Chem*, 94: 103370.

Long CL, Li R, 2004. Ethnobotanical studies on medicinal plants used by the Red-headed Yao People in Jinping, Yunnan Province, China. *J Ethnopharmacol*, 90: 389–395.

Lyles JT, Negrin A, Khan SI, et al., 2014. *In vitro* antiplasmodial activity of benzophenones and xanthones from edible fruits of *Garcinia* species. *Planta Med*, 80: 676–681.

Ma T, Sima YK, Ma HF, et al., 2013. A study on the geographic distribution and floristic characteristics of *Garcinia* Linn. in Yunnan Province. *J Yunnan Univ*, 35: 99–107.

Manohar S, Naik P, Patil L, et al., 2014. Chemical composition of *Garcinia xanthochymus* seeds, seed oil, and evaluation of its antimicrobial and antioxidant activity. *J Herbs Spices Med Plants*, 20: 148–155.

Murmu P, Kumar S, Patra JK, et al., 2016. Ethnobotanical, nutritional, phytochemical and antimicrobial studies of *Garcinia xanthochymus* fruit extracts. *Biotechnol J Int*, 13: 1–11.

Na Z, Hu HB, Xu YK, 2013a. Cytotoxic caged xanthones from the fruits of *Garcinia bracteata*. *Chem Nat Compd*, 49: 505–506.

Na Z, Song QS, Hu HB, 2013b. A new prenylated xanthone from latex of *Garcinia cowa* Roxb. *Rec Nat Prod*, 7: 220–224.

Na Z, Xu YK, 2009. Chemical constituents from twigs of *Garcinia xipshuanbannaensis*. *China J Chin Mater Med*, 34: 2338–2342.

Nazre M, Newman M, Pennington R, et al., 2018. Taxonomic revision of *Garcinia* section *Garcinia* (Clusiaceae). *Phytotaxa*, 373: 1–52.

Negi J, Bisht V, Singh P, et al., 2013. Naturally occurring xanthones: chemistry and biology. *J Appl Chem*, 2013: 621459.

Niu SL, Li DH, Li XY, et al., 2018. Bioassay- and chemistry-guided isolation of scalemic caged prenylxanthones from the leaves of *Garcinia bracteata*. *J Nat Prod*, 81: 749–757.

Obolskiy D, Pischel I, Siriwatanametanon N, et al., 2009. *Garcinia mangostana* L.: a phytochemical and pharmacological review. *Phytother Res*, 23: 1047–1065.

Ovalle-Magallanes B, Eugenio-Perez D, Pedraza-Chaverri J, 2017. Medicinal properties of mangosteen (*Garcinia mangostana* L.): a comprehensive update. *Food Chem Toxicol*, 109: 102–122.

Owona BA, Abia WA, Moundipa PF, 2020. Natural compounds flavonoids as modulators of inflammasomes in chronic diseases. *Int Immunopharmacol*, 84: 106498.

Pal S, Nirmal S, Borhade P, et al., 2005. Antiinflammatory activity of various extracts of leaves of *Garcinia xanthochymus*. *Indian J Pharm Sci*, 67: 394.

Panda SK, Mohanta YK, Padhi L, et al., 2016. Large scale screening of ethnomedicinal plants for identification of potential antibacterial compounds. *Molecules*, 21: 293.

Panthong K, Hutadilok-Towatana N, Panthong A, 2009. Cowaxanthone F, a new tetraoxygenated xanthone, and other anti-inflammatory and antioxidant compounds from *Garcinia cowa*. *Can J Chem*, 87: 1636–1640.

Panthong K, Pongcharoen W, Phongpaichit S, et al., 2006. Tetraoxygenated xanthones from the fruits of *Garcinia cowa*. *Phytochemistry*, 67: 999–1004.

Patil MM, Appaiah KA, 2015. *Garcinia*: bioactive compounds and health benefits. In: Martirosyan DM (ed.) *Introduction to Functional Food Science*, Dallas: Food Science Publisher, pg. 110–125.

Payamalle S, Joseph KS, Bijjaragi SC, et al., 2017. Anti-diabetic activity of *Garcinia xanthochymus* seeds. *Comp Clin Pathol*, 26: 437–446.

Pereañez JA, Patiño AC, Núñez V, et al., 2014. The biflavonoid morelloflavone inhibits the enzymatic and biological activities of a snake venom phospholipase A2. *Chem-Biol Interact*, 220: 94–101.

Phukhatmuen P, Raksat A, Laphookhieo S, et al., 2020. Bioassay-guided isolation and identification of antidiabetic compounds from *Garcinia cowa* leaf extract. *Heliyon*, 6: e03625.

Pinto M, Sousa M, Nascimento M, 2005. Xanthone derivatives: new insights in biological activities. *Curr Med Chem*, 12: 2517–2538.

Pongphasuk N, Khunkitti W, Chitcharoenthum M, 2003. Anti-inflammatory and analgesic activities of the extract from *Garcinia mangostana* Linn. *III WOCMAP Congress on Medicinal and Aromatic Plants*, Chiang Mai, Thailand, Feb., 2003, 125–130.

Qin HN, Yang Y, Dong SY, et al., 2017. Threatened species list of China's higher plants. *Biodivers Sci*, 25: 696–744.

Sae-Lim P, Seetaha S, Tabtimmai L, et al., 2020. Chamuangone from *Garcinia cowa* leaves inhibits cell proliferation and migration and induces cell apoptosis in human cervical cancer *in vitro*. *J Pharm Pharmacol*, 72: 470–480.

Sahu A, Das B, Chatterjee A, 1989. Polyisoprenylated benzophenones from *Garcinia pedunculata*. *Phytochemistry*, 28: 1233–1235.

Sarma R, Das M, Mudoi T, et al., 2016. Evaluation of antioxidant and antifungal activities of polyphenol-rich extracts of dried pulp of *Garcinia pedunculata* Roxb. and *Garcinia morella* Gaertn. (Clusiaceae). *Trop J Pharm Res*, 15: 133–140.

Schobert R, Biersack B, 2019. Chemical and biological aspects of garcinol and isogarcinol: recent developments. *Chem Biodivers*, 16: e1900366.

Sharma PB, Handique PJ, Devi HS, 2015. Antioxidant properties, physico-chemical characteristics and proximate composition of five wild fruits of Manipur, India. *J Food Sci Tech Mys* 52: 894–902.

Shen J, Tian Z, Yang JS, 2007. The constituents from the stems of *Garcinia cowa* Roxb. and their cytotoxic activities. *Die Pharm*, 62: 549–551.

Shen K, Xi Z, Xie J, et al., 2016. Guttiferone K suppresses cell motility and metastasis of hepatocellular carcinoma by restoring aberrantly reduced profilin 1. *Oncotarget*, 7: 56650–56663.

Sui H, Tan H, Fu J, et al., 2020. The active fraction of *Garcinia yunnanensis* suppresses the progression of colorectal carcinoma by interfering with tumor associated macrophage-associated M2 macrophage polarization *in vivo* and *in vitro*. *FASEB J*. [published online ahead of print, 2020 Apr 13].

Tewtrakul S, Wattanapiromsakul C, Mahabusarakam W, 2009. Effects of compounds from *Garcinia mangostana* on inflammatory mediators in RAW264.7 macrophage cells. *J Ethnopharmacol*, 121: 379–382.

Trisuwan K, Boonyaketgoson S, Rukachaisirikul V, et al., 2014. Oxygenated xanthones and biflavanoids from the twigs of *Garcinia xanthochymus*. *Tetrahedron Lett*, 55: 3600–3602.

Tsai YF, Yang SC, Chang WY, et al., 2018. *Garcinia multiflora* inhibits FPR1-mediated neutrophil activation and protects against acute lung injury. *J Cell Physiol*, 51: 2776–2793.

Upegui Y, Robledo SM, Gil Romero JF, et al., 2015. *In vivo* antimalarial activity of α-mangostin and the new xanthone δ-mangostin. *Phytother Res*, 29: 1195–1201.

Vo HT, Ngo NT, Bui TQ, et al., 2015. Geranylated tetraoxygenated xanthones from the pericarp of *Garcinia pedunculata*. *Phytochem Lett*, 13: 119–122.

Wahyuni FS, Ali DAI, Lajis NH, 2017. Anti-inflammatory activity of isolated compounds from the stem bark of *Garcinia cowa* Roxb. *Pharmacogn J*, 9: 55–57.

Wang F, Ma H, Liu Z, et al., 2017a. α-Mangostin inhibits DMBA/TPA-induced skin cancer through inhibiting inflammation and promoting autophagy and apoptosis by regulating PI3K/Akt/mTOR signaling pathway in mice. *Biomed Pharmacother*, 92: 672–680.

Wang LP, Fu WW, Tan HS, et al., 2016. Chemistry of xanthones isolated from *Garcinia* species in China. *World Chin Med*, 11: 1154–1170.

Wang MH, Zhang KJ, Gu QL, et al., 2017b. Pharmacology of mangostins and their derivatives: a comprehensive review. *Chin J Nat Med*, 15: 81–93.

Wang TY, Li Q, Bi KS, 2018a. Bioactive flavonoids in medicinal plants: structure, activity and biological fate. *Asian J Pharm Sci*, 13: 12–23.

Wang X, Chen WT, 2012. Gambogic acid is a novel anti-cancer agent that inhibits cell proliferation, angiogenesis and metastasis. *Anti-Cancer Agents Med Chem*, 12: 994–1000.

Wang ZQ, Li XY, Hu DB, et al., 2018b. Cytotoxic garcimultiflorones K-Q, lavandulyl benzophenones from *Garcinia multiflora* branches. *Phytochemistry*, 152: 82–90.

Winata HS, Rosidah R, Sitorus P, 2018. Assessment of anti-inflammatory activity of ethanolic extract of Asam kandis (*Garcinia xanthochymus* Hook. f. ex T. Anderson) fruit. *Asian J Pharm Clin Res*, 11: 81–83.

Wu F, Zhao S, Yu B, et al., 2020. A new coronavirus associated with human respiratory disease in China. *Nature*, 579: 265–269.

Wu JH, Tung YT, Chyu CF, et al., 2008. Antioxidant activity and constituents of extracts from the root of *Garcinia multiflora*. *J Wood Sci*, 54: 383–389.

Wu SB, Long CL, Kennelly EJ, 2014. Structural diversity and bioactivities of natural benzophenones. *Nat Prod Rep*, 31: 1158–1174.

Wu ZQ, Guo QL, You QD, et al., 2004. Gambogic acid inhibits proliferation of human lung carcinoma SPC-A1 cells *in vivo* and *in vitro* and represses telomerase activity and telomerase reverse transcriptase mRNA expression in the cells. *Biol Pharm Bull*, 27: 1769–1774.

Xia Z, Zhang H, Xu D, et al., 2015. Xanthones from the leaves of *Garcinia cowa* induce cell cycle arrest, apoptosis, and autophagy in cancer cells. *Molecules*, 20: 11387–11399.

Xu DQ, Lao YZ, Xu NH, et al., 2015. Identification and characterization of anticancer compounds targeting apoptosis and autophagy from Chinese native *Garcinia* species. *Planta Med*, 81: 79–89.

Xu G, Feng C, Zhou Y, et al., 2008. Bioassay and ultraperformance liquid chromatography/mass spectrometry guided isolation of apoptosis-inducing benzophenones and xanthone from the pericarp of *Garcinia yunnanensis* Hu. *J Agric Food Chem*, 56: 11144–11150.

Xu J, Cui YY, Niu YF, et al., 2016. Chemical constituents and their anti-inflammatory activities of the barks of *Garcinia multiflora* Champ. *Chin Tradit Pat Med*, 38: 579–583.

Yang CY, Chen C, Lin CY, et al., 2019. Garcimultiflorone K inhibits angiogenesis through Akt/eNOS- and mTOR-dependent pathways in human endothelial progenitor cells. *Phytomedicine*, 64: 152911.

Yapwattanaphun C, Subhadrabandhu S, Sugiura A, et al., 2002. Utilization of some *Garcinia* species in Thailand. *Acta Hortic*, 575: 563–570.

Yoon KD, Lee JY, Kim TY, et al., 2019. *In vitro* and *in vivo* anti-hyperglycemic activities of taxifolin and its derivatives isolated from pigmented rice (*Oryzae sativa* L. cv. *Superhongmi*). *J Agri Food Chem*, 68: 742–750.

Yu LM, Zhao MM, Yang B, et al., 2007. Phenolics from hull of *Garcinia mangostana* fruit and their antioxidant activities. *Food Chem*, 104: 176–181.

Zahan N, 2017. *Antioxidant and Antimicrobial Investigations of Methanol Extract of Garcinia Cowa Stem*. Aftabnagar, Dhaka: East West University.

Zhang A, He W, Shi H, et al., 2016a. Natural compound oblongifolin C inhibits autophagic flux, and induces apoptosis and mitochondrial dysfunction in human cholangiocarcinoma QBC939 cells. *Mol Med Rep*, 14: 3179–3183.

Zhang BJ, Fang QW, Tan HS, et al., 2017a. Preparative isolation and purification of garcinol by high-speed countercurrent chromatography. *World Sci Technol/Modern Tradit Chin Med Mater Med*, 19: 254–259.

Zhang BJ, Fu WW, Wu R, et al., 2019a. Cytotoxic prenylated xanthones from the leaves of *Garcinia bracteata*. *Planta Med*, 85: 444–452.

Zhang H, Zhang DD, Lao YZ, et al., 2014. Cytotoxic and anti-inflammatory prenylated benzoylphloroglucinols and xanthones from the twigs of *Garcinia esculenta*. *J Nat Prod*, 77: 1700–1707.

Zhang J, Zheng Z, Wu M, et al., 2019b. The natural compound neobractatin inhibits tumor metastasis by upregulating the RNA-binding-protein MBNL2. *Cell Death Dis*, 10: 554.

Zhang KJ, Gu QL, Yang K, et al., 2017b. Anticarcinogenic effects of α-mangostin: a review. *Planta Med*, 83: 188–202.

Zhang L, Kong SY, Zheng ZQ, et al., 2016b. Nujiangexathone A, a novel compound derived from *Garcinia nujiangensis*, induces caspase-dependent apoptosis in cervical cancer through the ROS/JNK pathway. *Molecules*, 21: 1360.

Zheng D, Zhang H, Zheng CW, et al., 2017. Garciyunnanimines A-C, novel cytotoxic polycyclic polyprenylated acylphloroglucinol imines from *Garcinia yunnanensis*. *Org Chem Front*, 4: 2102–2108.

Zheng Z, Wu M, Zhang J, et al., 2019. The natural compound neobractatin induces cell cycle arrest by regulating E2F1 and Gadd45α. *Front Oncol*, 9: 654.

9

Pu-er Tea from *Camellia sinensis* (L.) Kuntze and Related Plants

Na Li
State Key Laboratory of Phytochemistry and Plant Resources in West China, Kunming Institute of Botany, Chinese Academy of Sciences, Kunming, Yunnan, People's Republic of China

University of Chinese Academy of Sciences, Beijing, People's Republic of China

Hong-Tao Zhu
State Key Laboratory of Phytochemistry and Plant Resources in West China, Kunming Institute of Botany, Chinese Academy of Sciences, Kunming, Yunnan, People's Republic of China

Yunnan Key Laboratory of Natural Medicinal Chemistry, Kunming Institute of Botany, Chinese Academy of Sciences, Kunming, Yunnan, People's Republic of China

Wang Dong
State Key Laboratory of Phytochemistry and Plant Resources in West China, Kunming Institute of Botany, Chinese Academy of Sciences, Kunming, Yunnan, People's Republic of China

Yunnan Key Laboratory of Natural Medicinal Chemistry, Kunming Institute of Botany, Chinese Academy of Sciences, Kunming, Yunnan, People's Republic of China

Chong-Ren Yang
State Key Laboratory of Phytochemistry and Plant Resources in West China, Kunming Institute of Botany, Chinese Academy of Sciences, Kunming, Yunnan, People's Republic of China

Ying-Jun Zhang*
State Key Laboratory of Phytochemistry and Plant Resources in West China, Kunming Institute of Botany, Chinese Academy of Sciences, Kunming, Yunnan, People's Republic of China

Yunnan Key Laboratory of Natural Medicinal Chemistry, Kunming Institute of Botany, Chinese Academy of Sciences, Kunming, Yunnan, People's Republic of China

* Corresponding author.

CONTENTS

9.1 Introduction

Tea is one of the most ancient and popular beverages consumed around the world. It is commonly made from the tender young leaves and buds of two widely cultivated *Camellia* species (Theaceae), *Camellia sinensis* (L.) O. Kuntze var. *sinensis* and *C. sinensis* var. *assamica* (Masters) Kitamura (Theaceae) (Wang et al., 2000). Tea can be classified into three major categories according to the manufacturing process and degree of fermentation: non-fermented tea (e.g. green tea), pre-fermented tea (e.g. oolong tea and black tea), and post-fermented tea (e.g. Pu-er tea and Fuzhuan brick tea) (Yang et al., 2006). It should be noted that strictly, this is not a true fermentation, there is no alcohol produced; however, there is breakdown due to solid-state "fermentation" by microorganisms. It is also an oxidative process, but for the remainder of this chapter, the term fermentation is used.

Among the various teas, Pu-er tea is the most representative and distinctive tea in Yunnan Province of China. It has attracted more attention due to its special flavor and taste as well as its various beneficial health effects (Kuo et al., 2005, Sano et al., 1986, Hou et al., 2009, Oi et al., 2012, Cao et al., 2011, Xu et al., 2012, Huang et al., 2013, Zhao et al., 2011a, Jie et al., 2005). This tea is normally made from the leaves of *C. sinensis* var. *assamica* in the southern and western area of Yunnan Province (Figure 9.1).

Investigations of chemical components and pharmacological activities of Pu-er tea and its original plants, *C. sinensis* var. *assamica*, have been carried out for about

FIGURE 9.1 Pu-er tea and its original plants *Camellia sinensis* var. *assamica* (photos by Tao Yu and Hong-Tao Zhu).

25 years, leading to the isolation and identification of more than 140 compounds (Li et al., 2019). These compounds belong to different types including hydrolysable tannins, flavan-3-ols and their derivatives, flavonols, flavones and their derivatives, simple phenolic acids, and alkaloids. In the present review, we summarize the current knowledge on Pu-er tea from the aspects of botany, development, and phytochemistry, to pharmacological activities, hoping to provide a comprehensive understanding of this tea.

9.2 Botany

Tea belongs to Sect. *Thea* (L.) Dyer, Genus *Camellia* L., Family Theaceae. According to Min Tianlu, there are 12 species and 5 varieties determined in Sect. *Thea* (Min, 1992, 2000). Usually, *C. sinensis* and its variety *C. sinensis* var. *assamica* are the original plants of drinkable tea. The material for making Pu-er tea is usually *C. sinensis* var. *assamica*. It is mainly distributed in the southwest of China, particularly Yunnan province and its neighboring areas, such as Guizhou, Guangxi, Hainan Province, and in Vietnam, Laos, Thailand, and the north of Burma. This species has also been introduced into Sri Lanka, India, and Africa. *C. sinensis* var. *assamica* is usually a small tree with large leathery leaves, growing in the evergreen broad-leaved forests with an altitude of 100–1500 m (Min, 1992). As an important material of the popular commercial tea products, it has been widely used to produce green tea, black tea, and white tea.

In addition, there are some varieties and closely related species of *C. sinensis* used as tea by local people. Among them, *C. taliensis* (W. W. Smith) Melchior, named *Da-Li-Cha* in Chinese, has been cultivated for making tea from long ago because of its pleasant flavor. *C. taliensis* is similar to the widely cultivated *C. sinensis* var. *assamica* in morphology. The obvious difference is a leaf-absent pubescence and five-locular ovary with tomenta for *C. taliensis*. It is an evergreen tree about 10–20 m high, distributed mainly from the west and southwest of Yunnan province in China to northern Myanmar, in the mountains of evergreen forests with altitudes of 1500–2400 m (Min, 1992). In China, it is mostly found in the west area of Ai-Lao mountains, e.g. Dali, Lincang, Dehong, Pu'er, and Xishuangbanna, and parts of the eastern area of the Ai-Lao mountain, like Yuanjiang (Jiang et al., 2009, Meng et al., 2019).

9.3 The Development and Manufacture Process of Pu-er Tea

As a well-known tea from ancient times, Pu-er has evolved with rich historical and cultural accumulation. Initially, Pu-er tea referred to the tea drunk by local Pu people (collective name of ancient tribes). Since the Tang dynasty, tea was introduced to Tibet by horse riders through a long and rough trade route known later as the famous Chinese Tea-horse Road. Pu-er tea changed gradually from loose sun-dried green tea into compressed tea in certain shapes for easy transport and storage. During this long journey tea had undergone a series of biotransformations ("natural fermentation"). In the process of tea trading, people noticed that, due to hot and humid conditions, some tea stored in improper environments was fermented, which accidentally produced the earliest "artificial" fermented Pu-er tea. The special mellow taste of fermented Pu-er tea and the concept of "aging" aroused public attention quickly. In the 1970s, commercial tea-makers in Yunnan Province successfully developed a method of making Pu-er ripened as a tea using a "piling up" process (i.e. solid-state fermentation), which greatly shortened the time of natural "aging." In the process of "piling up," the sun-dried tea leaves are moistened and piled up for a few weeks and turned over every two days to prevent excessively high temperatures. This technology of solid-state fermentation by microorganisms is widely used in the modern manufacture of Pu-er ripened tea, which has improved its manufacturing efficiency effectively. At present, Pu-er tea is protected by the National Geographical Indication Products, and the producing areas are confined to certain regions in Yunnan province, including 11 cities, 75 counties, and 639 townships (Lv et al., 2013, Shao et al., 1995, Zhang et al., 2011c).

According to the processing, the technical and the quality characteristics, Pu-er tea can be classified into Pu-er raw tea, Pu-er aging tea, and Pu-er ripened tea. Pu-er raw tea, a kind of green tea, is prepared from the sun-dried green tea of *C. sinensis* var. *assamica* by further steam softening and a compressing process. Pu-er ripened tea is normally made from the sun-dried green tea by microbial post-fermentation at higher temperature (about 50°C) and higher humidity conditions, leading to a series of oxidation, condensation, and degradation of polyphenols, and giving a special sensory quality (Lv et al., 2013). Pu-er aging tea is made from Pu-er raw tea after natural aging (fermentation) during long-term storage. Among them, Pu-er ripened tea is currently the most produced and widely consumed. Therefore, when the public mentions "Pu-er tea," it generally refers to the microbially fermented Pu-er ripened tea.

The post-fermentation process of Pu-er ripened tea involves many kinds of microorganisms, including molds, yeasts, bacteria, actinomycetes, etc. There are differences in the composition of the microorganisms among the "piles" of Pu-er ripened tea from different places. *Aspergillus* spp., e.g. *A. niger, A. gloucus, A. foetidus*, especially *A. niger*, are commonly regarded as the dominant microorganisms involved in the fermentation process (Abe et al., 2008, Zhao et al., 2010, Zhao and Zhou, 2005, Chen et al., 2006). Moreover, microorganism strains belonging to *Penicillium, Rhizopus, Talaromyces, Trichoderma, Candida, Saccharomyces, Bacillus, Lactobacillus, Actinomyces*, and *Streptomyces* are also isolated from different origins, fermentation periods, and storage periods of Pu-er ripened tea (Abe et al., 2008, Zhang et al., 2016, Mo et al., 2005, Gong et al., 1993, Jeng et al., 2007, Yang et al., 2011, Li et al., 2012, Zhao and Zhou, 2005, Fang et al., 2008). These microorganisms play a crucial role in the fermentation process and quality formation of the Pu-erh tea, such as infusion of color, taste, and aroma (Chen et al., 2009a, Hwang et al., 2003, Zhou et al., 2015, Zhou et al., 2004, Kang et al., 2015, Chen et al., 2008, Chen et al., 2009b). Xu et al. deduced that some of the volatile compounds in Pu-er ripened tea were derived from the action of microbial enzymes and hypothesized that *A. niger* plays a decisive role in the development of the volatile compounds (Xu et al., 2005). Jeng et al. found that Pu-er tea short-term fermented with *Streptomyces bacillaris* or *S. cinereus* enhances the color and content of statin, GABA, and polyphenols that might have health benefits (Jeng et al., 2007). Therefore, the selective inoculation of the certain microorganism strains for post-fermentation helps to shorten the fermentation period, and also helps to obtain certain bioactive compounds and to control the product quality of Pu-er ripened tea.

9.4 Phytochemistry

The chemical composition of Pu-er raw tea is very similar to that of green tea (Lv et al., 2013). Owing to the microbial post-fermentation, the chemical constituents of Pu-er ripened tea are different from those of Pu-er raw tea and its original materials. Gallic acid, gallocatechin, and caffeine are the main chemical components of Pu-er ripened tea (Xu et al., 2013). To date, 144 compounds have been isolated from Pu-er tea and its material involving one cultivated species, *C. sinensis* var. *assamica*, and one wild species, *C. taliensis*. These compounds can be divided into six categories according to their different chemical structures, hydrolysable tannins, flavan-3-ols and their derivatives, flavonols, flavones and their derivatives, alkaloids, simple phenolic compounds, and other compounds. Among them, tea polyphenols, consisting mainly of hydrolysable tannins and flavonoids, play an important role in bioactivities of Pu-er tea. Flavonoids are divided into three main subclasses, flavonols, flavones, and flavanols. Flavanols are usually referred to as flavan-3-ols. Flavonols and flavones have a similar structure with a double bond between C-2, three positions and an oxygen group at the C-4 position; the only difference is that the flavones lack a hydroxyl group at the C-3 position (Figure 9.2).

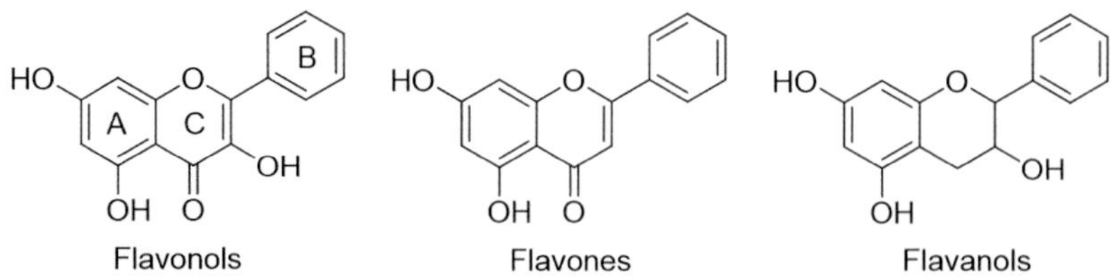

FIGURE 9.2 Three main subclasses of flavonoids.

9.4.1 Flavan-3-ols and Their Derivatives

Flavan-3-ols, e.g. catechins, proanthocyanidins, are the major components of green tea and the original tea plants. As the origin of the bitter and astringent taste in green tea, they show a significant anti-oxidative effect, owing to the efficacy of scavenging free radicals (Liu et al., 2009, Gao et al., 2008, Almajano et al., 2008).

Catechins, as the predominant components varying from 30% to 40% wt/wt of dissolved solids in green tea, are primarily responsible for the health benefits of tea plants and green tea (Jankun et al., 1997, Yang et al., 2002, Cao and Cao, 1999, Almajano et al., 2008). Similar to green tea, the main chemical components of Pu-er raw tea and its original plants are catechins (epicatechin (EC), catechin (C), epicatechin-3-*O*-gallate (ECG), gallocatechin (GC), gallocatechin-3-*O*-gallate (GCG), epigallocatechin (EGC), and epigallocatechin-3-*O*-gallate (EGCG)) (Figure 9.3), as well as its derivatives (e.g., epigallocatechin 3-*O*-caffeoate, epicatechin 3,5-di-*O*-gallate, epicatechin-3-*O*-benzoate). Of these, EGCG seems to elicit the most public attention (Calland et al., 2012, Brusselmans et al., 2003).

Proanthocyanidins also play an important role in the taste of green tea because of their strong astringency. Proanthocyanidins isolated from *C. sinensis* var. *assamica* form a series of four–eight linked catechin dimers, e.g. procyanidin B-2, prodelphinidin B-4, and trimer, procyanidin C-1 (Figure 9.4). To date, only ten proanthocyanidins have been reported from Pu-er raw tea and its original plants (Li et al., 2019).

In addition, the fresh leaves of *C. sinensis* var. *assamica* contain chalcan-flavan dimers (e.g., assamicains A-C), theasinensins (e.g., theasinensin A, desgalloyl theasinensin F), etc. (Hashimoto et al., 1989) (Figure 9.5). Among them, assamicanins appear to be ring-opened products of proanthocyanidins. Flavo-alkaloids are not common in plants but often possess noticeable biological activities (Blair et al., 2017). Recently, one kind of flavo-alkaloid, with an *N*-ethyl-2-pyrrolidinone substituted at C-6/8 of flavan-3-ols, namely, etc-pyrrolidinones E–J, displayed substantial protection against high-glucose-induced cell senescence on human umbilical vein endothelial cells (HUVECs). These compounds were isolated from Xigui tea, a Pu-er raw tea produced from an old tea tree cultivar of *C. sinensis* var. *assamica* (Cheng et al., 2018).

Due to the decomposition by microbial enzymes and autoxidation during the post-fermentation process, the chemical constituents of Pu-er raw tea, especially the catechins, change dramatically. As a result, Pu-er ripened tea contains negligible amounts of catechins (Lin et al., 1998). Proanthocyanidins have not been isolated from Pu-er ripened tea, either. Meanwhile, some new and complex chemical constituents that contribute to the unique quality of Pu-er ripened tea are formed. Some catechin

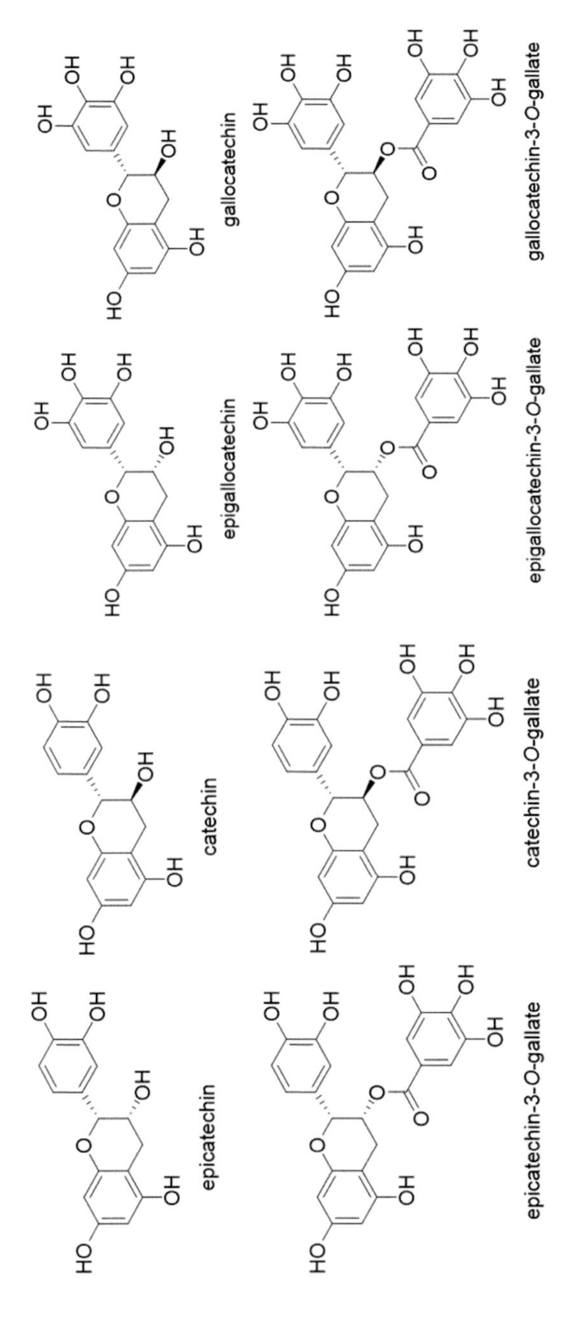

FIGURE 9.3 The most commonly existed catechins in tea.

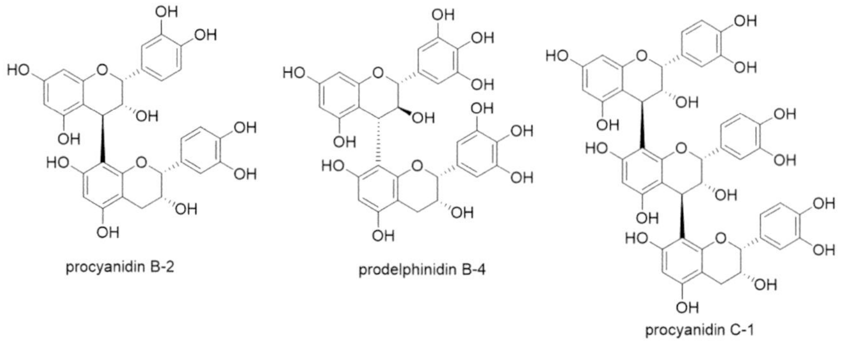

procyanidin B-2

prodelphinidin B-4

procyanidin C-1

FIGURE 9.4 Representative proanthocyanidins in Pu-er raw tea and *C. sinensis* var. *assamica*.

assamicain A

assamicain C

desgalloyl theasinensin F

theasinensin A

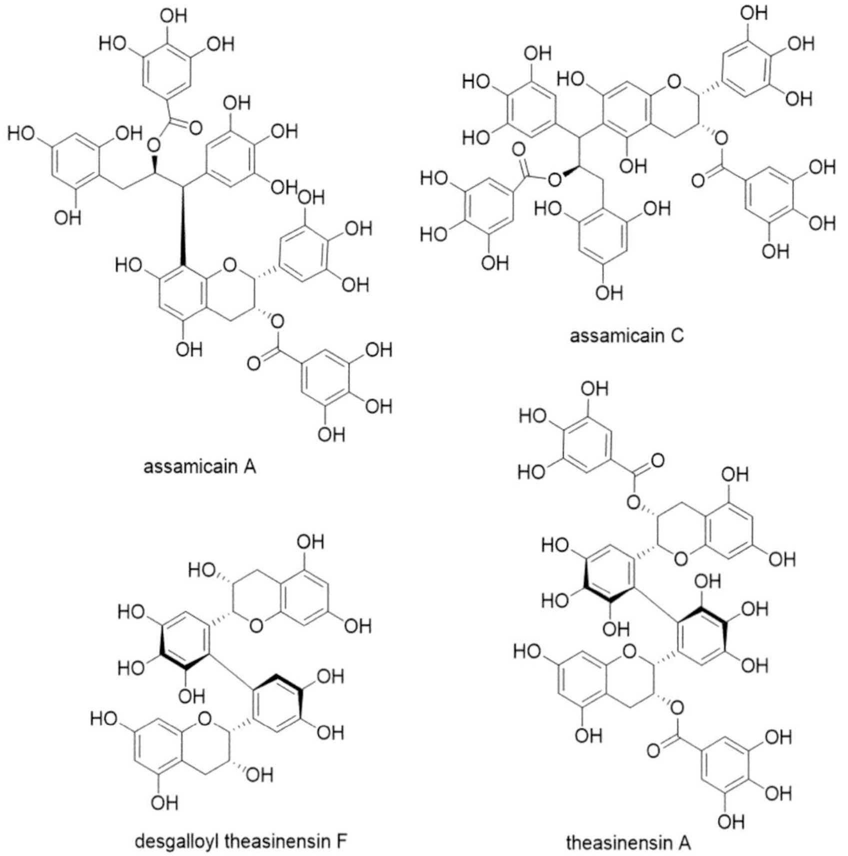

FIGURE 9.5 Representative flavan-3-ol dimers in Pu-er raw tea and *C. sinensis* var. *assamica*.

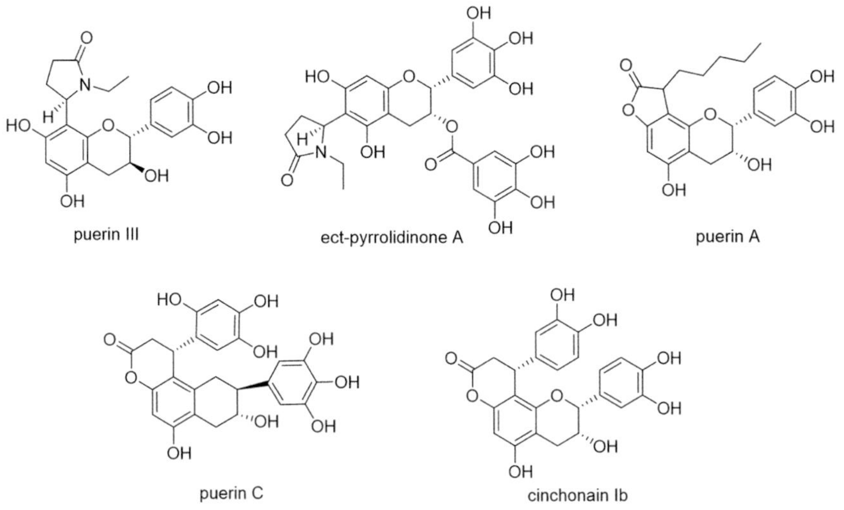

FIGURE 9.6 Representative flavan-3-ols in Pu-er ripened tea.

metabolites characteristic to Pu-er ripened tea have been reported (Figure 9.6), such as 8-*C* substituted flavan-3-ols (puerins A and B) (Zhou et al., 2005), cinchonain-type catechins (e.g. cinchonain Ib) (Zhou et al., 2005), and phenylpropanoid-substituted flavan-3-ols (puerins C-F) (Tao et al., 2014), as well as carboxymethyl- and carboxyl-catechins (Tian et al., 2014). Moreover, flavo-alkaloids (puerins I–VIII) were also isolated from Pu-er ripened tea (Wang et al., 2014). Since the galloyl groups of catechins (EGCG, ECG) tend to be removed during the post-fermented process, none of these flavo-alkaloids contains galloyl groups.

9.4.2 Hydrolysable Tannins

Hydrolysable tannins, the second major components in green tea and tea plants, are the main constituents contributing to the lasting sweet aftertaste in green tea. The study on tea tannins was started as early as 1958 (Meng et al., 2019). The content of hydrolysable tannins in Pu-er raw tea is much higher than that of the green tea made from *C. sinensis* and Pu-er ripened tea (Zhang et al., 2011b, Ku et al., 2010). Fourteen hydrolysable tannins were reported from Pu-er raw tea and its original plants. And 13 of them have been isolated from *C. taliensis*. HPLC analysis of *C. sinensis* var. *assamica* and *C. taliensis* indicated that hydrolysable tannins are characteristic constituents in the leaves of *C. taliensis* (Gao et al., 2008). Among them, 1,2-di-*O*-galloyl-4,6-(*S*)- hexahydroxydiphenoyl-*β*-D-glucopyranose was considered as a characteristic hydrolysable tannin of *C. taliensis*, of which the content in the dried leaves reached 2.44% (Gao et al., 2008) (Figure 9.7). It is noted that another hydrolysable tannin, 1-*O*-galloyl-4,6-*O*-(*S*)-hexahydroxyldiphenoyl-*β*-D-glucopyranose (strictinin), appeared to be the predominant major hydrolysable tannin (3.27%) in the leaves of *C. taliensis* collected from Yuanjiang county (Zhu et al., 2012). The hydrolysable tannins

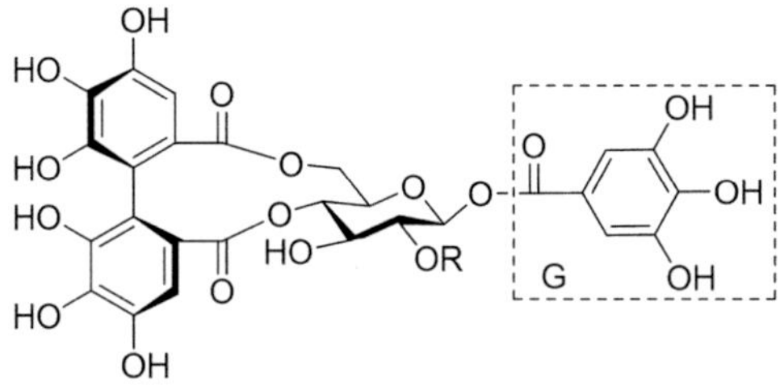

1,2-di-*O*-galloyl-4,6-(*S*)-HHDP-*β*-D-glucopyranose R = G
strictinin R = H

FIGURE 9.7 Representative hydrolysable tannins in Pu-er raw tea and its original plants.

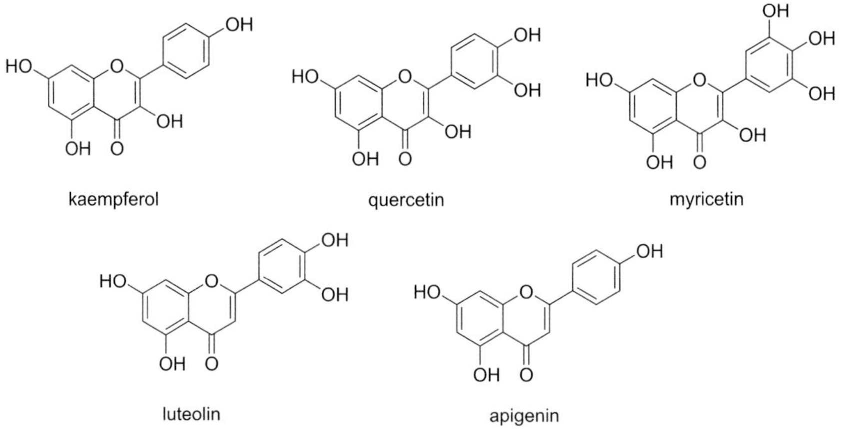

kaempferol quercetin myricetin

luteolin apigenin

FIGURE 9.8 The main types of flavonols and flavones in tea.

exhibited stronger anti-oxidant activities than the other types of compounds from *C. taliensis*, by DPPH and ABTS$^+$ assays.

9.4.3 Flavonols, Flavones, and Their Derivatives

The flavonols (kaempferol, quercetin, and myricetin), flavones (luteolin and apigenin), and their glycosides, which also affect the astringent taste and color of tea, are significant components in tea (Balentine et al., 1997, Hollman and Arts, 2000) (Figure 9.8). Flavonols and flavones are usually found in plants bound to sugars as O-glycosides: predominantly bound to C-3 positions in the parent flavonoids, although there are other positions. Flavones may also occur as C-glycosides. Different sugars, mainly glucose, rhamnose, and galactose, have been found to be bound to flavonoids in tea. Monosaccharides, disaccharides, and trisaccharides occur in these glycosides.

As a result, 27 flavonols, flavones, and their glycosides were isolated from Pu-er raw tea and ripened tea. The flavonol aglycones are not found in significant quantities in tea beverages (Balentine et al., 1997). Similar to other teas, the flavonols in Pu-er raw tea and ripened tea are also predominately present as glycosides rather

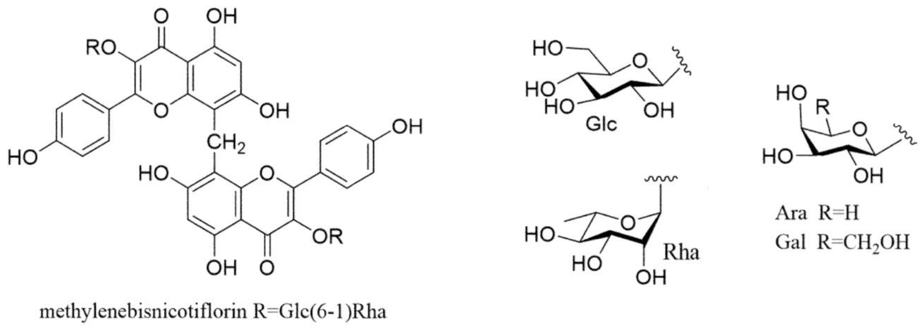

FIGURE 9.9 Representative dimeric flavonol glycoside in Pu-er ripened tea.

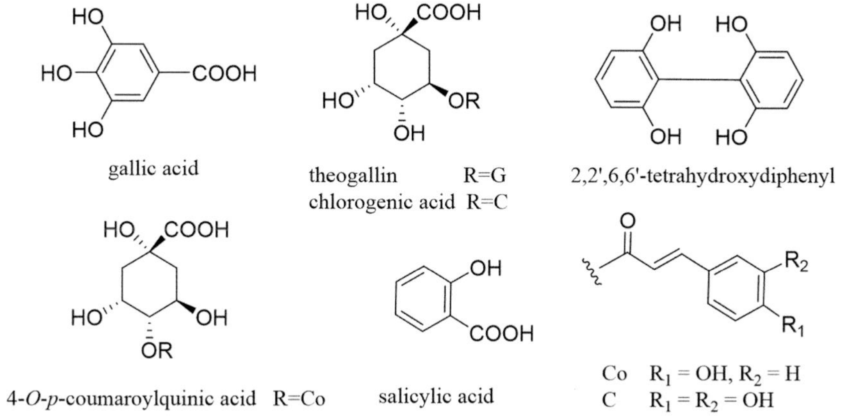

FIGURE 9.10 Representative simple phenols in Pu-er tea.

than as aglycones (Wang et al., 2000, Wu et al., 2012, Lv et al., 2013, Zhang et al., 2011a, Dong et al., 2008), such as kaempferol 3-*O*-β-D-glucoside, rutin, and kaempferol 3-*O*-rutinoside. A rare methylene-bridged dimeric flavonol glycoside, methylenebisnicotiflorin, was isolated from Pu-er ripened tea (Tao et al., 2016) (Figure 9.9). Flavone glycosides, such as isoschaftoside, luteolin, and 3′,4′,5-trihydroxy-7-methoxyflavone, have also been isolated from Pu-er ripened tea, but not from raw tea (Tian et al., 2014, Lv et al., 2010). Flavonols, flavones, and their glycosides possess various biological activities, such as antimicrobial, anti-viral, cytotoxic, anti-inflammatory, anti-oxidant, anti-hypertensive, and hypolipidemic activities (Verbeek et al., 2005, Formica and Regelson, 1995).

9.4.4 Simple Phenolic Compounds

As shown in Figure 9.10, these compounds consist of phenolic acids and their derivatives, such as gallic acid and its quinic acid ester, theogallin, coumaryl- and caffeoyl quinic acids, as well as simple phenols, e.g. pyrogallol. Besides gallic acid and theogallin, only a few kinds of caffeoyl quinic acids, including $3\alpha,5\alpha$-dihydroxycaffeoylquinic acid and chlorogenic acid, were isolated from Pu-er raw tea. Most of these phenolic compounds were identified in Pu-er ripened tea, such as salicylic acid, 4-*O-p* -coumaroylquinic acid, and 2,2',6,6'-tetrahydroxydiphenyl. Among them, gallic acid is the most important phenolic acid in both Pu-er raw and ripened tea with notable bioactivity. The amount of gallic acid increases during the post-fermentation owing to its liberation from catechin gallates and hydrolyzable tannins (Lin et al., 1998). Compared with Pu-er ripened tea, theogallin and chlorogenic acid are rich in Pu-er raw tea (Hou et al., 2009, Shao et al., 1995, Zhao et al., 2011c, Zhou et al., 2008).

9.4.5 Alkaloids and Their Derivatives

The tea alkaloids, comprising mainly of caffeine, theophylline, and theobromine, were also one kind of the important components in tea (Figure 9.11). Theobromine is described as bitter and metallic, and caffeine is bitter too (Ahmed et al., 2010). The contents of caffeine of Pu-er raw tea and ripened tea have the highest values of all of the alkaloids, followed by theobromine and theophylline (Zhao et al., 2011c, Lin et al., 1998). In addition, 8-oxocaffeine and pyrimidine alkaloids, such as deoxythymidine, thymine, and uracil, have been isolated from Pu-er ripened tea, but not from raw tea (She et al., 2007). The formation of 8-oxocaffeine likely arises from the biotransformation of caffeine during the post-fermentation process, and deoxythymidine might be derived from the combination of a pyrimidine alkaloid in the tea leaves and a microbial secondary metabolite. Thus, they were recognized as one of the characteristic components of Pu-er ripened tea.

9.4.6 Other Compounds

Some trace amounts of special components in Pu-er tea have also been investigated. Clinical trials have confirmed that statin decreases the incidence of major coronary and cerebrovascular events, and this may be due to its hypolipidemic and anti-inflammatory effects (Thavendiranathan et al., 2006, Houslay et al., 2007). Lovastatin, the first and only statin identified in Pu-er ripened tea, has cholesterol-lowering activity (Hwang et al., 2003). It was supposed to be produced by fungus during the post-fermentation process. A new amide, *N*-(3,4-dihydroxybenzoyl)-3,4-dihydroxybenzamide, was also isolated from Pu-er ripened tea. It is a very useful compound in preventing H_2O_2-induced cell death of HMEC (Zhang et al., 2011b). Theanine, also called γ-glutamylethylamide, was one of the important chemical bases for the fresh and sweet taste of tea (Zhang et al., 2007, Nakagawa, 1975) and it possesses important

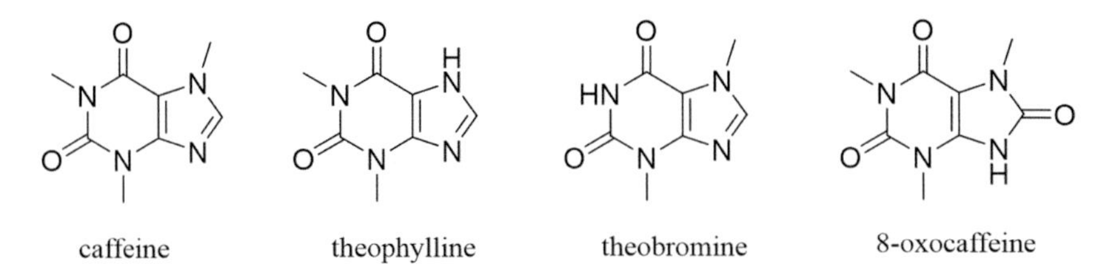

FIGURE 9.11 Representative alkaloids in Pu-er tea.

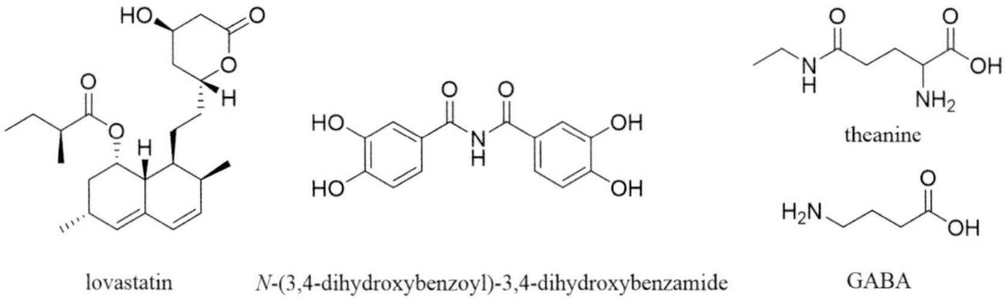

lovastatin N-(3,4-dihydroxybenzoyl)-3,4-dihydroxybenzamide theanine GABA

FIGURE 9.12 Other types of compounds in Pu-er tea.

bioactivities such as anti-hypertension, neural regulation, enhancing cognitive performance, and so on (Boros et al., 2016, Eschenauer and Sweet, 2006, Kim et al., 2012, Lardner, 2014, Saeed et al., 2017). Moreover, γ-aminobutyric acid (GABA), detected in some Pu-er ripened tea, has received more attention because it acts as a major inhibitory neurotransmitter in the central nervous system, and a relaxant, and can boost immunity in humans (Abdou et al., 2006, Zhao et al., 2011b) (Figure 9.12).

However, the majority of Pu-er ripened tea's chemical composition remains unrevealed despite its importance. These unidentified water-soluble constituents are often referred to as theabrownins, which account for a large part of the solids in Pu-er ripened tea infusions and are, therefore, much more important with respect to coloration and taste. In addition to the oxidation and polymerization products of polyphenols, Pu-er tea theabrownins consist of a highly complex mixture of amino acids, nitrogen-containing substances, sugars, gallic acid, and caffeine. Despite several attempts and efforts, detailed chemical studies of theabrownins have been limited due to the high degree of polymerization, large molecular weight, and difficulties in isolation and structural identification.

9.5 Biological Activities

Pharmacological studies have expounded the medicinal value and biological activities of the extracts and the chemical constituents of Pu-er tea primarily involved in hypolipidemic and anti-obesity, anti-diabetic, anti-oxidant, anti-tumor, anti-inflammatory, neuroprotective effects, etc. The main research on Pu-er tea are on the hypolipidemic and anti-obesity and anti-diabetic effects.

9.5.1 Hypolipidemic and Anti-Obesity

Among the health beneficial effects of Pu-er tea, the hypolipidemic and anti-obesity effects in animals and humans have become a hot topic for food research and molecular nutrition. Pu-er ripened tea can lower the levels of triglyceride (TG) and total cholesterol (TC) more significantly and efficiently than the other teas (Sano et al., 1986, Kuo et al., 2005). Moreover, it can lower the serum low-density lipoprotein-cholesterol (LDL-C) and elevate the levels of high-density lipoprotein-cholesterol (HDL-C) in rat on hyperlipidemia model by elevating superoxide dismutase (SOD) and glutathione peroxidase (GSH-Px), as well as suppressing iNOS enzyme expression (Hou et al., 2009). The mechanisms of the hypolipidemic and anti-obesity effects of tea polyphenols have been studied, and the results revealed that the tea polyphenols inhibited lipogenesis in MCF-7 breast cancer cells by down-regulation of fatty acid synthase (FAS) gene expression in the nucleus and stimulation of cell energy expenditure in the mitochondria, while the molecular mechanisms of FAS gene suppression by tea polyphenols (EGCG, theaflavins) may be the down-regulation of EGFR/PI3K/Akt/Sp-1 signal transduction pathways (Yeh et al., 2003, Lin and Lin-Shiau, 2006).

9.5.2 Anti-Diabetic

In 2013, Huang et al. studied the bioactivity of aqueous extracts of Pu-er tea and its fractions, e.g. *in vitro* anti-oxidant activity, α-glycosidase inhibitory property, and their effect on postprandial hyperglycemia in diabetic mice (Huang et al., 2013). The results showed that a 95% ethanol precipitate exhibited remarkable inhibitory effect against α-glycosidase *in vitro* and showed a significant effect on postprandial hyperglycemia in diabetic mice as compared with a model group. Additionally, the effects were not significantly ($p > 0.05$) different from the positive control acarbose at the same dosage (50 mg/kg body weight), which indicates that these fractions could be developed as potential anti-diabetic agents. Zhou et al. investigated the effect and mechanism of tea polyphenols on cardiac function in rats with diabetic cardiomyopathy and revealed that tea polyphenols can alter the levels of autophagy to improve glucose and lipid metabolism in diabetes (Zhou et al., 2018). There are also many other studies suggesting that tea polyphenols could be beneficial as additional therapy in the management of diabetic nephropathy (Yokozawa et al., 2005, Renno et al., 2008).

9.5.3 Anti-Oxidation

Many research studies have evaluated the anti-oxidative effects of tea extracts and phenolics using different models, such as DPPH and $ABTS^+$ radicals scavenging assay, tyrosinase inhibition assay, and ferric reducing/anti-oxidant power (FRAP) assay. And the ability was found to be better than vitamins C and E. Lots of animal studies offer a unique opportunity to evaluate the contribution of the anti-oxidant properties of tea polyphenols to the physiological effects of tea administration in different models of oxidative stress (Frei and Higdon, 2003). Research indicated that the anti-oxidative effects of tea polyphenols are not only owing to their ability to scavenge superoxide (Nanjo et al., 1996), but also due to improved activity of anti-oxidant enzymes, e.g. catalase, superoxide dismutase, and glutathione peroxidase (Yokozawa et al., 1999). Pu-er ripened tea was also found to possess protective effects against hydrogen peroxide-induced damage in human fibroblast HPF-1 cells (Jie et al., 2006). Thousands of articles have reported the anti-oxidative activities of tea polyphenols, but most of them focus on the mixture of polyphenol extracts or the tea infusions; further research on the activities of the monomeric compounds is needed to clarify their mechanism of action (Hara, 1993, Almajano et al., 2008).

Research on monomeric compounds has indicated that the anti-oxidative effects of hydrolysable tannins and flavan-3-ols are apparently greater than other polyphenols.

For hydrolyzable tannins, the radical scavenging activities were dependent on both the type of phenolic groups and their numbers in the molecules. The activity of the phenolic group was in the order of hexahydroxydiphenoyl (HHDP) > galloyl; the tannins with many HHDP groups in the molecule exhibited stronger activity compared with tannins with galloyl groups only (Yoshida et al., 1989). For flavan-3-ols, the presence of at least a catechol B-ring was confirmed to be essential for the radical scavenging activities, and the existence of a galloyl group could considerably strengthen the activity (Cai et al., 2006). The more galloyl and B-ring hydroxy groups attached to the molecule, the stronger the anti-oxidative activity (Villaño et al., 2007), which was in accordance with the results of the quantum chemistry calculation method (Tang et al., 2012).

9.5.4 Anti-Tumor

The relationship between tea consumption and the risk of cancer was studied in an earlier stage by contrasting tea drinkers with non-drinkers (Setiawan et al., 2001, Nakachi et al., 2000). Zhao et al. reported that water extracts of Pu-er tea can inhibit tumor cell growth by down-regulating mutant p53 and inhibit the expression of HSP70 and HSP90 in tumor cells (Zhao et al., 2011a). In recent years, clarifying the mechanisms of anti-tumor activities of tea polyphenols gradually become a major focus for researchers. In 2010, Cheng et al. studied the beneficial role of EGCG in the molecular mechanism of human oral cancer, and revealed that the immunomodulatory protein, indoleamine 2,3-dioxygenase, is suppressed by (–)-EGCG via blocking of γ-interferon-induced JAK-PKC-delta-STAT1 signaling in oral cancer cells (Cheng et al., 2010). Lots of other mechanisms of anti-tumor activity of tea polyphenols, especially EGCG, have been illustrated, e.g. inhibition of NF-κB and HIF-1α to impair angiogenesis via VEGF gene down-regulation (Li et al., 2013), activation of p53, p21, and p27 to inhibit the activity of cyclin-dependent kinase (Liu et al., 2013), and inhibition of DNA methyltransferase (Henning et al., 2013). However, illustrating various other anti-tumor mechanisms of monomeric compounds in Pu-er tea is still a key task and a monumental challenge to be addressed.

9.5.5 Anti-Inflammatory

In 2011, Cavet et al. revealed the anti-inflammatory effect of EGCG in human corneal epithelial cells (Cavet et al., 2011). In 2014, Leong et al. studied the anti-inflammatory effect of tea polyphenol using a mouse post-traumatic osteoarthritis (OA) model, and revealed that EGCG could slow the progression of OA and relieve OA-associated pain (Leong et al., 2014). Hisanaga et al. investigated the anti-inflammatory activity and molecular mechanisms of theasinensin A in both cell and animal models, and concluded that in the cell model, it reduced the levels of pro-inflammatory mediators including inducible nitric oxide synthase protein (iNOS), TNF-α, IL-12, NO, and monocyte chemotactic protein-1 (MCP-1) induced by LPS; while in the animal model, it suppressed the production of IL-12, TNF-α, and MCP-1 and attenuated mouse paw edema induced by LPS. Pu-er ripened tea extracts also showed strong NO suppressing effect (Lin et al., 2003). The anti-inflammatory activities of rutin and quercetin were investigated in rats using the Mizushima model of acute and chronic inflammation (Hisanaga et al., 2014). Results showed that rutin was extremely effective in reducing edema, nodules, and ankylosis, and quercetin was less effective than rutin (Guardia et al., 2001).

9.5.6 Neuroprotective Effects

Many investigations have demonstrated that drinking tea can decrease the risk of neurodegenerative diseases such as Parkinson's disease and Alzheimer's disease. Recently, many other neuroprotective mechanisms of different components of tea were also discussed. Previous studies revealed that tea polyphenols can lower the morbidity of these diseases by reducing oxidative stress and regulating signaling pathways (Cong et al., 2016, Qi et al., 2017, Mandel et al., 2004). Theanine can inhibit the glutamate receptors and regulate the extracellular concentration of glutamine, exhibiting neuroprotective effects (Lardner, 2014), while the neuroprotective mechanisms of caffeine may contribute to the ability to antagonize the adenosine receptor AR (El-Missiry et al., 2018). Even more, the acetylcholinesterase inhibitory activity of green and white tea (Okello et al., 2012), as well as some monomeric catechins were also evaluated (Meng et al., 2018). All these studies suggested that some tea constituents could be clinically useful for treating neurodegenerative disorders.

9.5.7 Other Bioactivities

Catechins, such as EGCG and its derivatives, are inhibitors of HCV entry (Calland et al., 2012, Ciesek et al., 2011), and exerted anti-HBV effects in a stable HBV-transfected cell line (Ye et al., 2009). Pu-er tea extracts were also demonstrated to have anti-HBV ability and could be used as a potential treatment against HBV infection with an additional merit of low cytotoxicity (Pei et al., 2011). The water extracts of Pu-er tea have a potential inhibitory effect on direct-acting mutagen, 4-nitroquinoline-*N*-oxide, and on indirect-acting mutagen, aflatoxin B1. In addition, it showed an inhibitory effect on the growth of *Staphylococcus aureus* and *Bacillus subtilis* (Wu et al., 2007). The antimicrobial activities of 7 catechins, as well as the aqueous extracts of 36 commercial black, green, oolong, and white teas against *Bacillus cereus* were evaluated by Friedman et al., and the results demonstrated that ECG, CG, GCG, and EGCG showed good antimicrobial activities at nanomolar levels (Friedman et al., 2006). Many other effects of tea and tea polyphenols, e.g. hepatoprotective (Zhang et al., 2014), hypotensive (Imura, 1985), and gastrointestinal digestion (Tenore et al., 2015), were also investigated.

9.6 Conclusion

Pu-er tea has shown a broad spectrum of pharmacological effects, such as anti-oxidation, hypolipidemic, anti-obesity, anti-diabetic, anti-tumor, antimicrobial, and anti-inflammatory activities. Among these, anti-oxidation, anti-tumor, and antimicrobial effects are characteristic of Pu-er raw tea (or green tea). It is well-known that tea polyphenols are the major pharmacologically active ingredients of tea. However, because of the microbial participation during the post-fermentation process, the main chemical components in Pu-er raw tea, especially catechins, are reduced significantly in the final fermented product, Pu-er ripened tea, whereas some characteristic components such as theabrownins and other less abundant constituents are being formed. This leads the main bioactivity of Pu-er ripened tea to be hypolipidemic, anti-obesity, and anti-diabetic, which is different from Pu-er raw tea. Since most of the previous pharmacological studies on Pu-er ripened tea were focused on the crude extracts, studies on the pure chemical components are urgently needed to clarify the active mechanism of Pu-er ripened tea, and provide scientifically the theoretical basis and guidance for its market development and utilization.

REFERENCES

Abdou AM, Higashiguchi S, Horie K, et al., 2006. Relaxation and immunity enhancement effects of γ-aminobutyric acid (GABA) administration in humans. *Biofactors*, 26: 201–208.

Abe M, Takaoka N, Idemoto Y, et al., 2008. Characteristic fungi observed in the fermentation process for Puer tea. *Int J Food Microbiol*, 124: 199–203.

Ahmed S, Unachukwu U, Stepp JR, et al., 2010. Pu-erh tea tasting in Yunnan, China: correlation of drinkers' perceptions to phytochemistry. *J Ethnopharmacol*, 132: 176–185.

Almajano MP, Carbo R, Jimenez JAL, et al., 2008. Antioxidant and antimicrobial activities of tea infusions. *Food Chem*, 108: 55–63.

Balentine DA, Wiseman SA, Bouwens LCM, 1997. The chemistry of tea flavonoids. *Crit Rev Food Sci Nutr*, 37: 693–704.

Blair LM, Calvert MB, Sperry J, 2017. Flavoalkaloids—isolation, biological activity, and total synthesis. *Alkaloids Chem Biol*, 77: 85–115.

Boros K, Jedlinszki N, Csupor D, 2016. Theanine and caffeine content of infusions prepared from commercial tea samples. *Pharmacogn Mag*, 12: 75–79.

Brusselmans K, De Schrijver E, Heyns W, et al., 2003. Epigallocatechin-3-gallate is a potent natural inhibitor of fatty acid synthase in intact cells and selectively induces apoptosis in prostate cancer cells. *Int J Cancer*, 106: 856–862.

Cai YZ, Sun M, Xing J, et al., 2006. Structure–radical scavenging activity relationships of phenolic compounds from traditional Chinese medicinal plants. *Life Sci*, 78: 2872–2888.

Calland N, Albecka A, Belouzard S, et al., 2012. (-)-Epigallocatechin-3-gallate is a new inhibitor of hepatitis C virus entry. *Hepatology*, 55: 720–729.

Cao Y, Cao R, 1999. Angiogenesis inhibited by drinking tea. *Nature*, 398: 381.

Cao ZH, Gu DH, Lin QY, et al., 2011. Effect of pu-erh tea on body fat and lipid profiles in rats with diet-induced obesity. *Phytother Res*, 25: 234–238.

Cavet ME, Harrington KL, Vollmer TR, et al., 2011. Anti-inflammatory and anti-oxidative effects of the green tea polyphenol epigallocatechin gallate in human corneal epithelial cells. *Mol Vis*, 17: 533–542.

Chen CS, Chan HC, Chang YN, et al., 2009a. Effects of bacterial strains on sensory quality of Pu-erh tea in an improved pile-fermentation process. *J Sens Stud*, 24: 534–553.

Chen KK, Zhang XL, Zhu HT, et al., 2008. Effects of *Aspergillus* on post-fermentative process of Pu-er tea. *Acta Bot Yunnanica*, 30: 624–628.

Chen KK, Zhu HT, Wang D, et al., 2006. Isolation and identification of *Aspergillus Species* from the post fermentative process of Pu-er ripe tea. *Acta Bot Yunnanica*, 28: 123–126.

Chen XY, Chen S, Zheng H, et al., 2009b. Purification of predominant microbes from Pu'er tea and their effects on the quality of Pu'er tea. *Mod Food Sci Technol*, 25: 604–607.

Cheng CW, Shieh PC, Lin YC, et al., 2010. Indoleamine 2,3-dioxygenase, an immunomodulatory protein, is suppressed by (-)-epigallocatechin-3-gallate via blocking of gamma-interferon-induced JAK-PKC-delta-STAT1 signaling in human oral cancer cells. *J Agric Food Chem*, 58: 887–894.

Cheng J, Wu FH, Wang P, et al., 2018. Flavoalkaloids with a pyrrolidinone ring from Chinese ancient cultivated tea Xi-Gui. *J Agric Food Chem*, 66: 7948–7957.

Ciesek S, von Hahn T, Colpitts CC, et al., 2011. The green tea polyphenol, epigallocatechin-3-gallate, inhibits hepatitis C virus entry. *Hepatology*, 54: 1947–1955.

Cong L, Cao C, Cheng Y, et al., 2016. Green tea polyphenols attenuated glutamate excitotoxicity via antioxidative and antiapoptotic pathway in the primary cultured cortical neurons. *Oxid Med Cell Longev*, 2016**:** 2050435.

Dong F, Yang ZY, He PM, et al., 2008. Liquid chromatographic-mass spectrometric analysis of antioxidant compounds from Pu-erh tea. *J Chin Inst Food Sci Technol*, 8: 133–141.

El-Missiry MA, Othman AI, El-Sawy MR, et al., 2018. Neuroprotective effect of epigallocatechin-3-gallate (EGCG) on radiation-induced damage and apoptosis in the rat hippocampus. *Int J Radiat Biol*, 94: 1–11.

Eschenauer G, Sweet BV, 2006. Pharmacology and therapeutic uses of theanine. *Am J Health-Syst Pharm*, 63: 26–30.

Fang X, Chen D, Li JJ, et al., 2008. Identification of microbial species in Pu-erh tea with different storage time. *Mod Food Sci Technol*, 24: 105–108.

Formica JV, Regelson W, 1995. Review of the biology of quercetin and related bioflavonoids. *Food Chem Toxicol*, 33: 1061–1080.

Frei B, Higdon JV, 2003. Antioxidant activity of tea polyphenols *in vivo*: evidence from animal studies. *J Nutr*, 133: 3275S–3284S.

Friedman M, Henika PR, Levin CE, et al., 2006. Antimicrobial activities of tea catechins and theaflavins and tea extracts against *Bacillus cereus*. *J Food Prot*, 69: 354–361.

Gao DF, Zhang YJ, Yang CR, et al., 2008. Phenolic antioxidants from green tea produced from *Camellia taliensis*. *J Agric Food Chem*, 56: 7517–7521.

Gong Z, Watanabe N, Yagi A, et al., 1993. Compositional change of Pu-erh tea during processing. *Biosci, Biotechnol Biochem*, 57: 1745–1746.

Guardia T, Rotelli AE, Juarez AO, et al., 2001. Anti-inflammatory properties of plant flavonoids. Effects of rutin, quercetin and hesperidin on adjuvant arthritis in rat. *Il Farmaco*, 56: 683–687.

Hara Y, 1993. Antioxidative activities of tea polyphenols. *Kassei Sanso, Furi Rajikaru*, 4: 307–14.

Hashimoto F, Nonaka GI, Nishioka I, 1989. Tannins and related compounds. LXXVII: novel chalcan-flavan dimers, assamicains A, B and C, and a new flavan-3-ol and proanthocyanidins from the fresh leaves of *Camellia sinensis* L. var. *assamica* Kitamura. *Chem Pharm Bull*, 37: 77–85.

Henning SM, Wang P, Carpenter CL, et al., 2013. Epigenetic effects of green tea polyphenols in cancer. *Epigenomics*, 5: 729–741.

Hisanaga A, Ishida H, Sakao K, et al., 2014. Anti-inflammatory activity and molecular mechanism of Oolong tea theasinensin. *Food Funct*, 5: 1891–1897.
Hollman PCH, Arts ICW, 2000. Flavonols, flavones and flavanols – nature, occurrence and dietary burden. *J Sci Food Agric*, 80: 1081–1093.
Hou Y, Shao W, Xiao R, et al., 2009. Pu-erh tea aqueous extracts lower atherosclerotic risk factors in a rat hyperlipidemia model. *Exp Gerontol*, 44: 434–439.
Houslay ES, Sarma J, Uren NG, 2007. The effect of intensive lipid lowering on coronary atheroma and clinical outcome. *Heart*, 93: 149–151.
Huang Q, Chen S, Chen H, et al., 2013. Studies on the bioactivity of aqueous extract of pu-erh tea and its fractions: *in vitro* antioxidant activity and α-glycosidase inhibitory property, and their effect on postprandial hyperglycemia in diabetic mice. *Food Chem Toxicol*, 53: 75–83.
Hwang LS, Lin LC, Chen NT, et al., 2003. Hypolipidemic effect and antiatherogenic potential of Pu-erh tea. *Acs Symp Ser*, 859: 87–103.
Imura K, 1985. Hypotensive components in green tea. *J Urban Living Health Assoc*, 29: 163–172.
Jankun J, Selman SH, Swiercz R, et al., 1997. Why drinking green tea could prevent cancer. *Nature*, 387: 561.
Jeng KC, Chen CS, Fang YP, et al., 2007. Effect of microbial fermentation on content of statin, GABA, and polyphenols in Pu-erh tea. *J Agric Food Chem*, 55: 8787–8792.
Jiang HB, Wang YG, Tang YC, et al., 2009. Investigation of wild tea plant (*Camellia taliensis*) germ plasm resource from Yunnan, China. *Southwest China J Agric Sci*, 22: 1153–1157.
Jie G, Lin Z, Zhang L, et al., 2006. Free radical scavenging effect of Pu-erh tea extracts and their protective effect on oxidative damage in human fibroblast cells *J Agric Food Chem*, 54: 8058–8064.
Jie GL, He PM, Ding R, 2005. Abecedarian study on the antioxidatnt activity of Pu-er tea. *Tea*, 31: 162–165.
Kang YS, Huang W, Yuan W, et al., 2015. Effects of different exogenous enzymes and yeast on components during fermentation of Pu-erh tea. *J Yunnan Agric Univ (Nat Sci Educ)*, 30: 784–789.
Kim NH, Jeong HJ, Kim HM, 2012. Theanine is a candidate amino acid for pharmacological stabilization of mast cells. *Amino Acids*, 42: 1609–1618.
Ku KM, Kim J, Park HJ, et al., 2010. Application of metabolomics in the analysis of manufacturing type of Pu-erh tea and composition changes with different postfermentation year. *J Agric Food Chem*, 58: 345–352.
Kuo KL, Weng MS, Chiang CT, et al., 2005. Comparative studies on the hypolipidemic and growth suppressive effects of oolong, black, pu-erh, and green tea leaves in rats. *J Agric Food Chem*, 53: 480–489.
Lardner AL, 2014. Neurobiological effects of the green tea constituent theanine and its potential role in the treatment of psychiatric and neurodegenerative disorders. *Nutr Neurosci*, 17: 145–155.
Leong DJ, Choudhury M, Hanstein R, et al., 2014. Green tea polyphenol treatment is chondroprotective, anti-inflammatory and palliative in a mouse post-traumatic osteoarthritis model. *Arthritis Res Ther*, 16: 508–518.
Li CC, Lu J, Yang RJ, et al., 2012. Isolation and identification of thermophilic bacteria during pile-fermentation of Pu'er tea. *J Beijing Univ Chem Technol Nat Sci Ed*, 39: 74–78.

Li N, Zhang YJ, Zhu HT, et al., 2019. Chemical and dominant microbial studies on Pu-er tea and its original plants. *Modernization Tradit Chin Med Materia Medica--World Sci Technol*, 21: 1173–1188.

Li X, Feng Y, Liu J, et al., 2013. Epigallocatechin-3-gallate inhibits IGF-I-stimulated lung cancer angiogenesis through downregulation of HIF-1α and VEGF expression. *J Nutrigenet Nutrigenomics*, 6: 169–178.

Lin JK, Lin CL, Liang YC, et al., 1998. Survey of catechins, gallic acid, and methylxanthines in green, oolong, pu-erh, and black teas. *J Agric Food Chem*, 46: 3635–3642.

Lin JK, Lin-Shiau SY, 2006. Mechanisms of hypolipidemic and anti-obesity effects of tea and tea polyphenols. *Mol Nutr Food Res*, 50: 211–217.

Lin YS, Tsai YJ, Tsay JS, et al., 2003. Factors affecting the levels of tea polyphenols and caffeine in tea leaves. *J Agric Food Chem*, 51: 1864–1873.

Liu Q, Zhang YJ, Yang CR, et al., 2009. Phenolic antioxidants from green tea produced from *Camellia crassicolumna* var. *multiplex*. *J Agric Food Chem*, 57: 586–590.

Liu S, Wang XJ, Liu Y, et al., 2013. PI3K/AKT/mTOR signaling is involved in (-)-epigallocatechin-3-gallate-induced apoptosis of human pancreatic carcinoma cells. *Am J Chin Med*, 41: 629–642.

Lv HP, Lin Z, Zhong QS, et al., 2010. Study on the chemical component of E8 fraction from Pu-erh tea. *J Tea Sci*, 30: 423–428.

Lv HP, Zhang YJ, Lin Z, et al., 2013. Processing and chemical constituents of Pu-erh tea: a review. *Food Res Int*, 53: 608–618.

Mandel S, Weinreb O, Amit T, et al., 2004. Cell signaling pathways in the neuroprotective actions of the green tea polyphenol (-)-epigallocatechin-3-gallate: implications for neurodegenerative diseases. *J Neurochem*, 88: 1555–1569.

Meng XH, Li N, Zhu HT, et al., 2019. Plant resources, chemical constituents, and bioactivities of tea plants from the genus *Camellia* section *Thea*. *J Agric Food Chem*, 67: 5318–5349.

Meng XH, Zhu HT, Yan H, et al., 2018. C-8 *N*-Ethyl-2-pyrrolidinone-substituted flavan-3-ols from the leaves of *Camellia sinensis* var. *pubilimba*. *J Agric Food Chem*, 66: 7150–7155.

Min TL, 1992. A revision of *Camellia* sect. *Tea*. *Acta Bot Yunnanica*, 14: 115–132.

Min TL, 2000. *Monograph of the Genus Camellia*. Kunming: Yunnan Science and Technology Press.

Mo HZ, Xu XQ, Yan MC, et al., 2005. Microbiological analysis and antibacterial effects of the indigenous fermented Puer tea. *Agro Food Industry Hi-Tech*, 16: 16–18.

Nakachi K, Matsuyama S, Miyake S, et al., 2000. Preventive effects of drinking green tea on cancer and cardiovascular disease: epidemiological evidence for multiple targeting prevention. *Biofactors*, 13: 49–54.

Nakagawa M, 1975. Contribution of green tea constituents to the intensity of taste element of brew. *J Jpn Soc Food Sci Technol*, 22: 59–64.

Nanjo F, Goto K, Seto R, et al., 1996. Scavenging effects of tea catechins and their derivatives on 1,1-diphenyl-2-picrylhydrazyl radical. *Free Radic Biol Med*, 21: 895–902.

Oi Y, Hou IC, Fujita H, et al., 2012. Antiobesity effects of Chinese black tea (Pu-erh tea) extract and gallic acid. *Phytother Res*, 26: 475–481.

Okello EJ, Leylabi R, McDougall GJ, 2012. Inhibition of acetylcholinesterase by green and white tea and their simulated intestinal metabolites. *Food Funct*, 3: 651–661.

Pei SB, Zhang Y, Xu H, et al., 2011. Inhibition of the replication of hepatitis B virus *in vitro* by Pu-erh tea extracts. *J Agric Food Chem*, 59: 9927–9934.

Qi G, Mi Y, Wang Y, et al., 2017. Neuroprotective action of tea polyphenols on oxidative stress-induced apoptosis through the activation of the TrkB/CREB/BDNF pathway and Keapl/Nrf2 signaling pathway in SH-SY5Y cells and mice brain. *Food Funct*, 8: 4421–4432.

Renno WM, Abdeen S, Alkhalaf M, et al., 2008. Effect of green tea on kidney tubules of diabetic rats. *Br J Nutr*, 100: 652–659.

Saeed M, Naveed M, Arif M, et al., 2017. Green tea (*Camellia sinensis*) and L-theanine: medicinal values and beneficial applications in humans - A comprehensive review. *Biomed Pharmacother*, 95: 1260–1275.

Sano M, Takenaka Y, Kojima R, et al., 1986. Effects of pu-erh tea on lipid metabolism in rats. *Chem Pharm Bull*, 34: 221–228.

Setiawan VW, Zhang ZF, Yu GP, et al., 2001. Protective effect of green tea on the risks of chronic gastritis and stomach cancer. *Int J Cancer*, 92: 600–604.

Shao W, Powell C, Clifford MN, 1995. The analysis by HPLC of green, black and Pu'er teas produced in Yunnan. *J Sci Food Agric*, 69: 535–540.

She GM, Chen KK, Zhang YJ, et al., 2007. The occurrence of 8-oxocaffeine and pyrimidine alkaloids in Pu-er ripe tea. *Acta Bot Yunnanica*, 29: 713–716.

Tang H, Cai L, Li H, et al., 2012. The quantum chemistry calculation on the antioxidative activity of polyphenols in tea. *South China Agric*, 6: 7–13.

Tao MK, Xu M, Zhang H, et al., 2016. Methylenebisnicotiflorin: a rare methylene-bridged bisflavonoid glycoside from ripe Pu-er tea. *Nat Prod Res*, 30: 776–782.

Tao MK, Xu M, Zhu HT, et al., 2014. New phenylpropanoid-substituted flavan-3-ols from Pu-er ripe tea. *Nat Prod Commun*, 9: 1167–1170.

Tenore GC, Campiglia P, Giannetti D, et al., 2015. Simulated gastrointestinal digestion, intestinal permeation and plasma protein interaction of white, green, and black tea polyphenols. *Food Chem*, 169: 320–326.

Thavendiranathan P, Bagai A, Brookhart MA, et al., 2006. Primary prevention of cardiovascular diseases with statin therapy: a meta-analysis of randomized controlled trials *Arch Intern Med*, 166: 2307–2313.

Tian LW, Tao MK, Xu M, et al., 2014. Carboxymethyl- and carboxyl-catechins from ripe Pu-er tea. *J Agric Food Chem*, 62: 12229–12234.

Verbeek R, Van Tol EAF, Van Noort JM, 2005. Oral flavonoids delay recovery from experimental autoimmune encephalomyelitis in SJL mice. *Biochem Pharmacol*, 70: 220–228.

Villaño D, Fernández-Pachón MS, Moyá ML, et al., 2007. Radical scavenging ability of polyphenolic compounds towards DPPH free radical. *Talanta*, 71: 230–235.

Wang H, Provan GJ, Helliwell K, 2000. Tea flavonoids: their functions, utilisation and analysis. *Trends Food Sci Technol*, 11: 152–160.

Wang W, Zhang L, Wang S, et al., 2014. 8-C *N*-ethyl-2-pyrrolidinone substituted flavan-3-ols as the marker compounds of Chinese dark teas formed in the post-fermentation process provide significant antioxidative activity. *Food Chem*, 152: 539–545.

Wu C, Xu H, Héritier J, et al., 2012. Determination of catechins and flavonol glycosides in Chinese tea varieties. *Food Chem*, 132: 144–149.

Wu SC, Yen GC, Wang BS, et al., 2007. Antimutagenic and antimicrobial activities of pu-erh tea. *LWT -- Food Sci Technol*, 40: 506–512.

Xu M, Tao MK, Dong HZ, et al., 2013. HPLC-DAD-MS analysis of chemical compositions of Pu-er ripe tea. *Nat Prod Res Dev*, 25: 1212–1217.

Xu P, Chen H, Wang Y, et al., 2012. Oral administration of puerh tea polysaccharides lowers blood glucose levels and enhances antioxidant status in alloxan-induced diabetic mice. *J Food Sci*, 77: H246–H252.

Xu X, Yan M, Zhu Y, 2005. Influence of fungal fermentation on the development of volatile compounds in the Puer tea manufacturing process. *Eng Life Sci*, 5: 382–386.

Yang CR, Chen KK, Zhang YJ, 2006. Categories of tea and the definition of Pu-erh tea. *Tea Sci Technol*, 2: 37–38.

Yang CS, Maliakal P, Meng X, 2002. Inhibition of carcinogenesis by tea. *Annu Rev Pharmacol Toxicol*, 42: 25–54.

Yang RJ, Lv J, Yan L, et al., 2011. Isolation and identification of thermophilic fungi during the fermentation of Puer tea. *J Tea Sci*, 31: 371–378.

Ye P, Zhang S, Zhao L, et al., 2009. Tea polyphenols exerts anti-hepatitis B virus effects in a stably HBV-transfected cell line. *J Huazhong Univ Sci Technol Med Sci*, 29: 169–172.

Yeh CW, Chen WJ, Chiang CT, et al., 2003. Suppression of fatty acid synthase in MCF-7 breast cancer cells by tea and tea polyphenols: a possible mechanism for their hypolipidemic effects. *Pharmacogenomics J*, 3: 267–276.

Yokozawa T, Nakagawa T, Lee KI, et al., 1999. Effects of green tea tannin on cisplatin-induced nephropathy in LLC-PK1 cells and rats. *J Pharm Pharmacol*, 51: 1325–1331.

Yokozawa T, Nakagawa T, Oya T, et al., 2005. Green tea polyphenols and dietary fibre protect against kidney damage in rats with diabetic nephropathy. *J Pharm Pharmacol*, 57: 773–780.

Yoshida T, Mori K, Hatano T, et al., 1989. Studies on inhibition mechanism of autoxidation by tannins and flavonoids: V: radical-scavenging effects of tannins and related polyphenols on l,l-diphenyl-2-picrylhydrazyl radical. *Chem Pharm Bull*, 37: 1919–1921.

Zhang L, Li N, Ma ZZ, et al., 2011a. Comparison of the chemical constituents of aged pu-erh tea, ripened pu-erh tea, and other teas using HPLC-DAD–ESI–MSn. *J Agric Food Chem*, 59: 8754–8760.

Zhang L, Ma ZZ, Che YY, et al., 2011b. Protective effect of a new amide compound from Pu-erh tea on human micro-vascular endothelial cell against cytotoxicity induced by hydrogen peroxide. *Fitoterapia*, 82: 267–271.

Zhang W, Yang R, Fang W, et al., 2016. Characterization of thermophilic fungal community associated with pile fermentation of Pu-erh tea. *Int J Food Microbiol*, 227: 29–33.

Zhang X, Wu Z, Weng P, 2014. Antioxidant and hepatoprotective effect of (-)-epigallocatechin 3-*O*-(3-*O*-methyl) gallate (EGCG3″Me) from Chinese oolong tea. *J Agric Food Chem*, 62: 10046–10054.

Zhang Y, Du X, Wang XS, 2007. Research progress of theanine in tea (*Camellia Sinensis*). *Food Res Dev*, 28: 170–174.

Zhang YJ, Chen KK, Yang CR, 2011c. Talk about Pu-er tea. *China Nature*, 6: 56–60.

Zhao L, Jia S, Tang W, et al., 2011a. Pu-erh tea inhibits tumor cell growth by down-regulating mutant p53. *Int J Mol Sci*, 12: 7581.

Zhao LF, Zhou HJ, 2005. Study on the main microbes of Yunnan puer tea during pile-fermentation process. *J Shangqiu Teach Coll*, 21: 129–133.

Zhao M, Ma Y, Wei ZZ, et al., 2011b. Determination and comparison of γ-aminobutyric acid (GABA) content in Pu-erh and other types of Chinese tea. *J Agric Food Chem*, 59: 3641–3648.

Zhao Y, Chen P, Lin LC, et al., 2011c. Tentative identification, quantitation, and principal component analysis of greenpu-erh, green, and white teas using UPLC/DAD/MS. *Food Chem*, 126: 1269–1277.

Zhao ZJ, Tong HR, Zhou L, et al., 2010. Fungal colonization of Pu-erh tea in Yunnan. *J Food Saf*, 30: 769–784.

Zhou CB, Chen WP, Wu ZL, et al., 2015. Research on function and safety of *Penicillium oxalicum*, a preponderant fungus during the fermentative process of Pu'er tea. *J Food Sat Qual*, 40: 63–66.

Zhou H, Chen Y, Huang SW, et al., 2018. Regulation of autophagy by tea polyphenols in diabetic cardiomyopathy. *J Zhejiang Univ Sci B*, 19: 333–341.

Zhou HJ, Li JH, Zhao LF, et al., 2004. Study on main microbes on quality formation of Yunnan Puer tea during pile-fermentation process. *J Tea Sci*, 24: 212–218.

Zhou S, Wu XD, Liu J, et al., 2008. Comparison of chlorogenic acid content in different kinds of tea. *West China J Pharm Sci*, 2: 190–192.

Zhou ZH, Zhang YJ, Xu M, et al., 2005. Puerins A and B, two new 8-C substituted flavan-3-ols from Pu-er tea. *J Agric Food Chem*, 53: 8614–8617.

Zhu LF, Xu M, Zhu HT, et al., 2012. New flavan-3-ol dimer from green tea produced from *Camellia taliensis* in the Ai-Lao mountains of Southwest China. *J Agric Food Chem*, 60: 12170–12176.

10

Yunnan *Coffea arabica* L.

Ming-Hua Qiu,* Gui-Lin Hu, Xiao-Yuan Wang, and Xia Wang
State Key Laboratory of Phytochemistry and Plant Resources in West China, Kunming Institute of Botany, Chinese Academy of Sciences, Kunming, Yunnan, People's Republic of China

CONTENTS

* Corresponding author.

10.1 Introduction

Coffee belongs to the botanical genus *Coffea* in the family Rubiaceae. There are nearly one hundred coffee species, of which two species are major in the coffee trade, *Coffea arabica* and *Coffea robusta*. Due to sensory superiority, the *Coffea arabica* species has the greatest commercial value and represents more than 65% of global coffee production (Mussatto et al., 2011).

It is believed that the first coffee plant introduced to Yunnan was brought by a French Catholic missionary in 1902, and planted in Zhukula, Binchuan, for his own drinking. Another recorded time of introduction to Yunnan is in 1908, when the Jingpo border people from Myanmar to Nong Xian village in Ruili city used as a garden ornamental planting. Then, coffee gradually spread in Yunnan Province.

Due to the natural conditions of high altitude and large temperature differences between day and night in Yunnan, these factors are especially suitable for the growth of arabica coffee; thus Yunnan Province is the largest coffee production and export base in China. The coffee beans grown in Yunnan have the characteristics of small granules, symmetrical face, full-bodied, rich taste and are famous for "strong but not bitter, fragrant but not strong, slightly fruity taste," which can be compared to one of the world's best coffees—Blue Mountain coffee. YAC has become one of the necessary raw materials for Nestlé and other international famous enterprises. Moreover, it was once rated as a category No.1 product by the International Coffee Organization (ICO) (Zhou, 2010).

Some studies have shown that the raw beans and roasted beans of YAC have rich substances. In recent years, a series of bioactive components, such as new coffee diterpenoid glycosides, a new type of coffee diterpenoid lactones, a series of coffee diterpene lactamides, and damarane triterpenes were found in YAC. Therefore, the unique sensory qualities and the active substances of YAC are explored.

10.2 Yunnan *Coffea arabica* (Arabica Coffee, Catimor)

10.2.1 Characteristics of YAC

The *Coffea* plants grown in Yunnan Province are mainly Catimor varieties (YAC). The Catimor (YAC) was obtained as rust-resistant and high-yield varieties from one of the hybrids of the Bourbon and Timor varieties, while Timor coffee varieties were obtained from a hybrid between *Coffea canephora* (syn. *Coffea robusta*) and *Coffea arabica*, so 75% of the genetic material of Catimor grown in Yunnan (YAC) came from arabica and 25% from robusta. YAC grows as perennial small trees or shrubs, with an extensive root system concentrated in the 0–60 cm soil zone; it usually has a thick, short taproot and many developed lateral roots. The stem of coffee is also known as the trunk; the stem is straight, and the tender stem is slightly square and green, while after the suberification it becomes round and brown. Coffee leaves are thin and leathery, ovate-lanceolate or lanceolate, glabrous on both sides, the tip of the leaf grows acuminate and the entire edge is in a shallow slope. There are midrib bulges on both sides of the leaf blade. The stipules are broadly triangular. Hermaphrodite and aromatic flowers are emitted in inflorescences on the axiles of plagiotropic branches. The corolla is white; the top is generally five-lobed, rarely four- or six-lobed. *Coffea arabica* generally blooms in March and April, with the characteristic of repeatedly flowering and a concentrated flowering period. The anthers protrude from the crown tube, the stigma is two-lobed, and the flowering of *Coffea arabica* is strongly influenced by the region's latitude and climate, especially rainfall and temperature. High temperatures and drought cause abnormal budding: the bud does not open when the temperature is below 10°C, while temperatures above 13°C are good for blooming. For *Coffea arabica*, the fruits are oval and mature in 7–9 months; ripe fruits ("berries") are red or yellow, with orange ones indicating cross-pollination. Berries are broadly oval when ripe; each fruit usually contains two flat seeds, but there are also some with single and three seeds. The exocarp is epidural, the mesocarp is fleshy with a sweet taste, the endocarp (also known as seed shell) is a layer of keratin composed of stone cells, and the seeds include a seed coat (silver skin), endosperm, cotyledon, embryo stem, and other parts. The back of the seed is raised while the ventral is flat. The size of the fruit and its endosperm (the "bean") varies with the cultivation conditions and *Coffea* varieties (Figure 10.1). *Coffea* fruit takes a long time to develop, taking nearly 8–10 months; the fastest growth appears in 2–3 months after flowering. The size of the fruit tends to be stable after 4 months, and the dry matter accumulation gradually increases, while the dry matter grows fastest at 5–6 months after flowering (iPlant.cn, 1999).

FIGURE 10.1 Flowers and fruits of Yunnan *Coffea arabica* (photos by Hua Zhou).

10.2.2 Cultivars of YAC

Coffea arabica derived from Ethiopia or the Arabian Peninsula. The *Coffea* germplasm was first introduced into China in 1884. After 1980, with the continuous development of *Coffea* resources and planting industry, some research institutions in Yunnan and Hainan Province introduced a large number of foreign high-quality *Coffea* germplasm resources from Portugal (CIFC), Kenya, Colombia, Brazil, Malaysia, and other countries. Then abundant germplasm resources were formed in the long process of natural and artificial selection. For example, Dehong Tropical Agricultural Science Institute of Yunnan (called "Dehong Institute") preserved 642 copies of *C. arabica*, 34 of *C. canephora*, 14 of *C. liberica*, 3 of *C. eugenioides*, and 1 of *C. racemose*. In addition, 25 new *Coffea* varieties, which can adapt to different ecological systems, have been selected and cultivated independently by "Dehong Institute" through excellent germplasm identification, natural mutation selection, hybrid breeding, and other new techniques, which greatly promoted coffee variety structure adjustment and upgrading. Importantly, the Catimor CIFC7963 (namely YAC), a quality cultivar with traits of resistance to adverse conditions, dwarf cultivar, high productivity, and less attack by longicorn, making it suitable for planting in a shade environment and at high altitude; thus, its cultivated areas developed rapidly. In 1996, CIFC7963 was promoted and applied in domestic coffee production. In 2014, it was approved by the variety certification committee of the Ministry of Agriculture and became the main cultivar in Yunnan coffee production. Also, there are other excellent Arabica coffee cultivars planted in Yunnan such as S288, Typica, Bourbon, T series, Dere series, and others (Figure 10.2) (Zhao et al., 2018).

10.3 Resource Availability

10.3.1 Wildly Grown or Cultivated and Distribution

YAC is suitable for planting in an ecological environment of calm wind, cool temperatures, high humidity, and under shade or semi-shade. The hot regions and subtropical regions of Yunnan low-latitude plateau are mainly located in the south and southwest. The dry and hot river valleys, with high mountains and deep valleys, with distinct three-dimensional climate: cool climate, large temperature difference between day and night, good thermal conditions, and moderate rainfall, are the most suitable areas for the growth of YAC. Thus, the main planting areas of Arabica coffee in Yunnan Province are distributed in Lincang, Pu'er, Baoshan, Dehong, and Xishuangbanna Prefecture. For a long time, the productive cultivation of Arabica coffee in Yunnan Province was mainly distributed below 1000 m above sea level, generally no more than 1200 m. However, with the development of production, the elevation of Yunnan arabica coffee cultivation continues to rise. At present, the largest area of high-altitude coffee that has been put into production is Lujiangba plantation area in Baoshan city, in which a total of 1334 hectares of arabica coffee is grown at altitudes ranging from 1400 to 1670 m. The highest planting area in Pu'er is the coffee garden built by Yunnan Ruifeng Tea and Coffee Company at 1480 m above sea level in Maqadi, Menglang Town, Lancang County. Since 2012, Lincang Lingfeng Coffee Company has planted an area of 15,300 hectares of YAC, accounting for 50% of the area above

FIGURE 10.2 Some *Coffea arabica* cultivars. (A,C) Dere-48-1; (B) Catimor-CIFC7963; (D) Dere38; (E) Bourban; (F) Dere132 (photos by Bing Xiao).

1300 m, which makes it a demonstration area of specialty coffee in China. Professors Zhang Hong-bo and Huang Jia-xiong have concluded that the higher the altitude in the appropriate range, the better the cup quality (Huang, 2012, Zhang et al., 2012). The cup quality of Catimor coffee grown at high altitude was significantly improved. Thus, high altitude is an important condition to produce specialty coffee. Developing high-altitude coffee production can improve the efficiency of coffee production and management. Moreover, it is of practical significance to promote the sustainable development of YAC production in China.

In addition, shade is another important factor in the growth of arabica coffee. Coffee is native to the lower layer of tropical rain forest, and has formed a growth habitat of calm wind, shade or half shade, and a moist environment in the long-term phylogeny, and is sensitive to strong light. As an important measure of agricultural biodiversity, the agroforestry system has been widely applied and developed in the cultivation of YAC in recent years. The shaded cultivation of arabica coffee in Yunnan is beneficial to the balance of nutritional growth and reproductive growth, not only effectively controlling the occurrence of major diseases, pets, and weeds, but also conducive to the high and stable yield of coffee, improved quality, and reduced production costs (Li et al., 2012). There are various cultivation modes of coffee biodiversity in Yunnan, but they are mainly divided into three types: strip shade cultivation, dot shade cultivation, and rapid growth shade cultivation, represented by coffee-rubber, coffee-macadamia, and coffee-banana.

10.3.2 Sustainability

Nowadays, the Yunnan government is eager to develop arabica coffee as a characteristic industry, and this will effectively promote the industrialization of YAC. Benefitting from the unique climate conditions, a suitable cultivation environment, and strong support from government, since the early 1990s, arabica coffee-growing areas in these places have rapidly expanded. In 2018, coffee-planting areas and production in Yunnan Province increased to 120,000 hm^2 and 150,000 t, respectively, both contributing to about 98% of the China's coffee plant areas and yield. This rapid expansion has secured China in a position within the top 12 coffee-growing countries (Rigal et al., 2020). In 2019, YAC was listed in China's agricultural brand catalog, which adequately indicated its higher quality and great economic value. Furthermore, it has become not only one of the China's most important agricultural products in global market competition, but also one of the ten key agricultural industries in Yunnan and the largest earning foreign exchange export agricultural industry, surpassing tobacco and vegetables in Yunnan Province at the same time.

10.4 Plant Parts for Medicinal Uses

10.4.1 The Folklore and Ancient Uses

Coffee trees are native to the highlands of Ethiopia in southwestern Africa. It is said that more than a thousand years ago, a shepherd discovered that a sheep had eaten a plant and became excited and lively, and thus, this led to the discovery of the magic plant coffee. Local indigenous people in Ethiopia often grind the fruit of the coffee tree, mix it with animal fat, and knead it into ball-shaped structures. In fact, during the 4th–9th centuries, people of these indigenous tribes regarded these coffee balls as precious food for the soldiers in Ethiopia.

It was not until around the 11th century that people began to brew coffee with water as a beverage. In the 13th century, Ethiopian troops invaded Yemen and brought coffee to the Arab world. Since Islam prohibits religious people from drinking alcohol, some people in religious circles believe that this coffee drink is nerve-stimulating and violates Islamic teachings. Therefore, coffee shops were once banned and closed. However, the Sudanese of Egypt believed that coffee did not violate teachings and lifted the ban. Coffee beverages then quickly became popular in the Arab region.

Coffee was mainly used in medicine and religion in ancient times. Doctors and monks admitted that coffee had the functions of refreshing, strengthening the stomach, strengthening the body, and stopping bleeding. The use of coffee was documented in the early 15th century, and it was integrated into religion during this period.

It was also popularly used as a drink in religious ceremonies. Because alcohol is strictly prohibited in Islam, coffee became an important social drink at that time.

10.4.2 The Modern Uses

After entering the modern era, as the understanding of the chemical composition and biological activity of coffee has become deeper, coffee has gradually evolved into a functional beverage, which has a series of effects such as anti-oxidation, prevention of diabetes, prevention of cardiovascular disease, anti-depression, and anti-cancer. In addition to traditional coffee beverages, the pulps of fruit, silverskin, leaves, and flowers of coffee plant have gradually been developed and used as beverages or raw materials for health products because they are rich in caffeine, chlorogenic acids, trigonelline, and other biologically active components.

10.5 Phytochemistry of YAC

10.5.1 Volatile Compounds of YAC

Coffee is one of the most popular brewed beverages in the world because of its complex aroma and pleasant taste, as well as its stimulation effects from caffeine. More than 90 years of scientific studies exploring the complex volatile aroma composition of coffee have yielded over 1000 volatile compounds (Lee and Shibamoto, 2002). The composition of aroma components in coffee not only depends on the species and cultivars of coffee, climate and growing conditions, soil, and bean quality, but is also influenced by blend types, post-harvest processing, degree and speed of roasting, and storage methods (Dias et al., 2018, Pereira et al., 2019).

Green coffee beans offer a light special aroma. The aroma of coffee is related to the chemical composition of the green coffee beans, and typical coffee aromas are developed during the roasting process. Roasting is a highly complex process in which the coffee is exposed to temperatures of up to 300°C, and it leads to a series of complex chemical reactions such as Maillard reactions, Strecker degradation, caramelization, and oxidation (van Boekel, 2006, Moon and Shibamoto, 2009).

YAC has been widely accepted by coffee consumers due to its unique flavor and delightful taste. In order to identify the flavor characteristics of YAC (Table 10.1), many researchers have studied aromatic components of arabica coffee from different regions in Yunnan Province. Some compounds produce sensory effects to different degrees in the human nose and thus the capacity to impact YAC's aroma has been detected (Dong et al., 2015).

Cheong et al. (2013) analyzed volatile constituents in roasted coffee beans (*C. arabica* L. cv. Catimor) from the Pu'er district in Yunnan Province, China. Analytical results showed that, based on the total volatile compounds identified, Yunnan coffee had most of its volatile chemicals abundantly expressed, given its largest total peak area. The total concentration of volatile compounds in Yunnan arabica coffee identified was 1239.04 mg/L. These volatile compounds can be classified as furans, acids, furanones, ketones, lactones, phenols, pyrazines, pyridines, pyrroles, and sulfur-containing compounds. The extracts from Yunnan coffee beans were found to

TABLE 10.1

Representative Volatile Compounds of Roasted Yunnan Arabica Coffee Beans

No.	Odorous Compounds	Aroma Perception
1	2-Methylbutanal	Creamy
2	3-Methylbutanal	Creamy
3	2,3-Butanedione	Creamy
4	2,3-Pentanedione	Creamy
5	Furfuryl methyl ether	Roasted potatoes
6	2,5-Dimethypyrazine	Nutty
7	2,3-Dimethylpyrazine	Nutty
8	2-Ethyl-6-methylpyrazine	Earth
9	Acetic acid	Acidic
10	Furfuryl methyl sulfide	Roasted potatoes
11	3,5-Diethyl-2-methylpyrazine	Roasted
12	2-Acetylfuran	Roasted
13	6,7-Dihydro-5-methyl-5H-cyclopentapyrazine	Nutty
14	3-Methylbutanoic acid	Acidic
15	2-Acetyl-3-methylpyrazine	Roasted potatoes, nutty
16	Guaiacol	Phenolic, burnt
17	2-Acetylpyrrole	Muscats
18	Maltol	Sweet
19	4-Ethyl-2-methoxyphenol	Clove
20	2,5-Dimethyl-4-hydroxy-3(2H)furanone	Caramel

have higher amounts of 2-furfurylthiol, a sulfur-containing volatile compound that is known to be largely responsible for the roasted coffee aroma (Cannon et al., 2010). Due to low flash points and susceptibility to oxidative degradations, sulfur-containing volatile compounds are usually present in trace amounts (less than 0.01% of the total amount of volatile components) but play an important role in the freshness of roasted coffee. Pyrazines are the second dominant volatile components found in the Yunnan arabica coffee extracts, which impart roasts and earthy aroma in roasted ground coffee. The most abundant pyrazine is 2-methylpyrazine which has a nutty aroma, yet the most aromatic pyrazines are 2-ethylpyrazine, 2-ethyl-6-methylpyrazine, and 2-ethyl-3,6-dimethylpyrazine, which are present in relatively smaller amounts. Likewise, furanones also belong to an important group of volatile components in coffee aroma, of which furaneol or 2,5-dimethyl-4-hydroxy-3(2H)-furanone, conferring caramel, is one of the key aroma compounds most frequently reported. In addition, the highest concentration of phenolics is detected in the Yunnan coffee extract with *p*-vinylguaiacol as the major chemical. *P*-vinylguaiacol is a volatile phenolic derivative defined as another member of the potent odorant group. Besides, the presence of carboxylic acids such as acetic, propanoic, butanoic, and 3-methylbutanoicacids accounts for the sourness of YAC.

Dong et al. (2015) detected the odor and flavor compounds of roasted ground coffee beans with different degrees of roasting from four regions (Pu'er, Baoshan, Dehong, Lincang) in Yunnan Province. The numbers of volatile compounds produced by three roasting types (light, medium, dark) were 60, 65, and 67, respectively. The results showed that the volatile components of YAC are mainly classified as furans, acids, furanones, ketones, phenols, pyrazines, pyridines, pyrroles, sulfides, and esters. With the increase of roasting degree, the content of furans, pyridines, and sulfides increased gradually, while acids and furanones decreased accordingly.

Other studies which evaluated the influence of the degree of roasting (light, medium, and dark) and roasting speed on the volatile composition of YACs include the work of Bin Zhou and others (Zhou and Ren, 2014, 2015), Xiao-na Wang (Wang et al., 2017), and Miao Yu (Yu, 2017). Overall, their findings indicated that furfuryl alcohol, furfural, furan, pyrazine, pyrrole, acids, and phenols are the main volatile compounds. These complex compounds make up the distinctive aroma and flavor of YAC. The flavor of the same coffee bean is different depending on the roasting degree or speed. Increasing the degree of roasting produces drinks with different aromatic profiles, ranging from coffees with aromas rich in higher-volatility compounds, such as volatile acids and furans responsible for fresh and floral notes, to coffees with aromas rich in compounds such as pyrazines, furfuryl alcohol, furfural, and pyridines, responsible for the characteristic sweet, roasted, and earthy notes. From the aroma composition, content, and evaluation results, it was concluded that the YAC with medium roasting degree can best reflect its characteristics comprehensively.

Over the past few decades, researchers have put a lot of effort into the study of the chemical composition of coffee, and progress has been made. Apart from the most researched ingredients such as caffeine, chlorogenic acids, and trigonelline, the trace ingredients present in coffee have received great attention. In order to search for the bioactive ingredients in coffee, many phytochemical studies on coffee have been carried out in recent years, many of which are based on YAC.

10.5.2 Caffeine

Caffeine is a trimethylxanthine compound whose xanthine skeleton is derived from purine nucleosides (Figure 10.3). Caffeine may be synthesized through various routes by using xanthosine or xanthine as starting materials. The most widely considered synthetic route is: xanthosine→7-methylxanthine→theobromine→caffeine. Three methyltransferases from the salicylic acid, benzoic acid, theobromine methyltransferase (SABATH) family play key roles in caffeine synthesis (Huang et al., 2016). According to the result of Liu et al. (2012) using microwave-assisted extraction and HPLC analysis, caffeine values of YAC ranged from 0.875 to 1.121% in raw and roasted coffee. Hu et al. (2012b) reported the changing process of caffeine in the green bean and pulp of YAC at different development stages and found that the average content of caffeine and 5- caffeoylquinic acids in YAC green beans showed a slight upward trend with the increase of maturity. The caffeine present in the pulp was the highest in the early stage, with an average content of 0.81%, and decreased with the increase of maturity, with only 0.47% of caffeine presented in overripe YAC pulp (Hu et al., 2020b).

10.5.3 Chlorogenic Acids (CGAs)

Phenolic compounds account for about 6–10% of the total content of green coffee beans (Heeger et al., 2017). The main phenolic compounds are CGAs, among which the caffeoylquinic acids (CQAs), especially 5-CQA, dominate along with lower

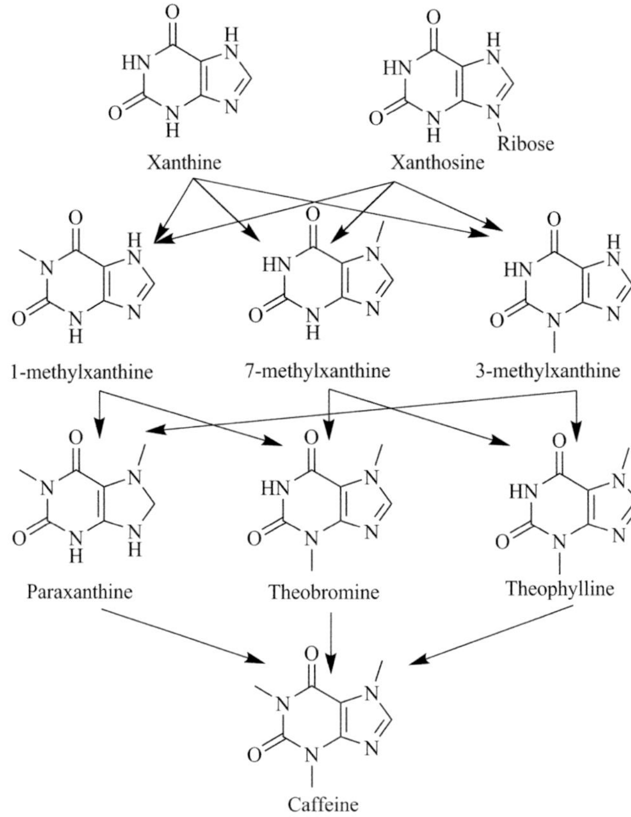

Xanthine

Xanthosine

Ribose

1-methylxanthine

7-methylxanthine

3-methylxanthine

Paraxanthine

Theobromine

Theophylline

Caffeine

FIGURE 10.3 The synthetic route of caffeine in coffee.

amounts of feruloylquinic acids and dicaffeoylquinic acids. Additionally, there are over 50 other structurally related cinnamoylquinic acids, some cinnamoyl-amino acid conjugates, and cinnamoyl-glycosides. CGAs can be mainly divided into CQAs with three isomers (3-, 4-, and 5-CQA), feruloylquinic acids (FQA) with three isomers (3-, 4-, and 5-FQA), and dicaffeoylquinic acids (diCQAs) with three isomers (3,4-diCQA; 3,5-diCQA; 4,5-diCQA) (Moeenfard et al., 2014) (Figure 10.4). According to Hu et al. (2020b), the content of 5-CQA in YAC stayed at a relatively low level in the initial stage of cherry development; in mid-stages, its content rose sharply, and was finally maintained between 6.0% and 8.0%. In pulp, the content of 5-CQA increased in early stages and decreased significantly in late stages, while in overripe pulp, the content of 5-CQA was only 0.32% (Hu et al., 2020b).

10.5.4 Trigonelline

Pyridine nucleotide cycle plays an important role in the formation of trigonelline: nicotinamide—nicotinic acid—nicotinic acid mononucleotide (NaMN)—nicotinic acid—denine dinucleotide (NaAD)—NAD—nicotinamide mononucleotide

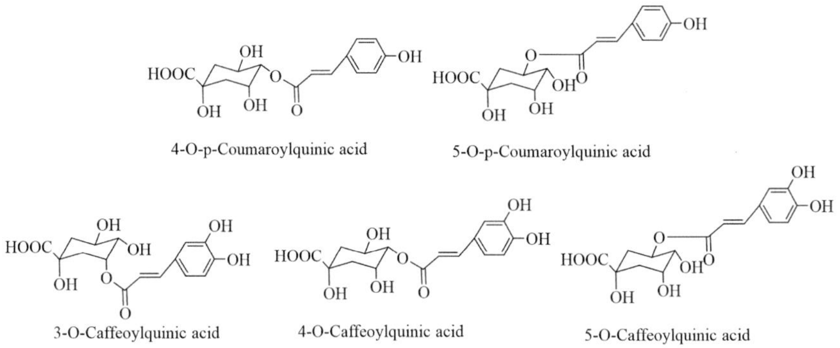

4-O-p-Coumaroylquinic acid 5-O-p-Coumaroylquinic acid

3-O-Caffeoylquinic acid 4-O-Caffeoylquinic acid 5-O-Caffeoylquinic acid

FIGURE 10.4 The main caffeoylquinic acids (CQAs) present in coffee.

(NMN)—nicotinamide. Trigonelline is finally synthesized from the nicotinic acid produced by this cycle (Zheng et al., 2004). Trigonelline in the early and middle stages of the YAC cherry development exhibited a slight downward trend, while the trigonelline in the coffee pulp accumulated rapidly. However, in the late stage, trigonelline in the beans began to accumulate, while the trigonelline in pulp decreased significantly (Hu et al., 2020b).

10.5.5 Diterpenes

Almost all diterpenes found in coffee are *ent*-kaurane types or its derivatives. They can be divided into several types like oxidized diterpenes, tetrahydrofuran-type diterpenes, lactone-type diterpenes, Δ4,18-type diterpenes, degraded-type diterpenes, Villanova-type diterpenes, and lactam-type diterpenes according to the skeleton (Figure 10.5) (Hu et al., 2019). By summarizing previous studies, it was found that YAC contains all these types of coffee diterpenes. Among them, lactone diterpenes and lactam diterpenes are only found in YAC (Hu et al., 2020a, Wang et al., 2018).

10.5.5.1 Oxidized-Type Diterpenes

The oxidized diterpenes and their glycosides that have been found in coffee have common oxidation positions at 16, 17, and 19. If C-16 is a chiral center, then 16-OH is in the alpha configuration. This kind of diterpene was found widely in both the cherries and roasted beans of YAC. Compounds **70**, **72**, and **74** belong to oxidized diterpenes, but will be introduced in the degradable diterpenes due to their structural characteristics (Figure 10.6).

10.5.5.2 Tetrahydrofuran-Type and Degraded-Type Diterpenes

Tetrahydrofuran-type diterpenes are the most abundant diterpenes in coffee (Figure 10.7). The structural characteristics of this type of diterpenes are based on the *ent*-kaurane skeleton, with carbon migration and rearrangement of methyl groups 18 and 19 to form a tetrahydrofuran ring. At the same time, the C-16 and 17-positions

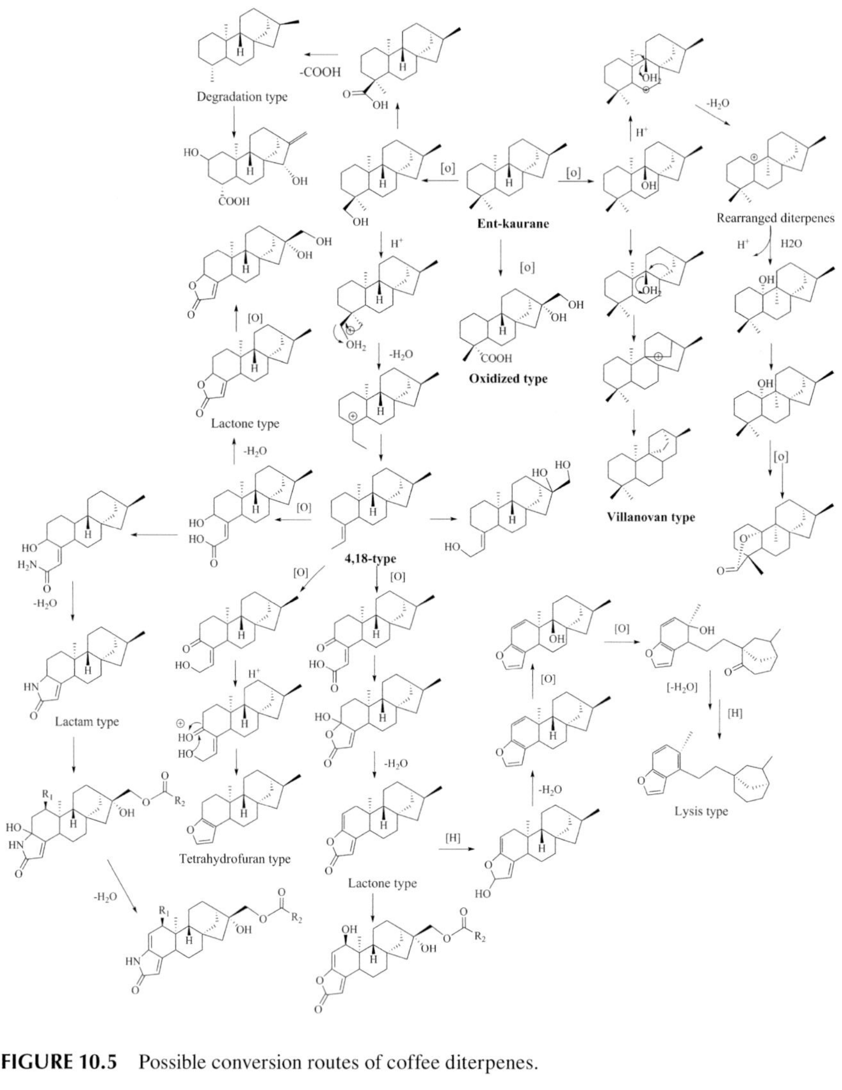

FIGURE 10.5 Possible conversion routes of coffee diterpenes.

of such compounds in coffee are easily oxidized, and then form coffee diterpene glycosides and coffee diterpene fatty acid esters with glucose or fatty acids, and all 16-OH are in the α configuration. Cafestol (**15**) and kahweol (**28**) are representative of this type of diterpenes, and are the most abundant. They can undergo dehydration and dehydrogenation reactions during roasting to form products such as dehydrocafestol (**24**), dehydrokahweol (**37**), cafestal (**27**), kahweal (**38**), isokahweol (**39**), and dehydro-isokahweol (**40**) (Dias et al., 2014). In addition to cafestol, kahweol, and their roasting products, a series of tetrahydrofuran-type diterpenes (**41–49**) with C-2 oxidized to carbonyl were also found in green beans and roasted beans of YAC (Shu et al., 2014).

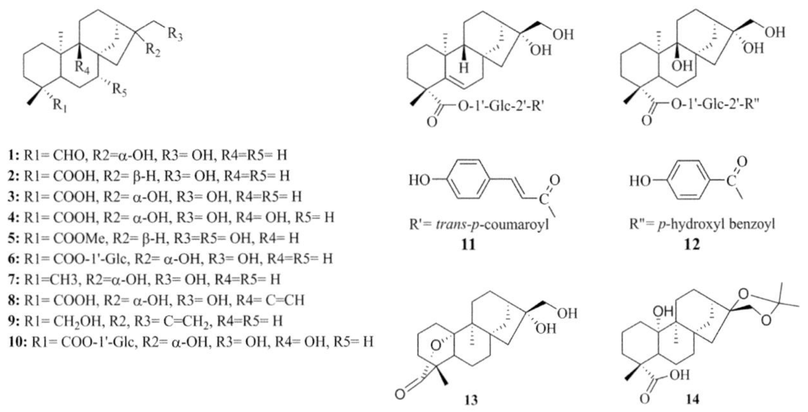

1: R1= CHO, R2=α-OH, R3= OH, R4=R5= H
2: R1= COOH, R2= β-H, R3= OH, R4=R5= H
3: R1= COOH, R2= α-OH, R3= OH, R4=R5= H
4: R1= COOH, R2= α-OH, R3= OH, R4= OH, R5= H
5: R1= COOMe, R2= β-H, R3=R5= OH, R4= H
6: R1= COO-1'-Glc, R2= α-OH, R3= OH, R4=R5= H
7: R1=CH3, R2=α-OH, R3= OH, R4=R5= H
8: R1= COOH, R2= α-OH, R3= OH, R4= C=CH
9: R1= CH_2OH, R2, R3= C=CH_2, R4=R5= H
10: R1= COO-1'-Glc, R2= α-OH, R3= OH, R4= OH, R5= H

R'= *trans-p*-coumaroyl
11

R"= *p*-hydroxyl benzoyl
12

13

14

FIGURE 10.6 Oxidized-type diterpenes present in coffee.

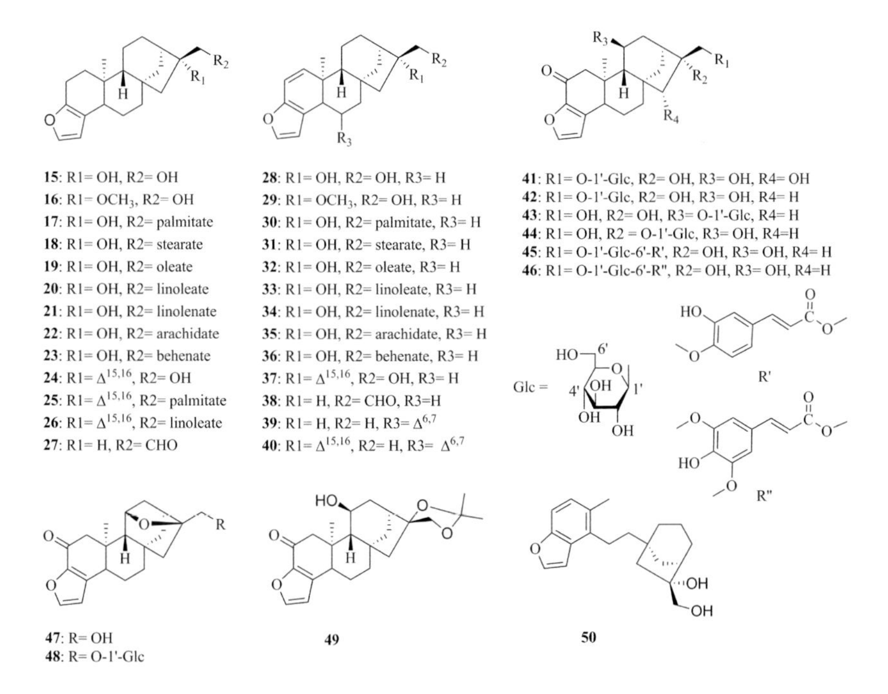

FIGURE 10.7 Tetrahydrofuran-type and degraded-type diterpenes present in coffee.

10.5.5.3 Lactone-Type Diterpenes

A series of diterpene esters whose methyl groups were rearranged to γ-lactones at C-18 and C-19 were isolated and identified from YAC green beans (Wang et al., 2018). The structural characteristics of this kind of diterpenes are similar to those of tetrahydrofuran diterpenes (Figure 10.8). The C-16 and C-17 positions are easily oxidized, and they are combined with fatty acids to form fatty acid esters.

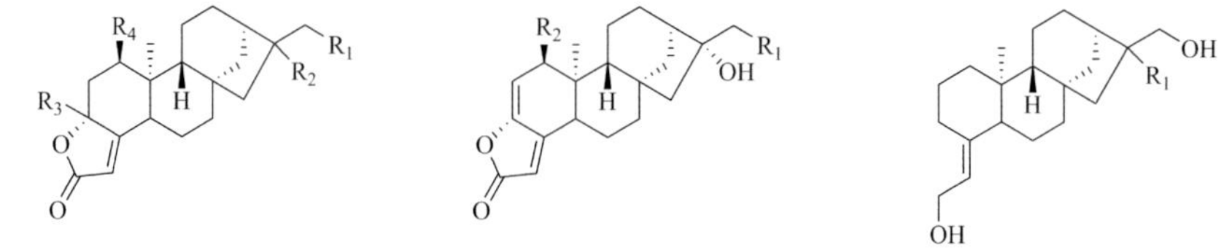

51: R1= OH, R2= α-OH, R3=R4= H
52: R1= OH, R2= α-OH, R3= OH, R4= H
53: R1= OH, R2= α-OH, R3= OCH_3, R4= H
54: R1= OH, R2= β-H, R3= OH, R4=H
55: R1= palmitate, R2= α-OH, R3=R4= H
56: R1= linoleate, R2= α-OH, R3=R4= OH,
57: R1= oleate, R2= α-OH, R3= OH, R4= H
58: R1= palmitate, R2= α-OH, R3=R4= OH
59: R1= palmitate, R2= α-OH, R3= OH, R4= OCH_3
60: R1= nonadecadienoate, R2= α-OH, R3= OH, R4= OCH_3

61: R1= OH, R2= H
62: R1= OH, R2= OCH_3
63: R1= palmitate, R2= OH
64: R1= eicosadienoate, R3=OH

65: R= β-H
66: R= α-OH

FIGURE 10.8 Lactone-type diterpenes present in coffee.

67: R1= β-COOH, R2= H, R3= OH
68: R1= α-COOH, R2= H, R3= OH
69: R1= α-COOH, R2= H, R3= O-1'-Glc
70: R1= α-COOH, R2= COOH, R3= O-1'-Glc

Atractyloside R= CH_3
Carboxyatractyloside R= COOH

71: R1= COOH, R2= H, R3= OH
72: R1= R2= COOH, R3= OH
73: R1= COOH, R2= H, R3= O-1'-Glc
74: R1= R2= COOH, R3= O-1'-Glc

75

76

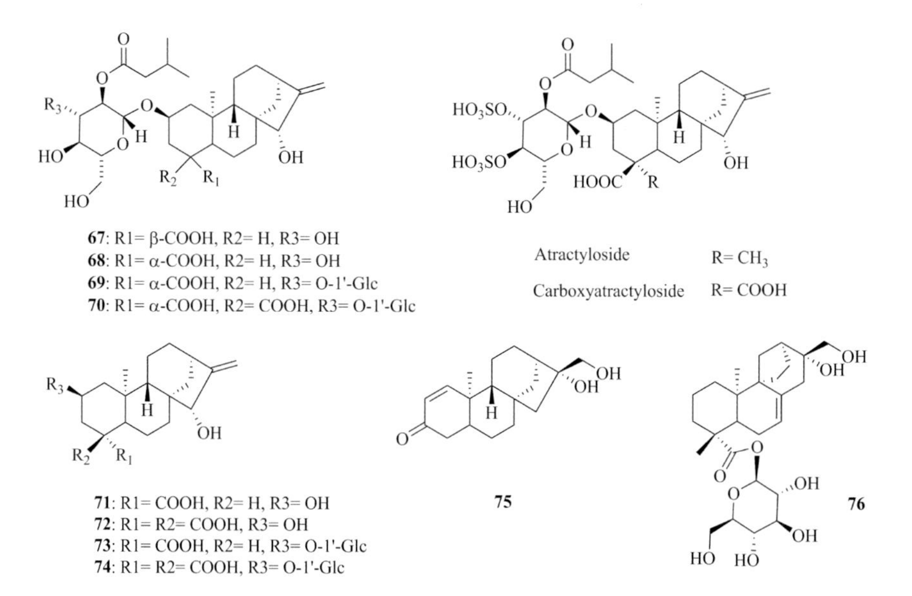

FIGURE 10.9 Degraded-type and Villanova-type diterpenes present in coffee.

10.5.5.4 Degraded-Type and Villanova-Type Diterpenes

A total of six degradable diterpenes were found in coffee. Their structural characteristics are the degradation of the methyl group at the C-18 or C-19. The C-16 and C-17 are usually double bonds, and the oxidation positions are usually at C-2, 15, and 19, where C-2 easily combines with sugars to form diterpene glycosides. This type of diterpenes is also found to be present in YAC roasted beans. Specially, a diterpene with both C-18 and C-19 degradation (**75**) was found to be present in the roasted beans of YAC (Shu et al., 2014) (Figure 10.9).

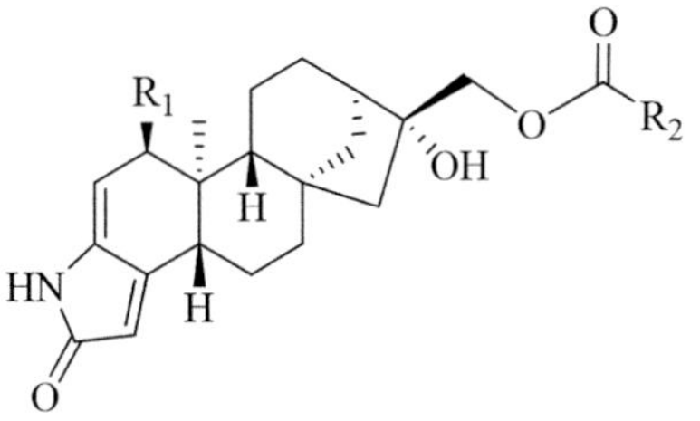

77: R1=H, R2=palmitate
78:R1=H, R2=stearate
79:R1=H, R2=Linoleate
80:R1=OH, R2=palmitate
81:R1=OH, R2=Linoleate

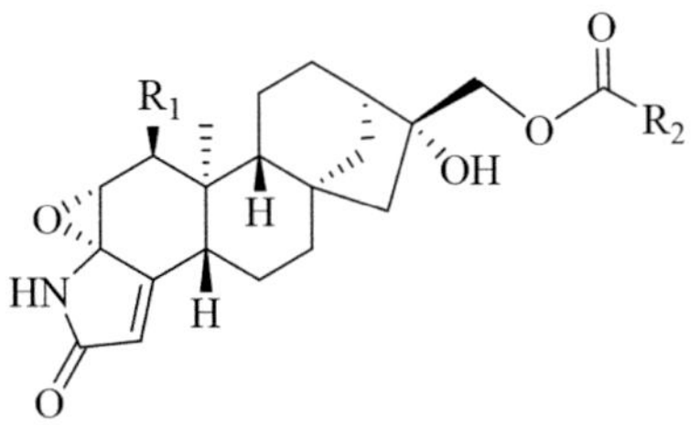

82:R1=OH, R2=palmitate
83:R1=OH, R2=Linoleate

FIGURE 10.10 Lactam-type diterpenes present in coffee.

Villanovane I (**76**) is a Villanova-type diterpene obtained from roasted beans of YAC. Different from the *ent*-kaurane diterpenes which has a C-8 bridged ring, the skeleton of this type of tetracyclic diterpenes has a C-9 bridged ring. This class of diterpenes is rarely reported in the literature; up till now, only one case has been reported in coffee plant (Shu et al., 2014).

10.5.5.5 Lactam-Type Diterpenes

Seven lactam diterpene esters (**77–83**) were recently found in roasted beans of YAC (Hu et al., 2020a) (Figure 10.10). All seven compounds showed the same carbon skeleton as tetrahydrofuran-type, γ-lactone-type, and Δ4,18-type *ent*-kaurane diterpenes. The difference is that there is a lactam fused with the A ring in these compounds. The A ring of such compounds is usually prone to oxidation. The content of lactam-type diterpene palmitate is the highest, followed by lactam linoleic acid diterpene ester.

10.6 Biological Activities Related to Medicinal Uses or Clinical Evidence

10.6.1 Caffeine

Caffeine can be distributed throughout the body via absorption by the stomach and intestines. Among all the known coffee components, caffeine exhibits the most significant activities against the nervous system. Once absorbed, caffeine will play a variety of roles in the body by inhibiting adenosine receptors, which include subtypes A1, $A2_A$, $A2_B$, and A3. The inhibition activities of caffeine against A1 and $A2_A$ subtype are the most prominent, and these inhibitory effects can be helpful to prevent Parkinson's disease (PD), alleviate hypertension, enhance cognitive function, relieve Alzheimer's disease, ease anxiety, and can also treat congestive heart failure and polar renal dysfunction. Apart from adenosine receptors, 5-hydroxytryptamine

(5-HT) receptor, cholinergic receptor, opioid receptor, and GABA receptor can also be affected by caffeine to different extents (Hu et al., 2019).

Caffeine may be the main active ingredient in coffee to prevent PD. It has been demonstrated that caffeine is resistant to dopaminergic neurotoxicity in animal models, and its mechanism of action may be due to the blockage of A2 adenosine receptors to stimulate dopamine release, thereby improving the function of the dopaminergic system (Trevitt et al., 2009). Another study showed that eicosanoyl-5-hydroxytryptamide, which was purified from coffee as an agent that leads to enhanced enzymatic activity of the specific phosphatase PP2A that dephosphorylates the pathogenic protein α-synuclein, works in synergy with caffeine in protecting against mouse models of PD and dementia with Lewy bodies. The mechanism of this synergy is also through enhancing PP2A, which is dysregulated in the brains of individuals with these α-synucleinopathies (Yan et al., 2018).

Extensive studies have shown that caffeine can effectively reduce the incidence of depression. By searching PubMed, Web of Science, China National Knowledge Infrastructure, and WANFANG DATA in English or Chinese from 1 January 1980 to 1 May 2015, the evidence of a linear association was found between coffee consumption and depression, and the risk of depression decreased by 8% (relative risk = 0.92, 95% confidence interval = [0.87, 0.97], $p = 0.002$) for each cup/day increment in coffee intake. A non-linear association was found between caffeine consumption and depression; the risk of depression decreased faster and the association became significant when the caffeine consumption was above 68 mg/day and below 509 mg/day (Wang et al., 2016). Kim and Kim (2018) investigated the associations of green tea, coffee, and caffeine consumption with self-report lifetime depression in the Korean population using data from the Korean National Health and Nutrition Examination Survey, and found that frequent green tea consumers (3 cups/week) had a 21% lower prevalence of depression. Additionally, a study conducted by Machado et al. (2020) investigated the potential of caffeine in preventing behavioral outcomes, neurodegeneration, and synaptic protein alterations in a mice model of agitated depression by bilateral olfactory bulbectomy, and confirmed that caffeine is effective in preventing neurodegeneration associated with memory impairment and may be considered as a promising therapeutic tool in the prophylaxis and/or treatment of depression.

10.6.2 Chlorogenic Acids

The CGAs have excellent anti-oxidant ability (Liang et al., 2016a,b, Mullen et al., 2011) which prevents free radical-induced diseases such as cardiovascular damage, nervous system damage, liver damage, and tumors (Park, 2013). In a study conducted by Liang et al. (2016b), principal component analysis was used to interpret the correlations between the physiochemical and anti-oxidant parameters of coffee, and found that the intracellular anti-oxidant capacity that best describes the potential health benefits of coffee positively corresponds with CGAs content.

CGAs can also effectively inhibit inflammatory factors such as TNF-α, IL-6, and IL-1β, thus preventing certain diseases caused by inflammation, such as acute liver injury, gastrointestinal disease, and endothelial dysfunction (Tajik et al., 2017, Chauhan et al., 2012, Krakauer, 2002).

Studies have shown that CGAs also play an important role in regulating sugar metabolism, which is mainly achieved by regulating enzyme activities related to sugar metabolism (Bassoli et al., 2015), reducing insulin resistance, or stimulating insulin secretion. Glucose-6-phosphatase (Glc-6-Pase) is a multicomponent system that exists primarily in the liver and catalyzes the terminal step in gluconeogenesis and glycogenolysis. Hydrolysis of glucose 6-phosphate (Glc-6-P) was measured in the presence of Svetol or chlorogenic acid in intact human liver microsomes. Svetol significantly inhibited the hydrolysis of Glc-6-P in intact human liver microsomes in a competitive manner, and suggested that caffeoylquinic acid and dicaffeoylquinic acid (CQAs) are the main compounds that mediate this activity (Henry-Vitrac et al., 2010).

The function of CGAs to promote ester metabolism makes them potentially effective in inhibiting obesity and cardiovascular diseases. For instance, the oxidative modification of low-density lipoprotein (LDL) is an important risk factor in the pathogenesis of arteriosclerosis, while an *in vivo* study in 11 healthy male students found that CGAs can prevent cardiovascular-related diseases by lowering the level of LDL-cholesterol in the body (Yukawa et al., 2004).

10.6.3 Diterpenes

10.6.3.1 Oxidized-Type Diterpenes

The degree of oxidation of these compounds has an effect on the intensity of activity, especially the degree of oxidation of the methyl group at the 19 position. For example, 16α, 17-dihydroxy-*ent*-kauran-19-al (**1**) appears to have a strong *in vitro* inhibitory activity effect on human ovarian cancer cells A2780 (IC_{50}: 0.38 μM) and human lung cancer cells 95-D (IC_{50}: 19.38 μM). However, 16α, 17-dihydroxy-*ent*-kauran-19-oic acid (**3**) only showed certain *in vitro* inhibitory effect on 95-D (IC_{50}: 39.83 μM) (Zhou et al., 2013). Compound **3** has certain anti-platelet aggregation effect, but compound **1** does not show this activity (Yang et al., 2002, Wang et al., 2011). Besides, 16-*ent*-kauren-19-ol (**9**) and its oxidation product, 16-*ent*-kauren-19-oic acid, were screened for Na^+, K^+ ATPase inhibitors, and the latter had 22 times the inhibitory activity of **9** (Ngamrojnavanich et al., 2003). Although the coffee diterpenes that have been found are all in the 16α-OH or 16β-H configuration, there are a large number of their isomers (16β-OH or 16α-H) in nature. The configuration has an influence on the activity intensity of this kind of diterpenes. For example, 16α, 17-dihydroxy-*ent*-kauran-19-oic acid (**3**) does not inhibit the replication of HIV virus *in vitro*, but its isomer 16β, 17-dihydroxy-*ent*-kauran-19-oic acid showed strong inhibitory effect (EC_{50}: 0.8 μg/ml) (Wu et al., 1996). Both 16α, 17-diol-*ent*-kaurane (**7**) and its isomer 16β, 17-diol-*ent*-kaurane have strong activities in inhibiting NO production, but the latter has better activity (Nhiem et al., 2015).

10.6.3.2 Tetrahydrofuran-Type and Degraded-Type Diterpenes

As the most studied diterpenes in coffee, cafestol and kahweol exhibit various biological activities; in particular they can increase the activity of cholesterol ester transfer protein (CETP), which in turn accumulates extracellular cholesterol to increase serum cholesterol levels (De Roos et al., 2000). The content of these two diterpenes in coffee drinks is directly affected by the brewing method. Unfiltered coffee such as espresso, French coffee, and Turkish coffee usually contains more caffeine (6–12 mg/cup), while in filtered coffee and instant coffee, the content of coffee is difficult to pass through the filter paper, and is reduced to 0.2–0.6 mg/cup (Gross et al., 1997).

There is a series of coffee diterpene fatty acid esters in coffee, which are an important part of coffee lipids. Diterpene esters of cafestol and kahweol with various fatty acids (C16: 0, C18: 0, C18: 1, C18: 2, C18: 3, C20: 0, C22: 0) (**17–23**, **30–36**). Cafestol palmitate (**17**) and kahweol palmitate (**30**) showed various similar biological activities, but kahweol palmitate has better anti-angiogenesis effect and activated glutathione S-transferase activity (Moeenfard et al., 2016, Lam et al., 1982, Huber et al., 2004).

10.6.3.3 Lactone-Type Diterpenes

Studies have shown that tricalysiolide A (**51**) and tricalysiolide C (**53**) exhibited certain *in vitro* inhibitory activities on human oral epidermal-like cancer cells KB at a concentration of 50 μM (Tamaki et al., 2008). Furthermore, the diterpenes caffarolides C–E (**57–59**) can activate platelet aggregation *in vitro* at a concentration of 30 mg/mL (Wang et al., 2018).

10.6.3.4 Degraded-Type Diterpenes

Compounds **69/70**, **71/72**, and **73/74** are three groups of diterpenoid glycosides with similar structures, of which **70**, **72**, and **74** are oxidized diterpenes. They are similar in structure to the phytotoxins atractyloside and carboxyatractyloside in *Atractylis gummifera*. The latter can strongly inhibit the translation of nucleic acids, and has strong toxicity to the liver. Therefore, Lang et al. (2013) screened the six compounds for adenine nucleotide shiftase inhibitory activities in mitochondria after the compounds **69–74** were found in the crude coffee extract. All showed inhibitory effects *in vitro*, with compound **70** having the strongest activity (Lang et al., 2013). This indicates that both the carboxyl group and the glycosyl group at position 18 have effects on this type of activity. Fortunately, these compounds will gradually degrade as the degree of roasting increases (Lang et al., 2014).

10.6.3.5 Lactam-Type Diterpenes

Compounds **77**, **78**, **79**, **81**, and **83** showed moderate glucosidase inhibitory activities with compounds **77** (8.28 ± 0.62 μM), **81** (12.44 ± 1.37 μM), and **83** (22.2 ± 5.34 μM) having IC_{50} values lower than positive control ursolic acid (25.96 ± 1.42 μM), indicating that the lactam *ent*-kaurane diterpenoids are potential components that exert hypoglycemic effect in coffee (Hu et al., 2020a).

10.7 Commercial Products from YAC

YAC has formed a highly prosperous industry, and there are various YAC products on the market. The most common ones are instant coffee, “hanging ear” coffee, and capsule coffee. Additionally, functional Chinese coffee products like *Cordyceps* coffee, *Ganoderma* coffee, *Chrysanthemum* coffee, *Hovenia* coffee, and *Dendrobium* coffee can also be found everywhere in the market.

Instant coffee is produced by extracting effective ingredients from roasted and ground coffee beans and then drying them. During this process, some of the aromatic substances will be lost and the flavor and taste of the finished product will not be as strong and pure as that of roasted and ground coffee. "Hanging ear" coffee is a kind of portable coffee, with the grounded coffee beans sealed in filter bags. After tearing the bag, spread the paper plywood on both sides and hang it on the cup, and slowly brew it with hot water. The coffee brewing is completed by drip filtering, and the acidity, sweetness, bitterness, and aroma of the coffee are perfectly reflected. Capsule coffee is coffee beans ground into coffee powder, and then packed into aluminum capsules, to avoid the problem of ordinary coffee beans or coffee powder becoming acidic and oxidized after contact with air.

10.8 Conclusion

Due to some limited factors, studies on volatile compounds of YAC have been mainly related to roasting degree and different geographical regions in Yunnan Province recently. In fact, many other aspects still remain to be fully elucidated, such as the effects of geographical origin and post-harvest processing, as well as other aspects that are better understood including the influence of species, seed quality, and storage conditions. Therefore, the key to the highly desirable aromatic character of YAC is still not fully understood. Another issue that has received very little attention is the question of the innumerable cultivars of coffee plants within YAC, which we believe merits further attention, given the substantial potential influence on drink quality. Furthermore, with the growing technological development in the areas of science and agriculture, many other aspects remain to be discovered (or rediscovered), and the theme of YAC quality and flavor offers an excellent opportunity for scientific investigation.

Furthermore, YAC-based phytochemical studies have shown that there are a variety of connotative substances in coffee. Although these substances may exist in small amounts, they may still affect the taste of the coffee beverage. Therefore, it is necessary to clarify the composition of the micro-components of *Coffea arabica* in more depth. In addition, the kind of biological activity these trace amounts of contained substances possess and how they affect human health are still worthy of more extensive and in-depth research.

REFERENCES

Bassoli BK, Cassolla P, Borba-Murad GR, et al., 2015. Instant coffee extract with high chlorogenic acids content inhibits hepatic G-6-Pase *in vitro*, but does not reduce the glycaemia. *Cell Biochemistry and Function*, 33: 183–187.

Cannon RJ, Trinnaman L, Grainger B, et al., 2010. The key odorants of coffee from various geographical locations. *Flavors in Noncarbonated Beverages*, 1036: 77–90.

Chauhan PS, Satti NK, Sharma P, et al., 2012. Differential effects of chlorogenic acid on various immunological parameters relevant to rheumatoid arthritis. *Phytotherapy Research*, 26: 1156–1165.

Cheong MW, Tong KH, Ong JJM, et al., 2013. Volatile composition and antioxidant capacity of Arabica coffee. *Food Research International*, 51: 388–396.

De Roos B, Van Tol A, Urgert R, et al., 2000. Consumption of French-press coffee raises cholesteryl ester transfer protein activity levels before LDL cholesterol in normolipidaemic subjects. *Journal of Internal Medicine*, 248: 211–216.

Dias RCE, de Faria-Machado AF, Mercadante A, et al., 2014. Roasting process affects the profile of diterpenes in coffee. *European Food Research and Technology*, 239: 961–970.

Dias RCE, Valderrama P, Marco PH, et al., 2018. Quantitative assessment of specific defects in roasted ground coffee via infrared-photoacoustic spectroscopy. *Food Chemistry*, 255: 132–138.

Dong WJ, Zhang F, Zhao JP, 2015. Application of electronic nose system coupled with HS-SPME-GC/MS for characterization of aroma fingerprint of roasted coffee beans from different cultivation regions in Yunnan Province. *Chinese Journal of Tropical Crops*, 36: 1903–1911.

Gross G, Jaccaud E, Huggett AC, 1997. Analysis of the content of the diterpenes cafestol and kahweol in coffee brews. *Food and Chemical Toxicology*, 35: 547–554.

Heeger A, Kosinska-Cagnazzo A, Cantergiani E, et al., 2017. Bioactives of coffee cherry pulp and its utilisation for production of Cascara beverage. *Food Chemistry*, 221: 969–975.

Henry-Vitrac C, Ibarra A, Roller M, et al., 2010. Contribution of chlorogenic acids to the inhibition of human hepatic glucose-6-phosphatase activity *in vitro* by Svetol, a standardized decaffeinated green coffee extract. *Journal of Agricultural and Food Chemistry*, 58: 4141–4144.

Hu GL, Gao Y, Peng XR, et al., 2020a. Lactam ent-Kaurane diterpene: A new class of diterpenoids present in roasted beans of coffea arabica. *Journal of Agricultural and Food Chemistry*, 68: 6112–6121.

Hu GL, Peng XR, Wang X, et al., 2020b. Excavation of coffee maturity markers and further research on their changes in coffee cherries of different maturity. *Food Research International*, 132: 109121.

Hu GL, Wang X, Zhang L, et al., 2019. The sources and mechanisms of bioactive ingredients in coffee. *Food & Function*, 10: 3113–3126.

Huang JX, 2012. Preliminary study on the influence of different altitudes on the quality of coffee Arabica. *Tropical Agricultural Science and Technology*, 32: 4–7.

Huang RQ, O'Donnell AJ, Barboline JJ, et al., 2016. Convergent evolution of caffeine in plants by co-option of exapted ancestral enzymes. *Proceedings of the National Academy of Sciences of the United States of America*, 113: 10613–10618.

Huber WW, Teitel CH, Coles BF, et al., 2004. Potential chemoprotective effects of the coffee components kahweol and cafestol palmitates via modification of hepatic N-acetyltransferase and glutathione S-transferase activities. *Environmental and Molecular Mutagenesis*, 44: 265–276.

iPlant.cn., 1999. *Coffea arabica* [Online]. http://www.iplant.cn/info/Coffea%20arabica?t=z. [Accessed].

Kim J, Kim J, 2018. Green tea, coffee, and caffeine consumption are inversely associated with self-report lifetime depression in the Korean population. *Nutrients*, 10: 1201.

Krakauer T, 2002. The polyphenol chlorogenic acid inhibits staphylococcal exotoxin-induced inflammatory cytokines and chemokines. *Immunopharmacology and Immunotoxicology*, 24: 113–119.

Lam LKT, Sparnins VL, Wattenberg LW, 1982. Isolation and identification of kahweol palmitate and cafestol palmitate as active constituents of green coffee beans that enhance glutathione S-transferase activity in the mouse. *Cancer Research*, 42: 1193–1198.

Lang R, Fromme T, Beusch A, et al., 2013. 2-O-beta-D-glucopyranosyl-carboxyatractyligenin from *Coffea* L. inhibits adenine nucleotide translocase in isolated mitochondria but is quantitatively degraded during coffee roasting. *Phytochemistry*, 93: 124–135.

Lang R, Fromme T, Beusch A, et al., 2014. Raw coffee based dietary supplements contain carboxyatractyligenin derivatives inhibiting mitochondrial adenine-nucleotide-translocase. *Food and Chemical Toxicology*, 70: 198–204.

Lee KG, Shibamoto T, 2002. Analysis of volatile components isolated from Hawaiian green coffee beans (*Coffea arabica* L.). *Flavour and Fragrance Journal*, 17: 349–351.

Li JH, Zhang HB, Zhou H, et al., 2012. Effect of shade/non-shade farming system on cup quality of Arabica coffee in Yunnan. *Tropical Agricultural Science and Technology*, 31: 20–23.

Liang NJ, Lu XN, Hu YX, et al., 2016a. Application of attenuated total reflectance-fourier transformed infrared (ATR-FTIR) spectroscopy to determine the chlorogenic acid isomer profile and antioxidant capacity of coffee beans. *Journal of Agricultural and Food Chemistry*, 64: 681–689.

Liang NJ, Xue W, Kennepohl P, et al., 2016b. Interactions between major chlorogenic acid isomers and chemical changes in coffee brew that affect antioxidant activities. *Food Chemistry*, 213: 251–259.

Liu HC, Shao JL, Li QW, et al., 2012. Determination of trigonelline, nicotinic acid, and caffeine in Yunnan Arabica coffee by microwave-assisted extraction and HPLC with two columns in series. *Journal of Aoac International*, 95: 1138–1141.

Machado DG, Lara MVS, Dobler PB, et al., 2020. Caffeine prevents neurodegeneration and behavorial alternations in a mice model of agitated depression. *Progress in Neuro-Psychopharmacology & Biological Psychiatry*, 98: 109776.

Moeenfard M, Cortez A, Machado V, et al., 2016. Anti-angiogenic properties of cafestol and kahweol palmitate diterpene esters. *Journal of Cellular Biochemistry*, 117: 2748–2756.

Moeenfard M, Rocha L, Alves A, 2014. Quantification of caffeoylquinic acids in coffee brews by HPLC-DAD. *Journal of Analytical Methods in Chemistry*, 2014: 965353.

Moon JK, Shibamoto T, 2009. Role of roasting conditions in the profile of volatile flavor chemicals formed from coffee beans. *Journal of Agricultural and Food Chemistry*, 57: 5823–5831.

Mullen W, Nemzer B, Ou B, et al., 2011. The antioxidant and chlorogenic acid profiles of whole coffee fruits are influenced by the extraction procedures. *Journal of Agricultural and Food Chemistry*, 59: 3754–3762.

Mussatto SI, Machado EMS, Martins S, et al., 2011. Production, composition, and application of coffee and its industrial residues. *Food and Bioprocess Technology*, 4: 661–672.

Ngamrojnavanich N, Sirimongkon S, Roengsumran S, et al., 2003. Inhibition of Na+,K+-ATPase activity by (-)-ent-kaur-16-en-19-oic acid and its derivatives. *Planta Medica*, 69: 555–556.

Nhiem NX, Hien NTT, Tai BH, et al., 2015. New ent-kauranes from the fruits of *Annona glabra* and their inhibitory nitric oxide production in LPS-stimulated RAW264.7 macrophages. *Bioorganic & Medicinal Chemistry Letters*, 25: 254–258.

Park JB, 2013. Isolation and quantification of major chlorogenic acids in three major instant coffee brands and their potential effects on H2O2-induced mitochondrial membrane depolarization and apoptosis in PC-12 cells. *Food & Function*, 4: 1632–1638.

Pereira GVD, Neto DPD, Magalhaes AI, et al., 2019. Exploring the impacts of postharvest processing on the aroma formation of coffee beans - A review. *Food Chemistry*, 272: 441–452.

Rigal C, Xu JC, Hu GL, et al., 2020. Coffee production during the transition period from monoculture to agroforestry systems in near optimal growing conditions, in Yunnan Province. *Agricultural Systems*, 177: 102696.

Shu Y, Liu JQ, Peng XR, et al., 2014. Characterization of diterpenoid glucosides in roasted puer coffee beans. *Journal of Agricultural and Food Chemistry*, 62: 2631–2637.

Tajik N, Tajik M, Mack I, et al., 2017. The potential effects of chlorogenic acid, the main phenolic components in coffee, on health: A comprehensive review of the literature. *European Journal of Nutrition*, 56: 2215–2244.

Tamaki N, Matsunami K, Otsuka H, et al., 2008. Rearranged ent-kauranes from the stems of *Tricalysia dubia* and their biological activities. *Journal of Natural Medicines*, 62: 314–320.

Trevitt J, Kawa K, Jalali A, et al., 2009. Differential effects of adenosine antagonists in two models of parkinsonian tremor. *Pharmacology, Biochemistry, and Behavior*, 94: 24–29.

Van Boekel MAJS, 2006. Formation of flavour compounds in the Maillard reaction. *Biotechnology Advances*, 24: 230–233.

Wang JP, Xu HX, Wu YX, et al., 2011. Ent-16 beta,17-dihydroxy-kauran-19-oic acid, a kaurane diterpene acid from *Siegesbeckia pubescens*, presents antiplatelet and anti-thrombotic effects in rats. *Phytomedicine*, 18: 873–878.

Wang LF, Shen XL, Wu YL, et al., 2016. Coffee and caffeine consumption and depression: A meta-analysis of observational studies. *Australian and New Zealand Journal of Psychiatry*, 50: 228–242.

Wang X, Meng QQ, Peng XR, et al., 2018. Identification of new diterpene esters from green Arabica coffee beans, and their platelet aggregation accelerating activities. *Food Chemistry*, 263: 251–257.

Wang XN, Wang SM, Yin JZ, et al., 2017. Analyzing volatile components in coffee melanoidins from different roasting degree - Yunnan Arabica coffee melanoidins. *Chinese Journal of Health Inspection*, 27: 1682–1685.

Wu YC, Hung YC, Chang FR, et al., 1996. Identification of ent-16 beta,17-dihydroxykauran-19-oic acid as an anti-HIV principle and isolation of the new diterpenoids annosquamosins A and B from *Annona squamosa*. *Journal of Natural Products*, 59: 635–637.

Yan R, Zhang J, Park HJ, et al., 2018. Synergistic neuroprotection by coffee components eicosanoyl-5-hydroxytryptamide and caffeine in models of Parkinson's disease and DLB. *Proceedings of the National Academy of Sciences of the United States of America*, 115: E12053–E12062.

Yang YL, Chang FR, Wu CC, et al., 2002. New ent-kaurane diterpenoids with anti-platelet aggregation activity from *Annona squamosa*. *Journal of Natural Products*, 65: 1462–1467.

Yu M, 2017. *Study on Flavor Quality of Coffee Beans in Dehong Area of Yunnan.* Heilongjiang Bayi Agricultural University. Daqing, China.

Yukawa GS, Mune M, Otani H, et al., 2004. Effects of coffee consumption on oxidative susceptibility of low-density lipoproteins and serum lipid levels in humans. *Biochemistry*, 69: 70–74.

Zhang HB, Guo TY, Bai XH, et al., 2012. Growing of *Coffea arabica* in high altitudes in China. *Chinese Journal of Tropical Agriculture*, 34: 21–28.

Zhao MZ, Guo TY, Bai XH, 2018. The collection and conservation status of coffee germplasm in the world. *Chinese Journal of Tropical Agriculture*, 38: 62–85.

Zheng XQ, Nagai C, Ashihara H, 2004. Pyridine nucleotide cycle and trigonelline (N-methylnicotinic acid) synthesis in developing leaves and fruits of *Coffea arabica*. *Physiologia Plantarum*, 122: 404–411.

Zhou B, Ren HT, 2014. Effect of different roast degree on aroma quality of coffee Arabica in Yunnan Province. *Food Research and Development*, 35: 65–73.

Zhou B, Ren HT, 2015. Effect of roasting time on volatile components in Yunnan Arabica coffee. *Modern Food Science & Technology*, 31: 236–244.

Zhou CX, Sun LR, Feng F, et al., 2013. Cytotoxic diterpenoids from the stem bark of *Annona squamosa* L. *Helvetica Chimica Acta*, 96: 656–662.

Zhou YF, 2010. Operation of industry chain to speed up Yunnan coffee industry development. *China Tropical Agriculture*, 5: 27–30.

Section 4

Medicinal Plants among Ethnic Groups

11

Medicinal Plants Used by Ethnic Groups in Yunnan

Ruifei Zhang
College of Life and Environmental Sciences, Minzu University of China, Beijing, People's Republic of China

Key Laboratory of Ethnomedicine (Minzu University of China), Ministry of Education of China, Beijing, People's Republic of China

Yuehu Wang
Kunming Institute of Botany, Chinese Academy of Sciences, Kunming, Yunnan, People's Republic of China

Jun Yang
Kunming Institute of Botany, Chinese Academy of Sciences, Kunming, Yunnan, People's Republic of China

Chun-Lin Long*
College of Life and Environmental Sciences, Minzu University of China, Beijing, People's Republic of China

Key Laboratory of Ethnomedicine (Minzu University of China), Ministry of Education of China, Beijing, People's Republic of China

Kunming Institute of Botany, Chinese Academy of Sciences, Kunming, Yunnan, People's Republic of China

CONTENTS

* Corresponding author.

11.1 Introduction

Ethnomedicine is the totality of health, knowledge, values, beliefs, skills, and practices of members of a society including all the clinical and non-clinical activities that relate to their health needs (Foster and Anderson, 1978). The practice of ethnomedicine, as an integral part of the culture of indigenous people, has interactions with the local economy, ecosystems, and religions (Anyinam, 1995). Folk knowledge is increasingly being fragmented and threatened by development pressures; however, traditional medicines still occupy a preferential place for a majority of the population in the developing countries and terminal patients in the developed areas, due to the low cost, easy access, and low side effects of ethnomedicines (Anyinam, 1995, Ramawat et al., 2009). Yunnan Province, having 18,000 higher plant species and 26 linguistic groups (Yang et al., 2004), is well-known for its rich biological and cultural diversity in China. Many ethnic groups, such as Dai (Huai and Pei, 2004), Lisu (Ji et al., 2004b), Miao (Huai and Pei, 2004), Yao (Long and Li, 2004), and Yi (Long et al., 2009), have developed their medicinal knowledge. The integration of modern science with traditional uses of ethnomedicines is the key to the use of indigenous knowledge; thus this chapter tries to bridge this gap by examining the botany, ethnobotany, phytochemistry, and pharmacology of medicinal plants used by ethnic groups in Yunnan. Eight ethnomedicines (Table 11.1), *Arundina graminifolia*, *Coptis teeta*, *Cynanchum otophyllum*, *Psammosilene tunicoides*, *Rodgersia sambucifolia*, *Winchia calophylla*, *Piper boehmeriaefolium*, and *Piper mullesua*, have been selected to present the research status of ethnomedicinal plants in Yunnan Province.

11.2 *Arundina graminifolia* (D. Don) Hochr.

Arundina graminifolia is a perennial flowering herbaceous plant in the family Orchidaceae (Chen and Stephan, 2009) (Figure 11.1A). It usually grows 40–100 cm tall. Stems are rigid, enclosed by leaf sheaths. Numerous leaves, 8–20 × 1–2 cm, leathery or papery, apex acuminate; sheaths 2–4 cm. Inflorescence 2–20 cm, racemose or one- or two-branched at base and paniculate, two–ten-flowered, flowers opening in succession; floral bracts broadly ovate-triangular, 3–5 mm, sheathing at base. Flowers white or pink, sometimes slightly tinged with purple; pedicel and ovary 1.5–3 cm. Sepals narrowly elliptic to narrowly elliptic-lanceolate, 25–40 × 7–9 mm. Petals ovate-elliptic, 25–40 × 13–15 mm; lip 25–40 × 12–24 mm, apical margin undulate; lateral lobes incurved, embracing column, rounded; mid-lobe sub-square, 8–16 × 10–16 mm, apex shallowly divided; disk with three (rarely five) lamellae. Column slightly arcuate, 20–25 mm. Capsule 28–35 × 8–15 mm. Flower and fruit June–November, sometimes January–April.

Arundina graminifolia is natively distributed in tropical and subtropical Asia, including south China, India, and other areas in southeast Asia (Chen and Stephan, 2009). It usually grows on slopes and grasslands, under the forest canopy, and by streams, with the altitude ranging from 400 to 2800 m, with good adaptability. However, due to the reduction of its natural habitats and an increasing demand for

TABLE 11.1

Traditional Uses of Ethnomedicines in Yunnan Province, China

Scientific Names	Local Names (Chinese)	Ethnic Groups	Parts Used	Traditional Uses	Classical Text (Chinese)	References
A. graminifolia	竹叶兰, 长杆兰, 山荸荠, 大叶寮刁竹, 文尚海, 百样解, 夕那格郎, 西剥罗, 西帮普兰	Blang, Dai, Han, Lisu, Wa	Rhizomes, the whole plant	Used as detoxification and to treat hepatitis, pneumonia, rheumatism, snake bites, tracheitis, and wounds and bruises.	《怒江药》, 《西双版纳傣药志》, 《中药大辞典》	Yan, 2017
Coptis teeta	云黄连, 云连, 滇连	Derung Han, Lisu, Nu	Rhizomes	Clearing heat and detoxification. Used to treat dysentery, furuncle, jaundice, and inflammation.	《滇南本草》, 《清宫医案》	Huang et al., 2005
Cynanchum otophyllum	青羊参, 闹狗药,小绿牛角藤, 牛尾参	Bai, Han, Yi	Rhizomes	Used to treat epilepsy, fibroids, low back pain, pulmonary tuberculosis, rheumatism, digest disorders, hydrophobia, and bites of venomous animals.	《滇南本草》, 《彝药志》	State Administration of TCM, 1999
P. tunicoides	金铁锁, 独鹿角姜, 独丁子, 对叶七, 金丝矮陀陀, 昆明沙参, 土人参, 蜈蚣七	Han, Miao, Yi	Roots	Used to treat wounds and bruises, rheumatism, furuncle, digest disorders, and bites of venomous animals.	《滇南本草》, 《植物名实图考》	Zhu and Yin, 2017

(*Continued*)

TABLE 11.1 (CONTINUED)
Traditional Uses of Ethnomedicines in Yunnan Province, China

Scientific Names	Local Names (Chinese)	Ethnic Groups	Parts Used	Traditional Uses	Classical Text (Chinese)	References
R. sambucifolia	岩陀, 红姜, 血三七, 毛头七, 红寒药, 牙勒街, 埃陀,破施, 赫贝, 绕才哼, 绕德补, 含优, 满优	Lisu, Bai, Naxi, Yi,	Rhizomes	Used to treat rheumatism, diarrhea, digest disorders, dysentery, irregular menstruation, metrorrhagia, and wounds.	《大理资志》, 《滇药录》, 《滇省志》, 《德宏药录》, 《德民志》, 《彝植药续》	Ma et al., 2013
W. calophylla	盆架树, 小灯台黑板树, 马灯盆, 山苦常, 白叶糖胶, 埋丁介	Dai	Barks, twigs and leaves, and resin	Used to treat asthma, bronchitis, cough, and wound.	《全国中草药汇编》, 《西双版纳傣药志》	You et al., 2020
Piper boehmeriaefolium	苎叶蒟，小麻叶，大麻疙瘩，芦子兰，歪叶子兰，芽帅样（Dai）	Dai	Stems, leaves	Used for traumatic injury, fracture, arthralgia, motor dysfunction, weakness and susceptible to diseases, limb numbness, dysmenorrhea, and amenorrhea.	《中国傣药志》	Ma et al., 2017
Piper mullesua	短蒟，芦子藤，质马此（Lisu）	Lisu	Whole plant	Used to treat rheumatic lumbocrural pain, arthritis, numbness of limbs, cold, and injury caused by falls.	《怒江中草药》	Nujiang Public Health Bureau, 1991

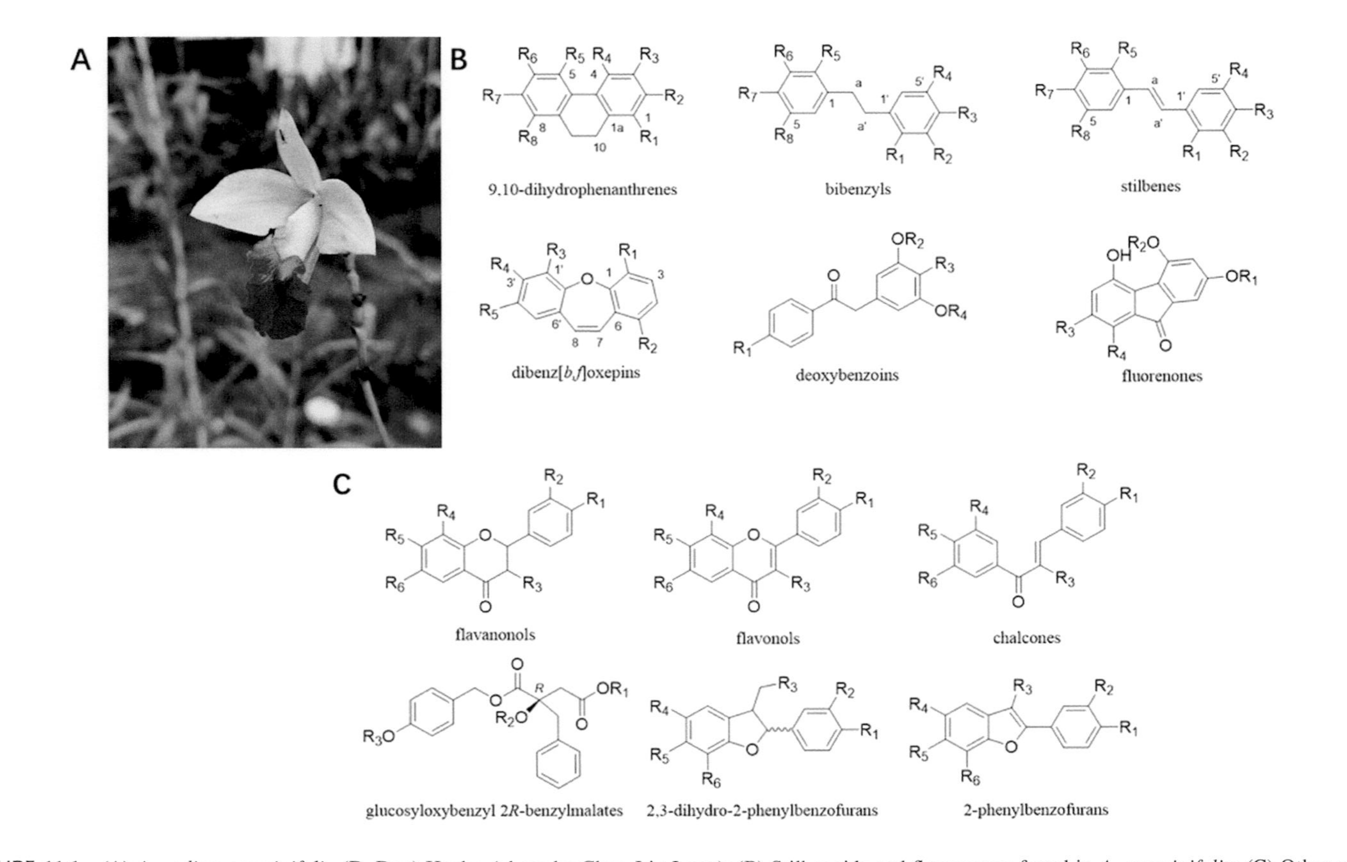

FIGURE 11.1 (A) *Arundina graminifolia* (D. Don) Hochr. (photo by Chun-Lin Long); (B) Stilbenoids and fluorenones found in *A. graminifolia*; (C) Other major phenolic compounds in *A. graminifolia*.

this herbal medicine, it has been recorded as a II-grade protected species in *State Key List of Protected Wild Plants* (Chinese Academy of Sciences, 1999).

The common English name for *A. graminifolia* is *bamboo-leaved orchid*. It also has many Chinese names, such as *Zhuye Lan* (bamboo leaf orchid), *Changgan Lan* (long stem orchid), *Caojiang* (ginger grass), and *Dayeliaodiaozhu* (big leaf bamboo) (Teoh, 2016). The rhizomes of *A. graminifolia*, named *Wenshanghai* or *Baiyangjie*, are one of the most well-known ethnomedicines used by the Dai ethnic group in Xishuangbanna, Yunnan (Peng et al., 2001, Lin Yanfang, 1995, Zhang et al., 2016). It is considered for detoxification, antiarthritis, and abirritation, and used as an ingredient in formulae such as *Dale Tang* (答勒汤), *Yajie Shaba* (雅解沙巴), and *Yajie* Tablets (雅解片) in Dai medicine. Blang people, calling it *Xinagelang* (夕那格郎), use the whole plant of *A. graminifolia* to treat lung abscess, pneumonia, and tracheitis (Yan, 2017). The Wa ethnic group in Yunnan name the whole plant of *A. graminifolia Xiboluo* (西剥罗) and use it to treat cough, phthisis, and tracheitis (Yan, 2017). The Lisu ethnic group call the whole plant of *A. graminifolia Xibangpulan* (西帮普兰) and use it to treat hepatitis, rheumatism, snake bites, and wounds and bruises (Chinese Academy of Sciences, 1999).

Liu et al. (2017a) and Teoh (2016) have summarized 115 phytochemicals isolated from *A. graminifolia* up to 2017, including stilbenoids, flavonols, fluorenone, glucosyloxybenzyl 2*R*-benzylmalate derivatives, phenylpropanoids, fatty acids, and other ketone and phenolic derivatives (Figure 11.1 B, C). The major constituents in *A. graminifolia* are stilbenoids, a group of plant phenolics containing the C6–C2–C6 (bibenzyl) unit in their structures (Ramawat and Mérillon, 2013). There are several well-known drugs or dietary supplements containing stilbenoids, such as resveratrol and tanshinone. However, most of the stilbenoids found in *A. graminifolia* have undergone few pharmacological studies. Stilbenoids and fluorenones isolated from *A. graminifolia* are listed in Table 11.2.

Numerous stilbenoids have been tested for their *in vitro* cytotoxicity against cancer cell lines. Gramideoxybenzoin D showed cytotoxicity against A549, SH-SY5Y, and PC3 cells with IC_{50} values of 2.2, 1.8, and 3.4 μM, respectively. Gramideoxybenzoin E showed cytotoxicity against NB4 and SH-SY5Y cells with IC_{50} values of 2.1 and 3.2 μM. Gramistilbenoid B was cytotoxic against NB4 and PC3 cells with IC_{50} values of 3.3 and 2.2 μM, respectively (Hu et al., 2013b). Graminibibenzyl A showed cytotoxicity against PC3 cells with an IC_{50} value of 3.6 μM, and graminibibenzyl B showed cytotoxicity against MCF7 cells with an IC_{50} value of 2.1 μM, respectively (Du et al., 2014). Gramniphenol H showed cytotoxicity against PC3 cells with an IC_{50} value of 3.5 μM. Gramniphenol I showed cytotoxicity against NB4 and PC3 cells with IC_{50} values of 3.6 and 3.8 μM, respectively (Li et al., 2013a). Gramistilbenoid L showed cytotoxicity against PC3 and SH-SY5Y cells with IC_{50} values of 5.5 and 3.6 μM, respectively (Yang et al., 2014). Gramniphenol J showed cytotoxicity against PC3 and SH-SY5Y cells with IC_{50} values of 8.2 and 7.6 μM (Gao et al., 2014). Gramniphenol K showed cytotoxicity against A549, SH-SY5Y, and MCF7 cells with IC_{50} values of 6.2, 7.5, and 4.7 μM, respectively (Meng et al., 2014). Gramniphenol H showed modest cytotoxicity against PC3 and SH-SY5Y cells with IC_{50} values of 6.2 and 4.3 μM, respectively (Niu et al., 2013). In summary, the stilbenoids in *A. graminifolia* likely have a broad spectrum of cytotoxicity. Moreover, 8 of the 11 active

TABLE 11.2

Stilbenoids and Fluorenones Isolated from *A. graminifolia*.

No.	Names	Class	Parts Used	References
1	Orchinol	Phenanthrenes	Rhizomes, aerial parts	Liu et al., 2005b, Auberon et al., 2016
2	4,7-Dihydroxy-2-methoxy-9,10-dihydrophenanthrene	Phenanthrenes	Rhizomes	Liu et al., 2005b
3	2,7-Dihydroxy-4-methoxy-9,10-dihydrophenanthrene	Phenanthrenes	Rhizomes	Liu et al., 2005b
4	7-Hydroxy-2-methoxyphenanthrene-1,4-dione	Phenanthrenes	Rhizomes	Liu et al., 2005b
5	7-Hydroxy-2-methoxy-9,10-dihydrophenanthrene-1,4-dione	Phenanthrenes	Rhizomes	Liu et al., 2005b
6	Arundinaol	Phenanthrenes	Rhizomes	Liu et al., 2005a
7	Blestriarene A	Phenanthrenes	Rhizomes	Yan et al., 2018
8	Shancidin	Phenanthrenes	Rhizomes	Yan et al., 2018
9	Densiflorol B	Phenanthrenes	Rhizomes, aerial parts	Yan et al., 2018, Auberon et al., 2016
10	Ephemeranthoquinone	Phenanthrenes	Rhizomes, aerial parts	Yan et al., 2018, Auberon et al., 2016
11	Coelonin	Phenanthrenes	Rhizomes, aerial parts	Yan et al., 2018, Auberon et al., 2016
12	Lusianthridin	Phenanthrenes	Rhizomes, aerial parts	Yan et al., 2018, Auberon et al., 2016
13	2,7-Dihydroxy-1-(*p*-hydroxylbenzyl)-4-methoxy-9,10-dihydrophenanthrene	Phenanthrenes	Rhizomes	Liu et al., 2012
14	4,7-Dihydroxy-1-(*p*-hydroxylbenzyl)-2-methoxy-9,10-dihydrophenanthrene	Phenanthrenes	Rhizomes	Liu et al., 2012
15	Arundigramin	Phenanthrenes	Aerial parts	Auberon et al., 2016
16	Arundiquinone	Phenanthrenes	Aerial parts	Auberon et al., 2016
17	7-Hydroxy-2,8-dimethoxy-phenanthrene-1,4-dione	Phenanthrenes	Whole plants	Liu et al., 2017a
18	7-Hydroxy-2,10-dimethoxy-phenanthrene-1,4-dione	Phenanthrenes	Whole plants	Liu et al., 2017a

(*Continued*)

TABLE 11.2 (CONTINUED)

Stilbenoids and Fluorenones Isolated from *A. graminifolia*.

No.	Names	Class	Parts Used	References
19	Flavanthrin	Phenanthrenes	Aerial parts	Auberon et al., 2016
20	Cucapitoside	Phenanthrenes	Whole plants	Gao et al., 2013a
21	Curcapital	Phenanthrenes	Whole plants	Gao et al., 2013a
22	Batatasin III	Bibenzyls	Rhizomes, aerial parts	Peng *et al.*, 2008, Auberon et al., 2016
23	3,3'-Dihydroxy-5-methoxybibenzyl	Bibenzyls	Rhizomes	Liu et al., 2012
24	Graminibibenzyl A	Bibenzyls	Whole plants	Du et al., 2014
25	Graminibibenzyl B	Bibenzyls	Whole plants	Du et al., 2014
26	5,12-Dihydroxy-3methoxybibenzyl-6-carboxylic acid	Bibenzyls	Whole plants	Du et al., 2014
27	5-Acetyloxy-12-hydroxy-3-methoxybibenzyl-6-carboxylic acid	Bibenzyls	Whole plants	Du et al., 2014
28	3-Hydroxy-5-methoxybibenzyl	Bibenzyls	Whole plants	Du et al., 2014
29	3,3'-Dihydroxy-5-methoxybibenzyl	Bibenzyls	Whole plants	Du et al., 2014
30	2,5,2',5'-Tetrahydroxy-3-methoxybibenzyl	Bibenzyls	Whole plants	Du et al., 2014
31	Gramniphenol H	Dibenz[*b*,*f*]oxepins	Whole plants	Li et al., 2013a
32	Gramniphenol I	Dibenz[*b*,*f*]oxepins	Whole plants	Li et al., 2013a
33	Gramniphenol J	Dibenz[*b*,*f*]oxepins	Whole plants	Gao et al., 2014
34	Gramniphenol K	Dibenz[*b*,*f*]oxepins	Whole plants	Meng et al., 2014
35	Bauhiniastatin D	Dibenz[*b*,*f*]oxepins	Whole plants	Li et al., 2013a
36	Pinosylvin	Stilbenes	Whole plants	Li et al., 2013a
37	3,5-Dihydroxy-stilbene-3-*O*-β-D-glucoside	Stilbenes	Whole plants	Li et al., 2013a
38	Rhapontigen	Stilbenes	Whole plants	Li et al., 2013a
39	3-Hydroxy-4,3,5-trimethoxy-trans-stilbene	Stilbenes	Whole plants	Li et al., 2013a

(Continued)

TABLE 11.2 (CONTINUED)

Stilbenoids and Fluorenones Isolated from *A. graminifolia.*

No.	Names	Class	Parts Used	References
40	2,3-Dihydroxy-3,5-dimethoxystilbene	Stilbenes	Whole plants	Li et al., 2013a
41	Gramistilbenoid L	Stilbenes	Whole plants	Yang et al., 2014
42	Gramistilbenoid A	Stilbenes	Whole plants	Hu et al., 2013b
43	Gramistilbenoid B	Stilbenes	Whole plants	Hu et al., 2013b
44	Gramistilbenoid C	Stilbenes	Whole plants	Hu et al., 2013b
45	Gramideoxybenzoin A	Doxybenzoins	Whole plants	Hu et al., 2013b
46	Gramideoxybenzoin B	Doxybenzoins	Whole plants	Hu et al., 2013b
47	Gramideoxybenzoin C	Doxybenzoins	Whole plants	Hu et al., 2013b
48	Gramideoxybenzoin D	Doxybenzoins	Whole plants	Hu et al., 2013b
49	Gramideoxybenzoin E	Doxybenzoins	Whole plants	Hu et al., 2013b
50	Gramideoxybenzoin F	Doxybenzoins	Whole plants	Hu et al., 2013b
51	Gramideoxybenzoin G	Doxybenzoins	Whole plants	Hu et al., 2013b
52	Gramideoxybenzoin H	Doxybenzoins	Whole plants	Hu et al., 2013b
53	Gramniphenol D	Fluorenones	Whole plants	Hu et al., 2013a
54	Gramniphenol E	Fluorenones	Whole plants	Hu et al., 2013a
55	Gramniphenol H	Fluorenones	Whole plants	Niu et al., 2013
56	Denchrysan A	Fluorenones	Whole plants	Niu et al., 2013
57	Dendroflorin	Fluorenones	Whole plants	Niu et al., 2013
58	Dengibsin	Fluorenones	Whole plants	Niu et al., 2013
59	1,4,5-Trihydroxy-7-methoxy-9*H*-fluoren-9-one	Fluorenones	Whole plants	Niu et al., 2013

compounds showed anti-proliferative activity against the PC3 cell line, indicating those stilbenoids may have a selective effect on prostate cancer.

In addition, 3,3'-dihydroxy-5-methoxy-bibenzyl and 7-hydroxy-2,8-dimethoxyphenanthrene1,4-dione exhibited anti-hepatic fibrosis activity against HSCT6 cells, with IC_{50} values of 61.9 μg/mL and 52.7 μg/mL, respectively (Liu et al., 2017b). Gamniphenols C, F, and G showed anti-tobacco mosaic virus (anti-TMV) activity, with IC_{50} values of 20.8, 40.8, and 57.7 μM, respectively. Gamniphenols B, D, and E displayed anti-HIV-1 activity with therapeutic index values above 100:1 (Hu et al., 2013a). Blestriarene A and shancidin had anti-bacterial activity against *Staphylococcus aureus*, *Bacillus subtilis*, and *Escherichia coli*, with MICs of 20–40 μg/mL (Yan et al., 2018). Bactericidal mechanisms may be related to the rupture of cell wall and membrane and leakage of nuclear mass. Moreover, blestriarene A and shancidin showed anti-hemolytic activity with an IC_{50} value of 16 μg/mL.

The phenylpropanoid derivatives also represent a major group of compounds in *A. gramnifolia*. Four 2-phenylbenzofurans, namely Gramniphenols F, G, and I and moracin M, have been isolated from the whole plant of *A. gramnifolia* (Hu et al., 2013a, Li et al., 2015b). Gramniphenol I exhibited notable anti-TMV activity with an inhibition rate of 16.8% (Li et al., 2015b). In 2015, four phenylpropanoids, arundina phenylpropanoid, coumaric acid, ω-hydroxypropioguaiacone, and 3-methoxy-4-hydroxy-Pr alc., were isolated from the whole plant of *A. graminifolia* (Dong et al., 2015). Among them, arundina phenylpropanoid exhibits potential anti-TMV activities with an inhibition rate of 22.6%. In another two studies (Gao et al., 2012, 2013a), three phenolic compounds, gramphenol A and gramniphenols A and B, were evaluated for their anti-TMV activity. Li et al. evaluated the anti-HIV-1 activity of five flavonoids isolated from *A. graminifolia*: gramflavonoid A showed moderate anti-HIV-1 activity and others also showed weak anti-HIV-1 activity (Li et al., 2013b). 3*S*,4*S*-3',4'-Dihydroxyl-7,8,-methylenedioxylpterocarpan, a flavonoid, showed high cytotoxicity against SH-SY5Y cell with IC_{50} values of 2.2 μM and moderate cytotoxicity with IC_{50} values of 5–10 μM for other four tested cell lines (A549, PC3, MCF7, and NB4 cells) (Shu et al., 2013).

Over the course of two years, Auberon et al. (Auberon et al., 2018, 2019) and Liu et al. (2019) independently isolated 17 glucosyloxybenzyl 2*R*-benzylmalate derivatives arundinosides A–Q from the whole plant of *A. graminifolia*. Both of them obtained arundinosides D–G, J, and K, although they used different chromatographic techniques. Auberon et al. (2018, 2019) used centrifugal partition chromatography, Sephadex LH-20 gel, and SPE cartridge; Liu et al. (2019) applied macroporous resin, C-18 reversed phase silica gel, and silica gel; and both groups finally purified these compounds using reverse-phase high performance liquid chromatography (HPLC). Also, Auberon obtained arundinosides D–F from both the aerial parts (Auberon et al., 2018) and underground parts (Auberon et al., 2019) of this plant. Neuroprotective, anti-hepatic fibrotic, α-glucosidase inhibitory, and anti-oxidant activity of these compounds were evaluated *in vitro*. Arundinosides A–G did not show neuroprotective ability to reduce the beta amyloid damage on PC12 cells at the maximum tested concentration of 100 μM (Auberon et al., 2018). Arundinosides D, F, H, I, and K showed inhibition against lipopolysaccharide (LPS)-induced damage in rat hepatic stellate HSC-T6 cells at a concentration of 100 μg/mL (Liu et al., 2019). Arundinosides D and O–Q exhibited α-glucosidase inhibitory activity (IC_{50} = 159.74, 22.06, 18.24, and

90.22 μg/mL, respectively) and $ABTS^+$ radical scavenging activity (IC_{50} = 4.98, 4.40, 2.96, and 6.75 μL/mL, respectively) (Auberon et al., 2019).

Xu et al. (2018) detected 54 constituents in the *A. graminifolia* essential oil, and 3 major components of the essential oil are hexadecanoic acid ethyl ester (25.33%), (Z,Z)-9,12-octadecadienoic acid methyl ester (23.09%), and (Z,Z,Z)-9,12,15-octadecatrienoic acid ethyl ester (17.67%).

A comparison of the anti-oxidant activity of leaf, root, stem, and whole plant extracts of *A. graminifolia* showed that rutin > root > leaf > whole plant > stem. Root extract is the strongest among these extracts but lower than rutin (Wang et al., 2009). In the rat liver cells and human red blood cells, the ethyl acetate extract of *A. graminifolia* could inhibit lecithin lipid peroxidation induced by carbon tetrachloride, with IC_{50} and maximum values of 0.098 mg/mL and 77.3%, respectively (Gao et al., 2013b). In the same study, scavenging effects of *A. graminifolia* on active oxygen species ·OH and O^{2-} generated by Fenton reaction and riboflavin photosensitization were firstly investigated by means of spectrophotometry. The damage to the DNA chain induced by the hydroxyl radical was observed by thiobarbituric acid spectrophotometry.

Residents of tropical Asia are familiar with *A. graminifolia* because it is a common, lowland to low montane, terrestrial orchid throughout the region; however, it is surprising that it has medicinal usage only in Yunnan and Hong Kong (Teoh Eng Soon, 2016). Ethnic groups in Yunnan have diverse medicinal uses for this plant, and the phytochemical and *in vitro* biological screens provide preliminary data for the anti-bacterial, anti-oxidant, anti-proliferative, and hepatoprotective activity. However, more *in vivo* and further pharmacological studies on the crude extracts and unique phytochemicals of *A. graminifolia* are needed.

11.3 *Coptis teeta* Wall.

Coptis teeta is a perennial flowering herb in the Ranunculaceae family (Wang et al., 2001) (Figure 11.2A). Petiole 8–19 cm, glabrous; leaf blade ovate-triangular, 6–12 × 5–9 cm, three-sect, papery, abaxially glabrous, adaxially sparsely puberulous on veins, base cordate; lateral segments subsessile to petiolulate, shorter than central one, obliquely ovate, unequally parted; central segment petiolulate, ovate-rhombic, pinnately divided; segments three–six pairs, remote, margin acute serrate, apex attenuate. Scapes 15–25 cm tall, glabrous. Inflorescences three–five-flowered; bracts elliptic, three-parted or pinnately divided. Sepals five, greenish yellow, elliptic, 7.5–8 × 2.5–3 mm, glabrous. Petals spatulate, 5.4–5.9 mm, glabrous, apex rounded to obtuse. Stamens 3–3.3 mm. Pistils 11–14. Follicles 7–9 mm, glabrous.

Coptis teeta is a narrow endemic restricted to the evergreen broad-leaved forests in the eastern Himalaya (Pandit and Babu, 2003), with an altitude ranging from 1500 to 2300 m. In Yunnan, it is mainly distributed throughout areas in the Gaoligong Mountains and west Biluoxue Mountains (Wang et al., 2001). In addition, the ethnic groups in this area have cultivated *C. teeta* for over 130 years (Huang et al., 2005). However, due to the long-term overuse and destruction of the natural habitats, *C. teeta* has been recorded as a II-grade vulnerable species in the *State Key List of Protected Wild Plants* in China (Chinese Academy of Sciences, 1999). Pandit and Babu (2003) have conducted a series of work on two subspecies *C. teeta* subs. *teeta* and *lohitensis*,

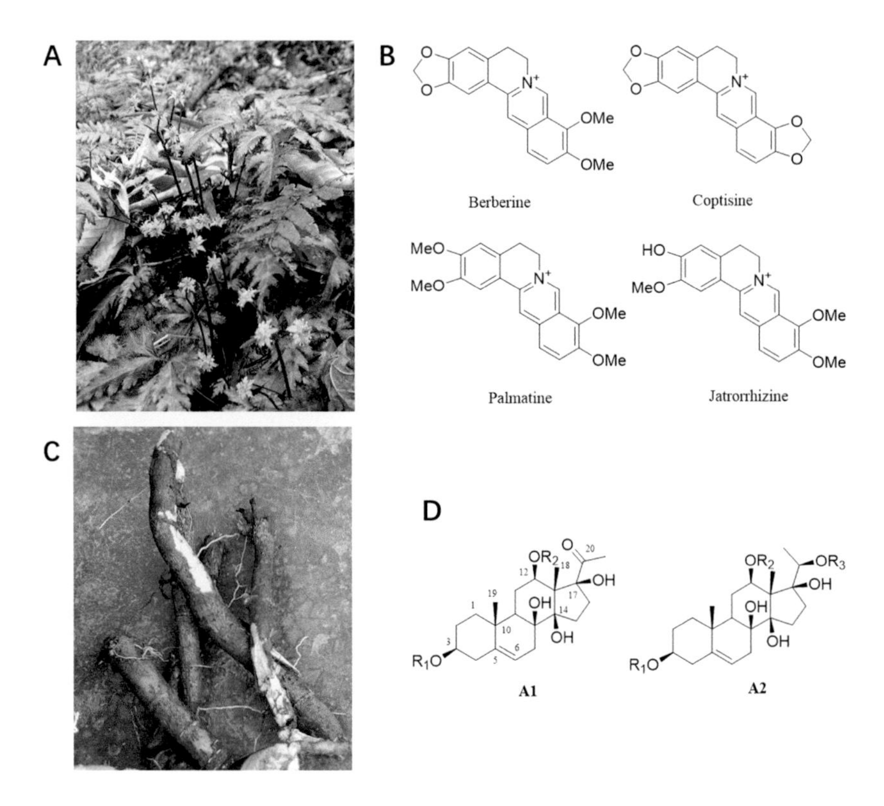

FIGURE 11.2 (A) *Coptis teeta* Wall. (photo by Chun-Lin Long); (B) Important active alkaloids in *C. teeta*; (C) Roots of *Cynanchum otophyllum* (photo by Zhuo Cheng); (D) Skeletons of the major C_{21} steroids in *C. otophyllum*.

including the taxonomy, cytogenetics, ecology, and reproductive biology. The reproductive biology of these two subspecies showed that ssp. *teeta* reproduces sexually and produces fewer ramets, compared with the ssp. *lohitensis*. And the seed setting is generally low, with a higher output in ssp. *teeta* and lower in ssp. *lohitensis*. The sexual reproduction does not contribute significantly towards population increase in *C. teeta,* which may be a cause of its vulnerable situation. Nevertheless, conservation strategies for this species need further studies.

The rhizomes of *C. teeta*, named *Yunlian*, are one of the botanical sources of a well-known traditional Chinese medicine *Huanglian* (黄连) (Tang et al., 2009). *C. teeta*, narrowly distributed in the eastern Himalaya, has been used for a long time as an ethnomedicine by local people, including the Derung, Lisu, and Nu ethnic groups in Yunnan (Huang et al., 2005). It is traditionally used to treat intestinal infections, such as acute gastroenteritis, bacillary dysentery, cholera, and malaria fever. Besides using *C. teeta* as a medicine, the Derung, Lisu, and Nu people in northwest Yunnan also trade *Yunlian* for increasing economic income (Long et al., 1999).

Phytochemical and pharmacological studies on *C. teeta* revealed that isoquinoline alkaloids, including berberine, coptisine, jatrorrhizine, and palmatine, are the major active compounds possessing antibacterial, anti-inflammatory, antineoplastic, and antiviral activities (Tang et al., 2009) (Figure 11.2B). In 2011, Shi et al. (2011) analyzed berberine hydrochloride, palmatine hydrochloride, and jatrorrhizine hydrochloride in *C. teeta* fibrous roots, leaves, and rhizomes, using a reverse phase HPLC method. Results showed that the total alkaloids in *C. teeta* rhizomes (10.74–12.80%) are the highest, compared with the fibrous roots (1.93–2.73%) and leaves (1.35–2.02%). In a further study, Li et al. (2011) analyzed the *C. teeta* total alkaloids during different harvest periods, using a reverse phase HPLC method. Finally, the optimal harvest time was determined to be October and November.

In 2019, the contents of 20 free amino acids in *C. chinensis* Franch., *C. deltoidei* C. Y. Cheng et Hsiao., and *C. teeta* were quantified, using an automatic amino acid analyzer (Li et al., 2019b). This study suggests that free amino acids may be used to identify these three *Coptis* species. Asparagine is the identifier of *C. chinensis*, *C. deltoidea*, and *C. teeta*; glutamine and arginine can distinguish *C. teeta* from *C. chinensis* and *C. deltoidea*. In another study, Qi et al. (2018) conducted a comprehensive quality assessment for these three *Coptis* species, using HPLC, Fourier transform near-infrared, and Fourier transform mid-infrared spectroscopy combined with multivariate statistical analysis. Results showed that berberine, epiberberine, groenlandicine, and magnoflorine were all identified as unique marker components.

Meng et al. (2013) found that the alkaloids in three botanical sources of *Huanglian* are similar, while the non-alkaloids are different. In the same study, 12 non-alkaloids were isolated from the rhizomes of *C. teeta* and identified as 3,5,7-trihydroxy-6,8-dimethylflavone, ferulic acid, *Z*-octadecyl caffeate, protocatechuic acid, methyl-3,4-dihydroxyphenyl lactate, 3,4-dihydroxyphenethyl alcohol, 3,5-dihydroxyphenethyl alcohol-3-*O*-β-D-glucopyranoside, (+)-lariciresinol, woorenosides I and II, longifolroside A, and (+)-syringaresinol-4-*O*-β-D-glucopyranoside.

Coptis teeta is an important botanical resource of bioactive alkaloids such as berberine and palmatine. Several modern technologies have been used in metabolomic studies of *C. teeta*; however, few studies of the pharmacological and toxicological effects of *Yunlian* have been conducted. An integrated study on phytochemistry and pharmacology may help to reveal the pharmacological mechanism related to the ethnomedicinal uses. Moreover, *in vitro* and *in vivo* studies on *C. teeta* are necessary in future.

11.4 *Cynanchum otophyllum* C.K. Schneid.

Cynanchum otophyllum is an endemic flowering plant in China, belonging to the Asclepiadaceae family (Li et al., 1995) (Figure 11.2C). Stems twining, to 2 m, pubescent along one line. Leaves opposite; petiole 1.5–5 cm, adaxially puberulent; leaf blade ovate-lanceolate, 4–11 × 2.5–8 cm, abaxially distinctly paler, membranous, minutely pilose, sometimes glabrescent, base strongly auriculate, basal lobes rounded, apex gradually acuminate. Inflorescences umbel-like or raceme-like; peduncle 2–4 cm, puberulent to subglabrous; cymules up to eight-flowered. Pedicel 3–5 mm, puberulent on one side. Sepals ovate-lanceolate, *ca.* 1 × 0.7 mm, puberulent; basal glands five. Corolla white, rotate; lobes oblong, 2–3 × *ca.* 1 mm, minutely puberulent inside. Corona slightly shorter than corolla, deeply five-lobed; lobes oblong-lanceolate, narrower toward base, apex rounded to subacute, sometimes slightly three-toothed, slightly fleshy, adaxially grooved with lateral longitudinal ridges, adaxial appendages minute or absent. Gynostegium short stipitate. Anther appendages ovate, erect; pollinia oblong. Stigma head slightly convex. Follicles lanceolate, 8–9 × *ca.* 1 cm, with two ridges or angles. Seeds ovate, *ca.* 6 × 3 mm; coma *ca.* 3 cm. Flowering in June to October, fruiting in August to December. It usually grows in thickets or open woods with an altitude ranging from 1000 to 3000 m, distributing in Guangxi, Guizhou, Hubei, Hunan, Sichuan, Xizang, and Yunnan provinces in China. The roots of *C. otophyllum*, documented in classical Chinese medical text *Diannan Bencao*, are used in folk medicine in Yunnan (State Administration of TCM, 1999). The Bai ethnic people use it to treat epilepsy, low back pain, rheumatism, digest disorders, hydrophobia, and dog bites. In Yi medicine, it is called *Roujiboqi* and used in the same manner as the Bai ethnic group. Due to the potent antiepileptic effect, it was clinically used in a Chinese prescription, *Qingyanghsen Tablet* (Yan Mingxing, 1983). In addition, C_{21} steroids are the major phytochemicals and active constituents in *Qingyanghsen Tablet* (Ming et al., 2006).

Numerous phytochemical studies have been conducted on the roots of *C. otophyllum*, leading to the isolation of a series of novel C_{21} steroidal glycosides (Zhan et al., 2019, Mu et al., 1986). Han et al. have summarized 74 C_{21} steroids found in *C. otophyllum* up to 2017 (Han et al., 2018). He et al. summarized the structural features of C_{21} steroids in the genus *Cynanchum*, classifying them into 17 groups with 6 skeletons (He et al., 2015); most C_{21} steroids in *C. otophyllum* possess the A1 or A2 skeleton (Figure 11.2D).

Over the past two years, several C_{21} steroid derivatives have been isolated from the roots of *C. otophyllum* in a series of studies. Dong et al. obtained 15 undescribed C_{21} steroids, namely cynanchins A–G and cynanotins A–H, together with 28 known analogues (Dong et al., 2018, 2019). Among them, 18 bear the A1 skeleton and 8 possess the A2 skeleton. Cynanotin F and 5α,6α-epoxycaudatin have an epoxide at $\Delta^{5,6}$ instead of a double bond (Dong et al., 2018). Zhan et al. (2019) isolated seven undescribed C_{21} steroids, cynotophyllosides P–V, along with three known analogs. Six of them have the A1 skeleton, and four bear the other skeleton. Although there are nearly 400 C_{21} steroids found in the genus *Cynanchum* (Han et al., 2018), showing a structural diversity, most analogs isolated from *C. otophyllum* belong to the A1 and A2 groups. These unique compounds may correlate with the anti-epileptic effect and the structure–activity relationship deserves further study.

Pharmacological studies reveal a broad spectrum of bioactivities of *C. otophyllum* and its C_{21} steroids, including neuroprotective (Dong et al., 2020), cytotoxic (Dong et al., 2018), hepatoprotective (Dong et al., 2019), immunomodulatory (Zhan et al., 2019), antivirus (Tao et al., 2011), and antimicrobial activity (Zhao et al., 2007). The most potent bioactivity is neuroprotective activity (Han et al., 2018).

Pretreatment with the total glucosides (*i.p.*, 50 mg/kg) of *C. otophyllum* roots did not affect glutamate decarboxylase and γ-aminobutyric acid (GABA) transaminase activities in the brain of normal mice; however, it reversed the decrease of glutamate decarboxylase in mice treated with thiosemicarbazide (Li et al., 1987). Otophylloside B (Ot B), a C_{21} steroidal glycoside bearing the A1 skeleton and protecting rats from audiogenic seizures with ED_{50} = 10.20 mg/kg, was isolated from the roots of *C. otophyllum* in 1986 (Mu et al., 1986). Yang et al. (2015) found that Ot B could extend the lifespan of *Caenorhabditis elegans*, delay the age-related decline of body movement, and improve the stress resistance. At the most active concentration of 50 μM, Ot B displayed the largest lifespan extension by up to 11.3%, by activating the FOXO transcription factor DAF-16. However, Ot B could not further extend the lifespan of long-lived mutant of insulin/IGF-1-like receptor (daf-2). In addition, Ot B also requires SIR-2.1 and CLK-1 which is an enzyme in ubiquinone synthesis, for lifespan extension. In another study on the Alzheimer's disease prevention of Ot B in *C. elegans* models, Yang et al. (2017) found that Ot B decreased the expression of *Aβ* at the mRNA level at a concentration of 50 μM. In the same study, further genetic analyses showed that Ot B increased the activity of heat shock transcription factor by up-regulating the expression of *hsf-1* and its target genes, *hsp-12.6*, *hsp-16.2*, and *hsp-70*. And Ot B also increased the expression of *sod-3* by partially activating DAF-16. Otophylloside N (Ot N), another neuroprotective steroidal glycoside bearing the A1 skeleton, was isolated form the roots of *C. otophyllum* in 2011 (Ma et al., 2011). Ot N exerted a neuroprotective effect against the pentylenetetrazol (PTZ)-induced neural damage in primary cortical neurons, mice, and zebrafish. Ot N treatment attenuated PTZ-induced morphology changes, cell death, LDH efflux in embryonic neuronal cells of C57BL/6J mice (PTZ = 30 mM, Ot N, 5 and 10 μg/mL), and convulsive behavior in zebrafish (PTZ = 10 mM, Ot N, 12, 25, and 50 μg/mL). In addition, Ot N reduced PTZ-induced cleavage of poly ADP-ribose polymerase and up-regulation of the Bax/Bcl-2 ratio and decreased the expression level of c-Fos in mouse cerebral cortex (PTZ = 30 mg/kg, Ot N, 2.5 and 5 μg/kg). *In vivo* antiepileptic studies showed that caudatin-3-*O*-*β*-D-cymaropyranosyl-(1→4)-*β*-D-cymaropyranosyl-(1→4) *β*-D-cymaropyranoside (10 μg/mL), and caudatin-3-*O*-*β*-D-oleandropyranosyl-(1→4)-*β*-D-oleandropyranosyl-(1→4)-*β*-D-cymaropyranosyl-(1→4)-*β*-D-cymaropyranoside (10 μg/mL), otophylloside F (300 μM), and rostratamine 3-*O*-*β*-D-oleandropyranosyl-(1→4)-*β*-D-cymaropyranosyl-(1→4)-*β*-D-cymaropyranoside (100 and 200 μM) all suppressed PTZ-induced seizure behaviors in larval zebrafish (Li et al., 2015a, 2015c). Preliminary structure–activity relation studies revealed that a pregnene skeleton with a C-12 ester group (ikemaoyl > cinnamoyl > hydroxy > *p*-hydroxybenzoyl) and a C-3 sugar chain consisting of three 2,6-dideoxysaccharide units is essential for this anti-epileptic activity in larval zebrafish (Li et al., 2015a). Cynanotosides A, B, and cynotophylloside H exhibited a dose-dependently (1–30 μM) protective activity against lutamate-, hydrogen peroxide-, and homocysteic acid-induced damage in hippocampal neuronal cell line HT22 (Zhao et al., 2013).

Thirteen of 23 C_{21} steroids found in the roots of *C. otophyllum* exerted cytotoxicity against 5 human tumor cell lines (HL-60, SMMC-7721, A-549, MCF-7, and SW480) at a concentration ranging from 11.4 to 37.9 μM (Dong et al., 2018). Ten of 20 C_{21} steroids obtained by Dong et al. (2019) at a concentration of 10 μM showed *in vitro* antiproliferation activity in hepatic stellate cells induced by transforming growth factor-β1 (TGF-β1). Zhan et al. (2019) studied the *in vitro* immunomodulatory activities against Con A- and LPS-induced proliferation in mice splenocytes. Cynotophylloside P and arcostin 3-*O*-*β*-D-cymaropyranosyl-(1→4)-*α*-L-digitoxopyranoside showed immunoenhancing activity (0.01–10 μg/mL), cynotophylloside Q showed immunosuppressive activity (0.01–10 μg/mL), and cynotophylloside R enhanced the proliferation of splenocytes at low concentrations (0.01 and 0.1 μg/mL) and suppressed immune cells at a concentration of 10 μg/mL.

Preliminary *in vitro* and *in vivo* studies on the crude extracts and on the C_{21} steroids from *C. otophyllum* provide pharmacological evidence for the anti-epileptic effect, offering scientific support to the ethnomedicinal use of *Qingyangshen*. In addition, *C. otophyllum* steroids possess a broad spectrum of cytotoxicity against tumor cells, indicating that these compounds may be explored as anti-tumor agents. Nevertheless, few toxicological studies have been conducted on *C. otophyllum*. More *in vivo* studies on the pharmacological mechanism, toxicology, and structure–activity relationship of unique C_{21} steroids from *C. otophyllum* are needed in future.

11.5 *Psammosilene tunicoides* W. C. Wu et C. Y. Wu

Psammosilene tunicoides is a perennial flowering plant in the family Caryophyllaceae (Lu et al., 2001) (Figure 11.3A). The genus *Psammosilene* is endemic to China and has only one species. In addition, this species is in danger of extinction at present and has been recorded as a II-grade vulnerable species in *State Key List of Protected Wild Plants* (China) (Chinese Academy of Sciences, 1999). Roots brown-yellow (Lu et al., 2001). Stems prostrate, purple-green, 20–35 cm, dichotomously branched, pubescent. Leaves subsessile, 1.5–2.5 × 1–1.5 cm, adaxially pilose, base rounded, rarely broadly

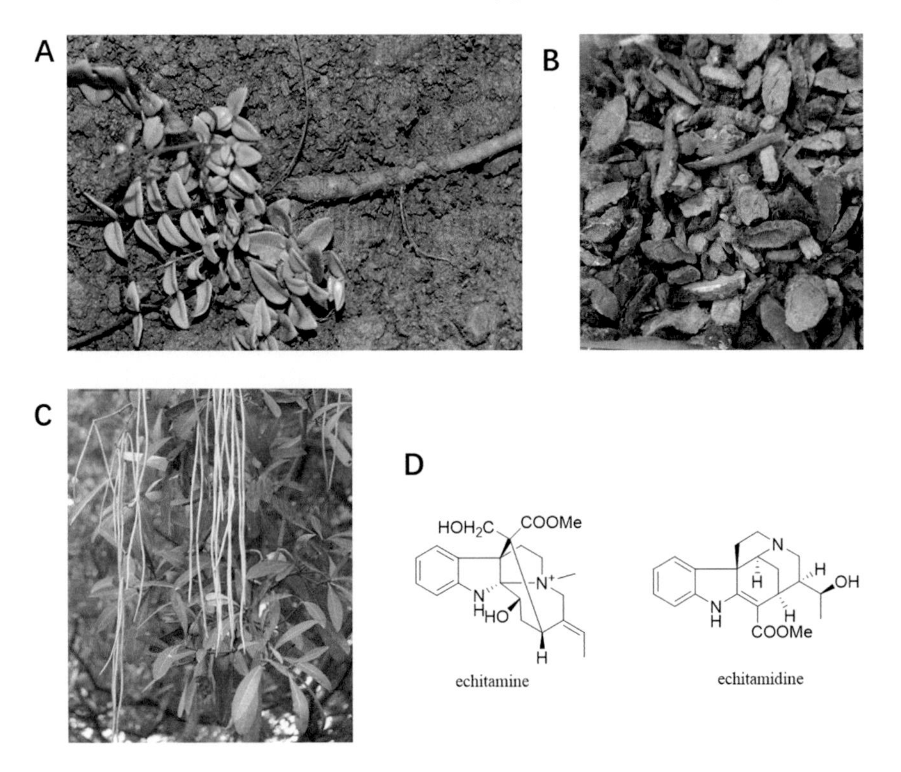

FIGURE 11.3 (A) *Psammosilene tunicoides* W.C. Wu et C.Y. Wu (photo by Chun-Lin Long); (B) *Yantuo* (*Rodgersia sambucifolia*) (photo by Chun-Lin Long); (C) *Winchia calophylla* A. DC. (photo by Chun-Lin Long); (D) Chemical structures of echitamine and echitamidine.

cuneate, apex entire, acute. Inflorescence a thyrse with a terminal, lax dichasium and two opposite, similarly lax and many-flowered dichasia proximally. Flowers 3–5 mm in diameter. Pedicel short or flowers sessile. Calyx tubular-campanulate, 4–6 mm, densely glandular pubescent, veins green; calyx teeth triangular-ovate, margin membranous, apex obtuse or acute. Petals 7–8 mm. Stamens exerted beyond calyx, 7–9 mm; anthers yellow. Ovary narrowly obovoid, *ca.* 7 mm; styles *ca.* 3 mm. Capsule *ca.* 7 mm. Seed brown, *ca.* 3 mm. Flowering in June to September, fruiting in July to October.

It is limited in distribution to warm and dry valleys along Jinsha Jiang and Yarlung Zangbo Jiang in Yunnan, SE Xizang, SW Sichuan, and West Guizhou (Lu et al., 2001). The habitats usually are rocky mountain slopes, dry pastures, calcareous rock crevices, and forests, with an altitude ranging from 900 to 3800 m.

The roots of *P. tunicoides* are documented in the traditional Chinese medicine text *Diannan Bencao* and widely used by the Miao, Yi, and Han ethnic groups in southwest China (Zhu and Yin, 2017). It is well known as *Jenb tieeg sox* in the Miao ethnic group and for treating rheumatism, bleeding, and wounds and bruises (Wang et al., 2006). As an external treatment, it treats furuncle and bites of venomous animals. Due to its potent anti-inflammatory and pain-relieving activity, it is used as an ingredient of several Chinese prescriptions, such as *Yunnan Baiyao*, *Jingulian* capsules, *Tongxuekang* capsules, and *Zhongzhuanggu* capsules (Li et al., 2016a).

Phytochemical studies reveal the major chemical constituents in *P. tunicoides* are triterpenoid saponins (Wen et al., 2020), cyclic peptides (Hou et al., 2020), carboline alkaloids (Wang et al., 2012), maltol glycosides (Qi et al., 2019), lignans (Tian, 2011), and volatile oils (Cao et al., 2009). These phytochemicals, especially the triterpenoid saponins, are considered as the active compounds in the roots of *P. tunicoides.*

The major triterpenoid saponins isolated from the roots of *P. tunicoides* are glycosides of quillaic acid (Li et al., 2016a, Tian, 2011). These saponins exhibited various biological activities, including anti-inflammatory (Wang et al., 2006), cytotoxic (Tian et al., 2012), and immunomodulatory activity (Zhang et al., 2012). The analgesic and anti-inflammatory effects of total saponins from *P. tunicoids* (TSPT) were evaluated in an adjuvant-induced arthritis (AA) rat model (Wang et al., 2006), by observing changes of the pain threshold and malondialdehyde (MDA) and cortisol levels in inflammatory-tissue soak of AA rats. TSPT showed analgesia effects and decreased the MDA level at the concentrations of 3.55 and 7.10 mg/kg (for 12 days) in rats, using the dexamethasone (2 mg/kg) and *Fengshining* (5 mg/kg) as standard drugs, indicating that the pharmacological mechanism analgesic and anti-inflammatory effects may be related to the decrease of MDA and regulation of cortisol. In another *in vivo* study, Song evaluated the anti-inflammatory activity and toxicology of TSPT (Song, 1981). In the behavioral experiment, the treatment of TSPT at a concentration of 5 mg/kg (subcutaneous injection) with mice raised the pain threshold in the hotplate test and lessened the frequency of writhing response induced by 0.6% acetic acid. In the same study, TSPT inhibited the croton oil-induced inflammation of the ears and granuloma caused by cotton in mice. *In vitro*, the saponins showed inhibitory activities against a variety of microorganisms; however, TSPT also affected hemolytic activity *in vitro.* The toxicological study showed that the LD_{50} (subcutaneous injection) of the TSPT in mice was 48.7 mg/kg. In 2012, Zhang et al. synthesized a group of oleanane-type triterpenoid saponins, TSA1–4, TSE1, and TSE2, originating from tunicosaponin A and tunicosaponin E found in the roots of *P. tunicoides* (Zhang et al., 2012). TSA2 (25 μg/mouse) enhanced the Con A-, LPS-, and ovalbumin-induced splenocyte proliferation, ovalbumin-specific antibody levels (IgG, IgG1, IgG2a, and IgG2b), and IFN-γ, TNF-α, IL-2, IL-4, and IL-5 levels in serum of ICR mice. A toxicological study showed that TSA2 did not cause any mortality and side effects when mice were administered subcutaneously at a dose up to 1.6 mg, and no significant hepatotoxic effect was observed in mice at doses ranging from 0.05 mg to 0.8 mg. Thus, TSA2 possesses higher adjuvant activities with less adverse effects and should be further explored as an immunomodulator for immune responses.

Cyclic peptides in *P. tunicoides* are a group of unique constituents in *P. tunicoides*. Tunicyclin A, possessing a diketopyrrolo[1,2-a] pyrazine unit, is the first tricyclic peptide obtained from a plant (Tian et al., 2009). Tunicyclin D showed a broad spectrum of antifungal activity against *Candida albicans* (SC5314), *C. albicans* (Y0109), *C. tropicalis*, *C. parapsilosis*, and *Cryptococcus neoformans* (BLS108) with MIC_{80} values of 4.0, 16.0, 0.25, 1.0, and 1.0 μg/mL, respectively (Tian et al., 2010). Two carboline alkaloids, tunicoidines F and G, were isolated from the aerial parts of *P. tunicoides* and were evaluated for their cytotoxicity against SMML-7721, A549, HCT116, and MCF-7 cell lines and anti-inflammatory activity in LPS-induced RAW264.7 macrophages (Wen et al., 2014). Results showed that tunicoidine G has weak cytotoxic activity against the HCT116 cell line at 20 μM. In another cytotoxic study on carboline alkaloids (Tian et al., 2012), 1-acetyl-3-carbomethoxyl-*β*-carboline was found to have cytotoxic activity against the HCT116 cell line with an IC_{50} value of 9.67 μg/mL. The volatile oil in *P. tunicoides* was extracted and analyzed by capillary column GC-MS (Cao et al., 2009). Ten major constituents were found as follows: hexadecanoic acid (19.25%), 3-octacosanone (15.93%), methyl salicylate (9.73%), 9,12-octadecadienoic acid (8.95%), 2-naphthylphenylamine (7.91%), 9-octadecynoic acid (6.50%), pentacosane (4.26%), pentadecanoic acid (3.81%), tetracosane (3.47%), and octacosane (2.41%).

In preliminary phytochemical and pharmacological studies, the anti-inflammatory and pain-relieving effects of the roots of *P. tunicoides* were examined, lending scientific support to the ethnomedicinal uses in Yunnan: the saponins are considered as the major compounds.

Cyclic peptides in *P. tunicoides* are unique constituents and deserve further study on their bioactivities. Several carboline alkaloids in *P. tunicoides* showed cytotoxicity against cancer cell lines, indicating the promising property to be used as antitumor agents. Nevertheless, due to the hemolytic activity of *P. tunicoides* extracts, *in vivo* toxicological studies are necessary.

11.6 *Rodgersia sambucifolia* Hemsl.

Rodgersia species are perennial flowering herbs distributed in East Asia and the Himalayas (Pan and James, 2001) (Figure 11.3B). Rhizomes are usually transversely elongating, thick, scaly. Leaves long petiolate, palmately, pinnately, or subpinnately compound; leaflets three–nine (or ten), base subsessile, margin doubly serrate, apex usually shortly acuminate. Inflorescence a paniculate cyme, ebracteate, many flowered. Sepals (four or)5(–seven), spreading, white, pink, or red. Petals usually absent, very rarely one, two, or five vestigial ones present. Stamens 10(–14). Ovary subsuperior, rarely semi-inferior, two- or three-loculed; placentation axile; ovules many; styles two or three. Capsule two- or three-valved.

The genus *Rodgersia* has five species and four of them are distributed in China (Pan and James, 2001). Among them, *R. sambucifolia* Hemsl. and *R. aesculifolia* Batalin are distributed in southwest China, and used by the local people in Yunnan. They usually grow in forests, forest margins, scrub, meadows, and rock clefts, with an altitude ranging from 1100 to 3800 m. In addition, environmental pollution, habitat destruction, and long-term over-exploitation resources of *Rodgersia* plants have dramatically reduced their resource reserves (Li et al., 2019a, Song et al., 2020). However, these plants are not currently protected species in China.

The rhizomes of *Rodgersia* plants, named *Yantuo* (岩陀often pronounced as *Aituo*) in Chinese, are used by several ethnic groups in Yunnan, including the Bai, Lisu, Naxi, and Yi ethnic groups (Ma et al., 2013). In Bai medicine, *Yantuo* is used by local people for rheumatism, cold and fever, diarrhea, digestive disorders, bacterial dysentery, irregular menstruation, metrorrhagia, and wounds. Similar traditional uses are also found in Lisu, Naxi, and Yi medicine. Meanwhile, *Yantuo* has different names in these ethnic groups in Yunnan (Ma et al., 2013, Chinese Academy of Sciences, 1999), including *sanrtyorxbeirx*, *nantyorx*, *maixyorx*, and *ssanebot* for *Bai* medicine, *Yelejie* for Lisu medicine, *Raocaiheng* for Naxi medicine, and *Poshi* for Yi medicine. *Yantuo* is also the major ingredient in some ethnomedicinal formulae, such as Yi medicine *Sechangzhixiesan* for enteritis (He et al., 2014), and *Yanlurukang* capsule for fibrous mastopathy, fibroids, and irregular menstruation (Cai and Zhang, 2017). Apart from medicinal uses, *Rodgersia* species have ornamental properties and contributes to water and soil conservation (Ma et al., 2013).

Four *Rodgersia* species in China all possess a potent botanical resource of bergenin (Li et al., 2019a), a trihydroxybenzoic acid glycoside, which has a broad spectrum of biological properties, such as anti-arrhythmic (Pu et al., 2002), antidiabetic (Kumar et al., 2012), antifungal (Prithiviraj et al., 1997), antimalarial (Liang et al., 2014), antinociceptive (de Oliveira et al., 2011), hepatoprotective (Lim et al., 2000), immunomodulatory (Nazir et al., 2007), and neuroprotective activities (Takahashi et al., 2003). Several chromatographic and spectroscopic methods for the extraction and analysis of bergenin in *Rodgersia* species have been conducted, including ultraviolet spectrophotometry, thin layer chromatography scanning, normal HPLC, and reversed phase HPLC (Li et al., 2019a). Chen et al. quantified the content of bergenin from *Rodgersia* species in Yunnan, using HPLC (Chen et al., 2011). Several phytochemical studies on *Rodgersia* plants led to the isolation of benzenoids, catechins, diterpenoids, flavonoids, isocoumarins, lignans, monoterpene glycosides, and steroids (Hu et al., 2007, Ji et al., 2004a). In 2002, three diterpene lactones, namely yantuines I–III, were isolated from the rhizomes of *R. sambucifolia* (Zheng et al., 2002). Hu et al. obtained two previously undescribed lignan diglycosides, hypophyllanthin-2a-*O*-*β*-apiofuranosyl-(1→6)-*O*-*β*-glucopyranoside and 8-methoxyhypophyllanthin-2a-*O*-*β*-apiofuranosyl-(1→6)-*O*-*β*-glucopyranoside, together with 5-(3-methylbutyl)-8-methoxy furanocoumarin, 1-acetoxyl-2e-piperonyl-6e-[6-methoxylpiperonyl]-3,7-dioxabicyclo-[3.3.0]-octane, *β*-rosasterol, muxiangrine III, and nirtetralin 7,7-methoxybergenin, from the ethanol extract of *R. sambucifolia* rhizomes (Hu et al., 2005). Also, from the roots of *R. sambucifolia*, Shi et al. determined four compounds, including bergenin, catechin, epicatechin-3-*O*-gallate, and oleanolic acid (Shi et al., 2008). From the root barks of *R.*

sambucifolia, an undescribed demethylacetovanillochromene glycoside, 2,2-dimethyl-2*H*-(8-hydroxy-6-acetyl)-benzopyran-8-*O*-*β*-D-apiofuranosyl-(1→6)-*β*-D-glucopyranoside was isolated, along with 23 known compounds (Hu et al., 2007). These known compounds include two benzenoids, two catechins, two diterpenoids, three flavonoids, two isocoumarins, four lignans, six monoterpene glycosides, and two steroids. From the rhizomes of *Rodgersia*, six monoterpene disaccharide glycosides were isolated and determined as (*E*)-3,7-dimethyl-1-*O*-[*α*-L-rhamnopyranosyl-(1→6)-*β*-D-glucopyranosyl]-oct-2-en-7-ol, (*E*)-3,7-dimethyl-1-*O*-[*α*-L-arabinofuranosyl-(1→6)-*β*-D-glucopyranosyl]-oct-2-en-7-ol,geranyl-1-*O*-*α*-L-arabinofuranosyl-(1→6)-*β*-D-glucopyranoside, geranyl-1-*O*-*α*-L-rhamnopyranosyl-(1→6)-*β*-D-glucopyranoside, geranyl-1-*O*-*β*-D-xylopyranosyl-(1→6)-*β*-D-glucopyranoside, and geranyl-1-*O*-*α*-L-arabinopyranosyl-(1→6)-β-D-glucopyranoside (Ji et al., 2004a). However, few biological screenings have been performed on the crude extracts and compounds from *Rodgersia*.

Although bergenin, the major active compound in *Yantuo*, has been widely studied for its pharmacological activities in *Rodgersia* species, more *in vivo* and *in vitro* pharmacological assays for this ethnomedicine are needed in future, to explain the pharmacological mechanisms related to traditional uses. Moreover, due to the systematic effects of various chemical constituents in *Yantuo*, the quality control considering unique compounds other than bergenin deserves a further study.

11.7 *Winchia calophylla* A. DC.

Winchia calophylla is a species in the Apocynaceae family (Li et al., 1995, 2007) (Figure 11.3C). Trees evergreen, glabrous, to 30 m tall. Branches greenish, angled when young. Leaves in whorls of three or four, rarely opposite; petiole 1–2 cm; leaf blade narrowly elliptic, 7–20 × 2.5–4.5 cm, thick papery, lustrous adaxially, paler abaxially, apex caudate, or acuminate; lateral veins 20–50 pairs. Cymes glabrous, *ca.* 4 cm; peduncle 1.5–3 cm. Pedicel to 3 mm. Corolla white, pubescent, tube 5–6 mm; lobes broadly ovate, 3–4 mm, overlapping to left. Disc absent. Ovaries connate. Follicles connate, 18–35 × 1–1.2 cm. Seeds narrowly elliptic; cilia brown-yellow, to 2 cm. Flowering in April to July, fruiting in August to December. *W. calophylla* distributes in the south and southeast Asia (Li et al., 1995). In China, it usually grows in monsoon or montane rain forests in south Yunnan and Hainan, with an altitude ranging from 300 to 1100 m.

The leaves and bark of *W. calophylla* and their resin are traditional Dai medicines in Yunnan, named *Maidingjie* by local people (You et al., 2020). The leaves and bark are used to treat cough and acute bronchitis, and the resin is used to stop bleeding. In addition, the wood of *W. calophylla* is used for making furniture and stationery (Li et al., 1995).

You et al. summarized 116 phytochemicals in *W. calophylla* up to 2020 (You et al., 2020), including indole alkaloids, iridoid glycosides, lignans, and triterpenoids. The 76 indole alkaloids are the unique and active constituents in *W. calophylla*. Among them, 18 echitamine and 18 echitamidine alkaloids are the two major groups (Figure 11.3D). The total alkaloid extract of *W. calophyll* inhibited bronchodilation in isolated smooth muscles of guinea-pig tracheal spirals, with an EC_{50} value of 0.27 mg/mL, using aminophylline as a positive control (EC_{50} = 0.024 and 0.016 mg/

mL for guinea-pig tracheal spirals and lung strips, respectively) (Zhu et al., 2005). Moreover, four alkaloids, cantleyine, loganin, *N*4-methyl akuammicine, and paeonol, exhibited relaxation effects on guinea-pig tracheal spirals (EC_{50} ranging from 0.13 to 0.75 mg/mL) and lung strips (EC_{50} ranging from 0.022 to 0.48 mg/mL). Numerous alkaloids from *W. calophylla* have been evaluated for their *in vitro* cytotoxicity in cancer cell lines (Cai et al., 2011, Li et al., 2012, Li et al., 2016b, Yuan et al., 2018, Zhong et al., 2017, Gan et al., 2006). However, most of them showed weak anti-proliferative activity. For example, Li et al. evaluated the cytotoxicity of alstilobanine C, calophyline A, *N*(4)-demethylechitamine, *N*(4)-demethylechitamine *N*-oxide, *Nb*-demethylalstogustine *N*-oxide, echitaminic acid, *N*-methyl aspidodasycarpine, and undulifolie in seven human cancer cell lines (A549 lung cancer, MCF-7 breast cancer, PC-3 prostate cancer, U87MG glioma, and U266, MM1.S, and MM1.R multiple myeloma cell lines) (Li et al., 2012). *N*-methyl aspidodasycarpine selectively inhibited the growth of PC-3 cell, with an IC_{50} value of 39.81 μM; alstilobanine C showed weak cytotoxicity against MCF-7, U87MG, and PC-3 cell lines, with IC_{50} values of 37.79, 39.00, and 44.42 μM; while other alkaloids showed no anti-proliferative activity up to a concentration of 50 μM. In 2016, Yang et al. (2016) tested the acetylcholinesterase enzyme inhibitory, anti-proliferative, and anti-fungal activities of ten alkaloids isolated from leaves and stems of *W. calophylla*. Deaceftylakummiline and *N*(4)-methyl-10-hydroxyl-desacetylakuammilin showed weak acetylcholinesterase enzyme inhibiting activity with IC_{50} values of 52.6 and 61.3 μmol/L. While deacetylakummiline showed inhibition against the plant pathogenic fungi *Penicillium italicum* and *Fusarium oxysporum* f. sp. *cubens* with IC_{50} values of 10.4 and 11.5 μM, respectively; and echitaminic acid inhibited *Rhizoctonia solani* with an IC_{50} value of 11.7 μM.

Although numerous phytochemicals have been isolated and identified from *W. calophylla*, few of them have undergone a deep pharmacological study. The *in vitro* inhibitory activity against bronchodilation provides preliminary data for the traditional use for cough and acute bronchitis. Nevertheless, further *in vivo* pharmacological and toxicological studies are necessary. In addition, although numerous alkaloids isolated from *W. calophylla* have been evaluated for cytotoxicity against cancer cells, most of them show weak anti-proliferative activity. More biological screening related to ethnomedicinal uses may help to explore the medicinal properties of those phytochemicals.

11.8 *Piper boehmeriifolium* (Miq.) Wall. ex C. DC.

Piper boehmeriifolium is the most commonly collected erect species of *Piper* in China and Indo-china (Tseng et al., 1999) (Figure 11.4A). Subshrubs erect, 1–3(5) m high, glabrous to uniformly hairy, dioecious, most parts usually drying black. Stems terete to thickly ridged when dry, minutely papillate to smooth, usually glabrous. Petiole (2)3–10 mm, glabrous or sometimes sparsely pubescent; leaf blades toward base of stem elliptic, narrowly elliptic, oblong, oblong-lanceolate, or ovate, (8)11–24 × (2.5)4–9.5 cm, papery to thinly papery, densely finely glandular, abaxially glabrous or occasionally puberulent, adaxially glabrous except sometimes for sparsely pubescent veins, base oblique, one side rounded, other side tapered and acute, bilateral difference 2–3 mm, apex acute to long acuminate; veins six–ten, usually one

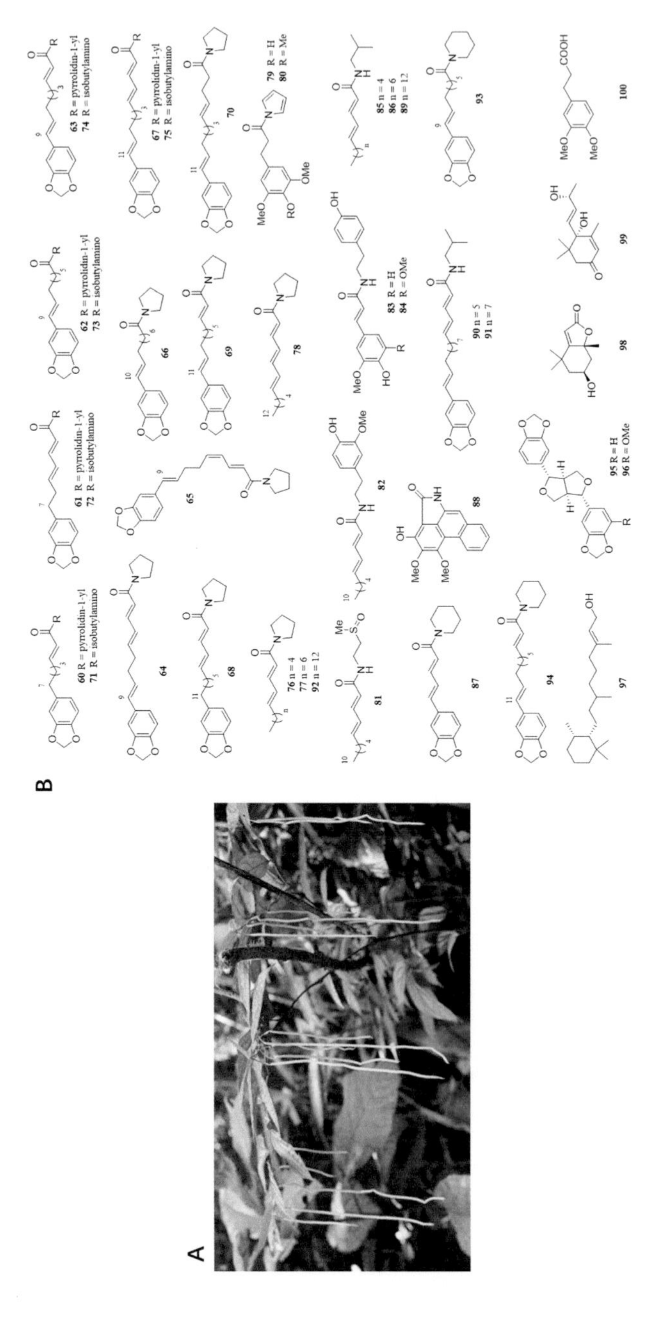

FIGURE 11.4 (A) *Piper boehmeriifolium* (Miq.) Wall. ex C. DC. (photo by Jun Yang); (B) Compounds of *Piper boehmeriifolium.*

more lateral vein on wider side, apical pair arising 1/3–1/2 way along midvein, alternate, reaching leaf apex, next pair often also above base; reticulate veins conspicuous, transversely oblong. Spikes mostly leaf-opposed, often terminal in male plants. Male spikes 10–16(23) cm × 2–3 mm; peduncle 1–3.5 cm; bracts orbicular, 1–2(2.5) mm wide, peltate, glabrous, obconical, shorter than wide, stamens two; filaments thick, short; anthers reniform. Female spikes 6–12 cm; peduncle and bracts as in male spikes; rachis sparsely pubescent; bracts 1–1.4 mm or slightly wider in diameter. Stigmas deciduous. Drupes densely clustered, subglobose, distinct, 1.2–3 mm in diam. Flowering in December to July. It usually grows in tropical rainforest or evergreen forests with an altitude ranging from 500 to 2200 m, distributed in Guangdong, Guangxi, Guizhou, and Yunnan provinces in China.

Piper boehmeriifolium is named *ya shuai yang* by the Dai people in Yunnan, China. Its stems and leaves are used to treat traumatic injury, fracture, arthralgia, motor dysfunction, weakness and susceptibility to diseases, limb numbness, dysmenorrhea, and amenorrhea. A water decoction or alcohol infusion of the stems and leaves (10–15 g) is taken orally; powders of stems and leaves or crushed fresh material are externally applied (Ma et al., 2017). *P. boehmeriifolium* roots are used in the treatment of tumors in India (Wang et al., 2014b).

From *P. boehmeriifolium*, *N*-containing compounds are 1-[(2*E*)-7-(3,4-methylenedioxyphenyl)-2-heptenoyl]pyrrolidine (**60**), 1-[(2*E*,4*E*)-7-(3,4-methylenedioxyphenyl)-2,4-heptadienoyl]pyrrolidine (**61**), 1-[(8*E*)-9-(3,4-methylenedioxyphenyl)-8-nonenoyl]pyrrolidine (**62**), 1-[(2*E*,8*E*)-9-(3,4-methylenedioxyphenyl)-2,8-nonadienoyl]pyrrolidine (**63**), 1-[(2*E*,4*E*,8*E*)-9-(3,4-methylenedioxyphenyl)-2,4,8-nonatrienoyl]pyrrolidine (**64**), 1-[(2*E*,4*Z*,8*E*)-9-(3,4-methylenedioxyphenyl)-2,4,8-nonatrienoyl]pyrrolidine (**65**), 1-[(9*E*)-10-(3,4-methylenedioxyphenyl)-9-decenoyl]pyrrolidine (**66**), 1-[(2*E*,4*E*,10*E*)-11-(3,4-methylenedioxyphenyl)-2,4,10-undecatrienoyl]pyrrolidine (**67**), 1-[(2*E*,4*E*)-11-(3,4-methylenedioxyphenyl)-2,4-undecadienoyl]pyrrolidine (**68**), 1-[(2*E*,10*E*)-11-(3,4-methylenedioxyphenyl)-2,10-undecadienoyl]pyrrolidine (**69**), 1-[(4*E*,10*E*)-11-(3,4-methylenedioxyphenyl)-4,10-undecadienoyl]pyrrolidine (**70**), (2*E*)-*N*-isobutyl-7-(3,4-methylenedioxyphenyl)hepta-2-enamide (**71**), (2*E*,4*E*)-*N*-isobutyl-7-(3,4-methylenedioxyphenyl)hepta-2,4-dienamide (**72**), (8*E*)-*N*-isobutyl-9-(3,4-methylenedioxyphenyl)nona-8-enamide (**73**), (2*E*,8*E*)-*N*-isobutyl-9-(3,4-methylenedioxyphenyl)nona-2,4-dienamide (**74**), retrofractamide B (**75**), 1-[(2*E*,4*E*)-2,4-decadienoyl]pyrrolidine (**76**), 1-[(2*E*,4*E*)-2,4-dodecadienoyl]pyrrolidine (**77**), 1-[(2*E*,4*E*,6*E*)-dodecatrienoyl]pyrrolidine (**78**), 3-(4-hydroxy-3,5-dimethoxyphenyl)propanoylpyrrole (**79**), 3-(3,4,5-trimethoxyphenyl)propanoylpyrrole (**80**), (2*E*,4*E*)-*N*-[2-(methylsulfinyl)ethyl]-2,4-decadienamide (**81**), (2*E*,4*E*)-*N*-[(4-hydroxy-3-methoxyphenyl)ethyl]-2,4-decadienamide (**82**), *N*-*trans*-feruloyltyramine (**83**), *N*-*trans*-sinapoyltyramine (**84**), pellitorine (**85**), (2*E*,4*E*)-*N*-isobutyl-2,4-dodecadienamide (**86**) (Tang et al., 2011), piperine (**87**) (Mahanta et al., 1974), piperolactam D (**88**) (Desai et al., 1990), *N*-isobutyl-2*E*,4*E*-octadecadienamide (**89**), guineensine (**90**), brachystamide B (**91**), *N*-pyrrolidyl-2*E*,4*E*-octadecadienamide (**92**), piperolein B (**93**), and piperchabamide B (**94**) (Liu et al., 2017c). Additionally, two lignans, (+)-sesamin (**95**) and (+)-5-methoxysesamin (**96**), three terpenoids, (+)-cassipourol (**97**), (−)-loliolide (**98**), and (+)-vomifoliol (**99**), as well as one phenylpropanoid, 3-(3,4-dimethoxyphenyl)propionic acid (**100**), have been reported (Liu et al., 2017c).

Compounds **59–62** and **66–69** are cytotoxic against several cancer cell lines (Wang et al., 2014a, Tang et al., 2011). Among them, 1-[(9*E*)-10-(3,4-methylenedioxyphenyl)-9-decenoyl] pyrrolidine (**66**) is the most active against human cervical carcinoma HeLa cells with an IC_{50} value of 2.7 μg/mL (Tang et al., 2011) (Figure 11.4B).

11.9 *Piper mullesua* Buch.-Ham. ex D. Don

Piper mullesua, is a climber in the Piperaceae family (Tseng et al., 1999) (Figure 11.5A). Climbers woody, glabrous except for rachis and bases of bracts. Stems slender, hard, basal part tuberculate. Prophylls very short; petiole 0.7–2 cm, slender; leaf blade elliptic or narrowly elliptic or ovate-lanceolate, 7.5–9 × 3–4 cm, papery to thinly leathery, without glands, base cuneate, symmetric or slightly oblique, apex caudate-acuminate; veins five(–seven), abaxially very prominent, apical pair arising 1–2.5 cm above base, usually alternate; reticulate veins conspicuous. Flowers bisexual. Spikes leaf-opposed, at apices of branchlets, subglobose, *ca.* 3 × 2.5–3 mm; peduncle 2–3 mm; rachis pubescent; bracts orbicular, *ca.* 1 mm wide, peltate, abaxially glabrous; stalk short. Stamens two; anthers reniform. Ovary obovoid; stigmas three or four, very small. Drupe obovoid, *ca.* 2.5 mm in diameter, partly immersed in rachis. Flowering in May to July. It usually grows on the edges of riverine valleys or slopes in tropical rainforests or evergreen forests with an altitude ranging from 800 to 2100 m, distributed in Hainan, Sichuan, Xizang, and Yunnan provinces in China.

Whole plant of *P. mullesua* (local name zhi ma ci) is used to treat rheumatic lumbocrural pain, arthritis, numbness of limbs, cold, and fall injury by Lisu people in Nujiang, Yunnan, China. A water decoction of the material (10–15 g) is taken orally (Nujiang Public Health Bureau, 1991).

Chemical constituents of this plant mainly are *N*-containing compounds, including pipermullesine A (**101**), pipermullesines B (**102**) and C (**103**), pipermullamides A–F (**104–109**), (+)-phenylalanine betaine (**110**), (–)-mangochinine (**111**), xylopinidine (**112**), (–)-oblongine (**113**), pellitorine (**114**), (2*E*,4*E*)-*N*-isobutyl-2,4-dodecadienamide (**115**), retrofractamide A (**116**), guineensine (**117**), brachystamide B (**118**), retrofractamide C (**119**), sarmentine (**120**), 3-(3,4-dimethoxyphenyl)propanoylpyrrole (**121**), *N-trans*-feruloyltyramine (**122**) (Xia et al., 2018), chingchengenamide A (**123**) (Zhang et al., 1998), *N*-isobutyl-16-phenylhexadeca-2*E*,4*E*-dienamide (**124**), and guineensine (**125**) (Srivastava et al., 2000b); lignans, including (–)-machilusin (**126**), galgravin (**127**), (–)-nectandrin A (**128**), nectandrin B (**129**), asarinin (**130**), sesamin (**131**), and fargesin (**132**) (Srivastava et al., 2000a, Xia et al., 2018, Zhang et al., 1998); phenylpropanoids, including myristicin (**133**) (Srivastava et al., 2000a) and methyl 3-(3,4-dimethoxyphenyl)propanoate (**134**) (Xia et al., 2018); arylalkenyl carboxylic acids and esters, including piperic acid (**135**), methyl piperate (**136**), methyl (2*E*,4*E*)-7-(1,3-benzodioxol-5-yl)hepta-2,4-dienoate (**137**), 1,3-benzodioxole-5-(2,4,8-triene-methyl nonoate) (**138**), and 1,3-benzodioxole-5-(2,4,8-triene-isobutyl nonaoate (**139**) (Srivastava et al., 2000a, Xia et al., 2018); terpenoids, including (–)-blumenol B (**140**), (–)-T-muurolol (**141**), *trans*-phytol (**142**), α-tocopherolquinone (**143**), and γ-tocopherol (**144**) (Xia et al., 2018); and steroids, including stigmast-4-ene-3,6-dione (**145**), (22*E*)-stigmasta-4,22-diene-3,6-dione (**146**), and (22*E*)-stigmasta-4,6,8(14),22-tetraen-3-one (**147**) (Xia et al., 2018).

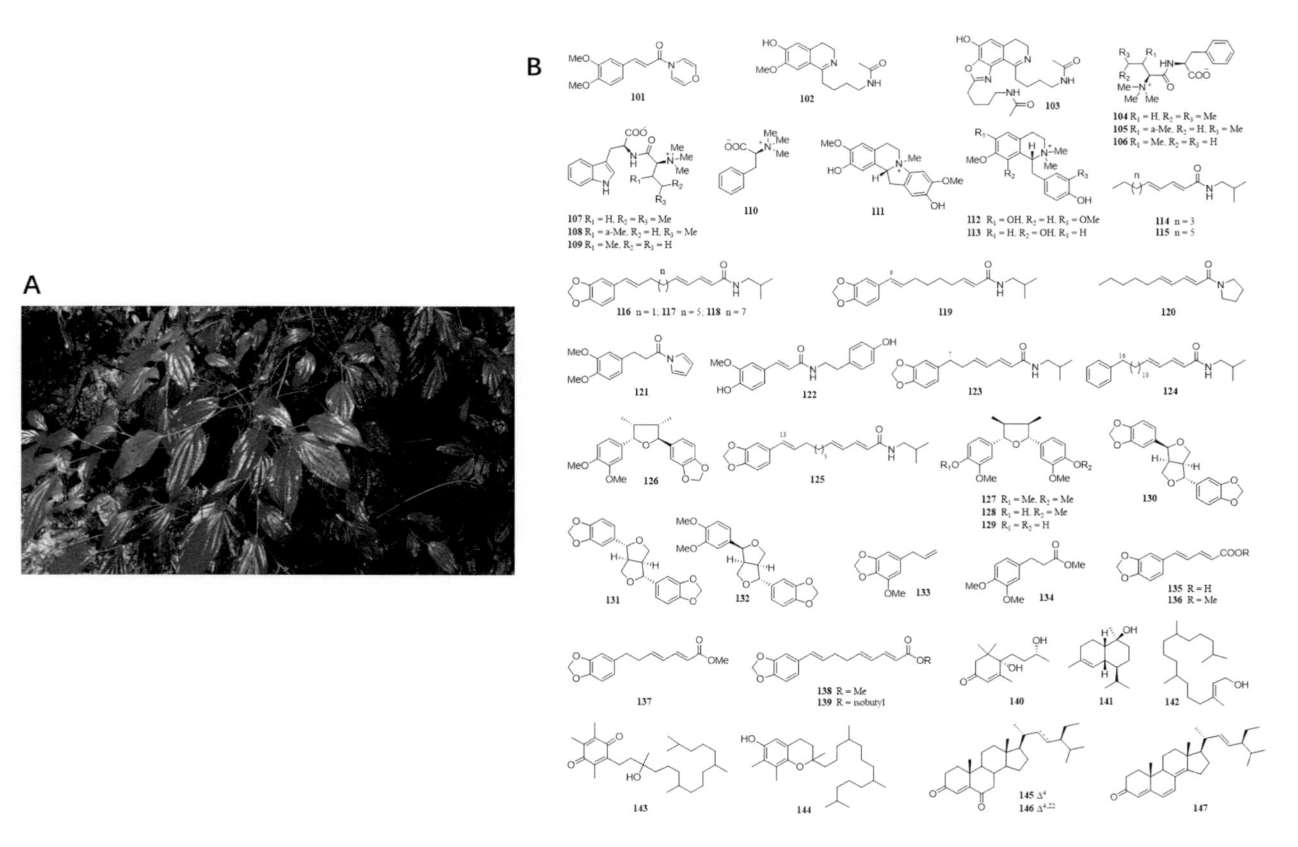

FIGURE 11.5 (A) *Piper mullesua* Buch.-Ham. ex D. Don (photo by Chun-Lin Long); (B) Compounds of *Piper mullesua*.

Alcoholic extracts of *P. mullesua* showed the activity against rabbit platelet aggregation induced by platelet-activating factor (PAF) with an IC_{50} value of 64.43 μg/mL (Shen et al., 1997). (–)-mangochinine (**111**), pellitorine (**114**), and (2*E*,4*E*)-*N*-isobutyl-2,4-dodecadienamide (**115**) from the plant showed weak inhibitory activity against rabbit platelet aggregation induced by PAF, with IC_{50} values of 470.3, 614.9, and 579.7 μg/mL, respectively (Xia et al., 2018) (Figure 11.5B).

11.10 Discussion and Conclusion

Eight ethnomedicinal plants have been examined for their traditional use by ethnic groups in Yunnan Province. Phytochemical studies reveal that the bioactive compounds include stilbenoids in *A. graminifolia*, alkaloids in *Coptis teeta*, C_{21} steroids in *Cynanchum otophyllum*, triterpenoids saponins in *P. tunicoides*, bergenin in *R. pinnata* and *R. sambucifolia*, and alkaloids in *W. calophylla*. Pharmacological studies on the antiepileptic, anti-inflammatory, antimicrobial, antinociceptive, and immunomodulatory activities of these ethnomedicines provide preliminary data for their traditional use by the Bai, Dai, Lisu, Miao, and Yi people. However, few of these pharmacological studies are conducted *in vivo*. In addition, several groups of phytochemicals from these medicinal plants may have side effects, such as alkaloids in *Coptis teeta* and *W. calophylla*, C_{21} steroids in *Cynanchum otophyllum*, and triterpenoid saponins in *P. tunicoides*. Since some of these ethnomedicines are already used in prescriptions, more toxicological studies on their crude extracts and compounds are necessary in future.

Although Yunnan has a high biodiversity and rich botanical resources, habitat destruction, increasing demand, and long-term overuse have threatened the biodiversity of these medicinal plants. Three ethnomedicinal plants in this chapter, *A. graminifolia*, *Coptis teeta*, and *P. tunicoides*, are recorded as II-grade protected species in the *State Key List of Protected Wild Plants* (China). The practice of ethnomedicine, as an integral part of the culture of indigenous people, has a close interface with local ecosystems. Ethnobotanical studies on these species are suggested to be conducted, to figure out how ethnic groups in Yunnan acquire and conserve these medicinal plants in the local ecosystem.

REFERENCES

Anyinam C, 1995. Ecology and ethnomedicine: exploring links between current environmental crisis and indigenous medical practices. *Soc Sci Med*, 40: 321–329.

Auberon F, Olatunji OJ, Antheaume C, et al., 2018. Arundinosides A–G, new glucosyloxybenzyl 2*R*-benzylmalate derivatives from the aerial parts of *Arundina graminifolia*. *Fitoterapia*, 125: 199–207.

Auberon F, Olatunji OJ, Krisa S, et al., 2016. Two new stilbenoids from the aerial parts of *Arundina graminifolia* (Orchidaceae). *Molecules*, 21: 1430.

Auberon F, Olatunji OJ, Waffo-Teguo P, et al., 2019. New glucosyloxybenzyl 2*R*-benzylmalate derivatives from the undergrounds parts of *Arundina graminifolia* (Orchidaceae). *Fitoterapia*, 135: 33–43.

Cai Q, Zhang G, 2017. Progress of clinical study of Yanlurukang capsule in China. *Chin J Fam Plan*, 25: 888–890.

Cai XH., Bao MF, Zhang Y, et al., 2011. A new type of monoterpenoid indole alkaloid precursor from *Alstonia rostrata*. *Org Lett*, 13: 3568–3571.
Cao G, Yang Z, Zhou X, et al., 2009. GC-MS determination of chemical constituents in volatile oil of *Psammosilene tunicoides*. *Physic Testing Chem Anal (Part B: Chemical Analysis)*, 45: 1276–1281.
Chen W, Meng Z, Yang S, et al., 2011. Content comparison of bergenin in Yunnan genus *Rodgersia*. *Med Plants*, 2: 26–29.
Chen XQ, Stephan WG, 2009. *Orchidaceae*. Beijing: Science Press & St. Louis: Missouri Botanical Garden Press.
Chinese Academy of Sciences, 1999. Subject database of China plant. http://www.plant.csdb.cn/protectlist (accessed May 24, 2020).
De Oliveira CM, Nonato FR, de Lima FO, et al., 2011. Antinociceptive properties of bergenin. *J Nat Prod*, 74: 2062–2068.
Desai SJ, Chaturvedi R, Mulchandani NB, 1990. Piperolactam D, a new aristolactam from Indian *Piper* species. *J Nat Prod*, 53: 496–497.
Dong J, Peng X, Li L, et al., 2018. C_{21} steroidal glycosides with cytotoxic activities from *Cynanchum otophyllum*. *Bioorg Med Chem Lett*, 28: 1520–1524.
Dong J, Peng X, Lu S, et al., 2019. Hepatoprotective steroids from roots of *Cynanchum otophyllum*. *Fitoterapia*, 136: 104171.
Dong J, Yue GGL, Lee JKM, et al., 2020. Potential neurotrophic activity and cytotoxicity of selected C_{21} steroidal glycosides from *Cynanchum otophyllum*. *Med Chem Res*, 29: 549–555.
Dong W, Zhou K, Wang YD, et al., 2015. A new phenylpropanoid from Dai Medicine *Arundina graminifolia* and its anti-tobaco mosaic virus activities. *Chin Tradit Herb Drugs*, 46: 2.
Du G, Shen Y, Yang L, et al., 2014. Bibenzyl derivatives of *Arundina graminifolia* and their cytotoxicity. *Chem Nat Compd*, 49: 1019–1022.
Foster GM, Anderson BG, 1978. *Medical Anthropology*. New York: John Wiley and Sons.
Gan LS, Yang SP, Wu Y, et al., 2006. Terpenoid indole alkaloids from *Winchia calophylla*. *J Nat Prod*, 69: 18–22.
Gao X, Yang L, Shen Y, et al., 2012. Phenolic compounds from *Arundina graminifolia* and their anti-tobacco mosaic virus activity. *B Korean Chem Soc*, 33: 2447–2449.
Gao Y, Jin Y, Yang S, et al., 2014. A new diphenylethylene from *Arundina graminifolia* and its cytotoxicity. *Asian J Chem*, 26: 3903.
Gao YT, Liu P, He ME, et al., 2013a. The inhibition effect of the extract of *Arundina graminifolia* on lipid peroxidation induced by carbon tetrachloride. *J Yunnan Univers Nationalities (Natural Sciences Edition*), 3: 5.
Gao Z, Xu S, Wei J, et al., 2013b. Phenolic compounds from *Arundina graminifolia* and their anti-tobacco mosaic virus activities. *Asian J Chem*, 25: 2747.
Han LX, Zhou M, Yang L, et al., 2018. Ethnobotany, phytochemistry and pharmacological effects of plants in genus *Cynanchum* Linn. (Asclepiadaceae). *Molecules*, 23: 1194.
He D, Yang S, Luo H, et al., 2015. Progress in steroid constituents of *Cynanchum* and pharmacological effects of the steroids. *J Southwest Univers Nationalities (Natural Science Edition*), 4: 423–431.
He Y, Zhang Q, Shi X, et al., 2014. Review on traditional chinese medicine prescription, composition and clinical application research for *Sechangzhixiesan*. *J Med Pharm Chin Minorities*, 3: 62–64.

Hou Y, Wang M, Sun C, et al., 2020. Tunicyclin L, a cyclic peptide from *Psammosilene tunicoides*: isolation, characterization, conformational studies and biological activity. *Fitoterapia*, 145: 104628.

Hu HB, Jian YF, Cao H, et al., 2007. Chemical constituents from the root bark of *Rodgersia sambucifolia*. *J Chin Chem Soc*, 54: 75–80.

Hu HB, Zheng SZ, Zheng XD, et al., 2005. Chemical constituents of *Rodgersia sambucifolia* Hemsl. *Indian J Chem Sect B*, 44: 2399–2403.

Hu QF, Zhou B, Ye YQ, et al., 2013a. Cytotoxic deoxybenzoins and diphenylethylenes from *Arundina graminifolia*. *J Nat Prod*, 76: 1854–1859.

Hu QF, Zhou B, Huang JM, et al., 2013b. Antiviral phenolic compounds from *Arundina gramnifolia*. *J Nat Prod*, 76: 292–296.

Huai HY, Pei SJ, 2004. Plants used medicinally by folk healers of the Lahu people from the autonomous county of Jinping Miao, Yao, and Dai in southwest China. *Econ Bot*, 58: S265–S273.

Huang J, Pei SJ, Wang YZ, 2005. Natural resources and conservation of *Coptis teeta*. *Chin Tradit Herb Drugs*, 36: 112–115.

Ji C, Tan N, Fu J, et al., 2004a. Monoterpene disaccharide glycosides from *Rodgersia pinnata*. *Yunnan Zhiwu Yanjiu*, 26: 465–470.

Ji H, Pei S, Long C, 2004b. An ethnobotanical study of medicinal plants used by the Lisu people in Nujiang, northwest Yunnan. *Chin Econ Bot*, 58: S253–S264.

Kumar R, Patel DK, Prasad SK, et al., 2012. Type 2 antidiabetic activity of bergenin from the roots of *Caesalpinia digyna* Rottler. *Fitoterapia*, 83: 395–401.

Li B, Li D, Yin L, et al., 2016a. Study advance on *Psammosilene tunicoides*. *Anhui Agri Sci Bull*, 22: 23–26.

Li H, Zang G, Zhang Z, et al., 2019a. Analysis on *Yantuo* resources and suitable environmental factors in Yunnan province. *Mod Chin Med*, 21: 1314–1333.

Li J, Zhao A, Li D et al., 2019b. Comparative study of the free amino acid compositions and contents in three different botanical origins of *Coptis* herb. *Biochem Syst Ecol*, 83: 117–120.

Li JL, Zhou J, Chen ZH, et al., 2015c. Bioactive C_{21} steroidal glycosides from the roots of *Cynanchum otophyllum* that suppress the seizure-like locomotor activity of zebrafish caused by pentylenetetrazole. *J Nat Prod*, 78: 1548–1555.

Li JY, Cai XL, Zhao TR, et al., 1987. Effect of total glucosides of *Qingyangshen* (*Cynanchum otophyllum*) on glutamate decarboxylase and GABA transaminase in mouse brain. *Chin Tradit Herb Drugs*, 18: 261–266.

Li L, Xu WX, Liu CB, et al., 2015a. A new antiviral phenolic compounds from *Arundina gramnifolia*. *Asian J Chem*, 27: 3525.

Li LM, Yang T, Liu Y, et al., 2012. Calophyline A, a new rearranged monoterpenoid indole alkaloid from *Winchia calophylla*. *Org Lett*, 14: 3450–3453.

Li PT, Gilbert MG, Stevens WD, 1995. *Asclepiadaceae*. Beijing: Science Press & St. Louis: Missouri Botanical Garden Press.

Li S, Zhang ZY, Liu LQ, et al., 2007. Karyomorphology of three species in *Alstonia* (Apocynaceae). *Yunnan Zhiwu Yanjiu*, 29: 434–438.

Li X, Zhang M, Xiang C, et al., 2015b. Antiepileptic C_{21} steroids from the roots of *Cynanchum otophyllum*. *J Asian Nat Prod Res*, 17: 724–732.

Li Y, Yang L, Shu L, et al., 2013b. Flavonoid compounds from *Arundina graminifolia*. *Asian J Chem*, 25: 4922.

Li YK, Zhou B, Ye YQ, et al., 2013a. Two new diphenylethylenes from *Arundina graminifolia* and their cytotoxicity. *B Korean Chem Soc*, 34: 3257–3260.

Li YT, Gao XD, Zhang W, et al., 2016b. Winchinines A and B, two unusual monoterpene indole alkaloids with a third nitrogen atom from *Winchia calophylla*. *RSC Adv*, 6: 59657–59660.

Li Z, Zhang T, Shi Y, et al., 2011. Determination of effective alkaloids contents in *Coptis teeta* Wall at different collecting period. *Southwest China J Agric Sci*, 24: 1294–1297.

Liang J, Li Y, Liu X, et al., 2014. *In vivo* and *in vitro* antimalarial activity of bergenin. *Biomed Rep*, 2: 260–264.

Lim HK, Kim HS, Choi HS, et al., 2000. Hepatoprotective effects of bergenin, a major constituent of *Mallotus japonicus*, on carbon tetrachloride-intoxicated rats. *J Ethnopharmacol*, 72: 469–474.

Lin Y, 1995. Dai medicine *Dale Tang* used to treat jaundice. *Yunnan Zhongyi Zhongyao Zazhi*, 16: 53–54.

Liu MF, Ding Y, Zhang DM, 2005a. Phenanthrene constituents from rhizome of *Arundina graminifolia*. *China J Chin Mater Med*, 30: 353–356.

Liu MF, Han Y, Xing DM, et al., 2005b. One new benzyldihydrophenanthrene from *Arundina graminifolia*. *J Asian Nat Prod Res*, 7: 767–770.

Liu M, Lü H, Ding Y, 2012. Antitumoral bibenzyl derivatives from tuber of *Arundina graminifolia*. *China J Chin Mater Med*, 37: 66–70.

Liu Q, Dai R, Lü F, et al., 2017a. Research progress in chemical constituents and pharmacological activities of *Baiyangjie*. *J Chin J Mod Appl Pharm*, 4: 618–624.

Liu Q, Wang H, Lin F, et al., 2017b. Study on the structures and anti-hepatic fibrosis activity of stilbenoids from *Arundina graminifolia* (D. Don) Hochr. *IOP Conference Series: Materials Science and Engineering*, 274: 012024.

Liu Q, Sun F, Deng Y, et al., 2019. HPLC-ESI-MS^n identification and NMR characterization of glucosyloxybenzyl 2*R*-benzylmalate derivatives from *Arundina graminifolia* and their anti-liver fibrotic effects *in vitro*. *Molecules*, 24: 525.

Liu T, Liang Q, Zhang XM, et al., 2017c. Chemical constituents from *Piper boehmeriifolium* (Miq.) Wall. ex C. DC. *Biochem Syst Ecol*, 75: 27–30.

Long C, Li R, 2004. Ethnobotanical studies on medicinal plants used by the red-headed Yao people in Jinping, Yunnan Province, China. *J Ethnopharmacol*, 90: 389–395.

Long C, Li H, Zhou Y, et al., 1999. Ethnobotanical studies in Gaoligong mountains: II. The Dulong ethnic group. *Acta Bot Yunnan*, Suppl XI: 137–144.

Long C, Li S, Long B, et al., 2009. Medicinal plants used by the Yi ethnic group: a case study in central Yunnan. *J Ethnobiol Ethnomed*, 5: 13.

Lu DQ, Lidén M, Oxelman B, 2001. *Psammosilene*. Beijing: Science Press & St. Louis: Missouri Botanical Garden Press.

Ma S, Sun Q, Gao Y, et al., 2013. Study on applications of the ethnomedicine *Rodgersia sambucifolia*. *Neimenggu Zhongyiyao*, (17): 61.

Ma XJ, Zhang LX, Lin YF, 2017. *Chinese Dai Medicines*. Beijing: People's Medical Publishing House.

Ma XX, Wang D, Zhang YJ, et al., 2011. Identification of new qingyangshengenin and caudatin glycosides from the roots of *Cynanchum otophyllum*. *Steroids*, 76: 1003–1009.

Mahanta PK, Ghanim A, Gopinath KW, 1974. Chemical constituents of *Piper syhaticum* (Roxb) and *Piper boehmerifoliurn* (Wall). *J Pharm Sci*, 63: 1160–1161.

Meng CY, Niu DY, Li YK, et al., 2014. A new cytotoxic stilbenoid from *Arundina graminifolia*. *Asian J Chem*, 26: 2411.

Meng F, Wang L, Zhang J, et al., 2013. Non-alkaloid chemical constituents from the rhizome of *Coptis teeta*. *J China Pharm Univ*, 44: 307–310.

Ming Q, Dong Y, Zou X, 2006. RP-HPLC assay for cynanchogenin in *Qingyanghsen* tablets. *Chin J Pharm Anal*, 26: 1156–1157.

Mu QZ, Lu JR, Zhou QL, 1986. Two new antiepilepsy compounds-otophyllosides A and B. *Scientia Sinica Series B – Chem Biol Agric Med Earth Sci*, 29: 295–301.

Nazir N, Koul S, Qurishi MA, et al., 2007. Immunomodulatory effect of bergenin and norbergenin against adjuvant-induced arthritis—A flow cytometric study. *J Ethnopharmacol*, 112: 401–405.

Niu DY, Han JM, Kong WS, et al., 2013. Antiviral fluorenone derivatives from *Arundina gramnifolia*. *Asian J Chem*, 25: 9514.

Nujiang Public Health Bureau, 1991. *Traditional Chinese Medicines in Nujiang*. Kunming: Yunnan Science & Technology Press.

Pan J, James C, 2001. *Rodgersia*. Beijing: Science Press & St. Louis: Missouri Botanical Garden Press.

Pandit MK, Babu CR, 2003. The effects of loss of sex in clonal populations of an endangered perennial *Coptis teeta* (Ranunculaceae). *Bot J Linn Soc*, 143: 47–54.

Peng X, He H, Mao M, et al., 2008. Studies on chemical constituents of *Arundina graminifolia* (D.Don) Hochr. *Yunnan Zhongyi Xueyuan Xuebao*, 31: 32–33.

Peng X, Zhao Y, Liao R, et al., 2001. Study on qualitative criterion of Daizu medical *Yajie* tablets. *J Med Pharm Chin Minorities*, 7: 28.

Prithiviraj B, Singh UP, Manickam M, et al., 1997. Antifungal activity of bergenin, a constituent of *Flueggea microcarpa*. *Plant Pathol*, 46: 224–228.

Pu HL, Huang X, Zhao JH, et al., 2002. Bergenin is the antiarrhythmic principle of *Fluggea virosa*. *Planta Med*, 68: 372–374.

Qi L, Ma Y, Zhong F, et al., 2018. Comprehensive quality assessment for *Rhizoma Coptidis* based on quantitative and qualitative metabolic profiles using high performance liquid chromatography, Fourier transform near-infrared and Fourier transform mid-infrared combined with multivariate statistical analysis. *J Pharmaceut Biomed*, 161: 436–443.

Qi X, Tian J, Shen Y, et al., 2019. Two new maltol glycosides from roots of *Psammosilene tunicoides*. *Chin Tradit Herb Drugs*, 50: 2513–2517.

Ramawat KG, Dass S, Mathur M, 2009. *Herbal Drugs: Ethnomedicine to Modern Medicine*. New York: Springer.

Ramawat KG, Mérillon JM eds., 2013. *Natural Products: Phytochemistry, Botany and Metabolism of Alkaloids, Phenolics and Terpenes*. Heidelberg: Springer.

Shen ZQ, Chen ZH, Wang DC, 1997. PAF antagonist screening from 12 alcohol extracts of *Piper* species native to Yunnan. *Acad J Kunming Med Coll*, 29: 23–25.

Shi G, Li D, Liu G, 2008. Studies on the chemical constituents of *Rodgersia sambucifolia*. *Dali Xueyuan Xuebao*, 7: 1–2.

Shi Y, Tian H, Zhang J, et al., 2011. Preliminary study on the individual differences of biomass and alkaloid contents in *Coptis Teeta*'s different parts. *J Yunnan Univers*, 33: 83–88.

Shu LD, Shen YQ, Yang LY, et al., 2013. Flavonoids derivatives from *Arundina graminifolia* and their cytotoxicity. *Asian J Chem*, 25: 8358–8360.

Song L, 1981. Pharmacological study of total saponins from *Psammosilene tunicoides*. *Acta Bot Yunnan*, 3: 287–290.

Song MF, Zhang LX, Zhang Y, et al., 2020. Effects of genetic variation and environmental factors on bergenin in *Rodgersia sambucifolia* Hemsl. *J Ethnopharmacol*, 247: 112201.

Srivastava S, Gupta MM, Tripathi AK, et al., 2000a. 1,3-Benzodioxole-5-(2,4,8-triene-methyl nonaoate) and 1,3-benzodioxole-5(2,4,8-triene-isobutyl nonaoate) from *Piper mullesua. Indian J Chem Sect B*, 39: 946–949.

Srivastava S, Verma RK, Gupta MM, et al., 2000b. Chemical constituents of *Piper mullesua. J Indian Chem Soc*, 77: 305–306.

State Administration of TCM, 1999. *Chinese Material Medica*. Shanghai: Shanghai Scientific and Technical Publishers.

Takahashi H, Kosaka M, Watanabe Y, et al., 2003. Synthesis and neuroprotective activity of bergenin derivatives with antioxidant activity. *Bioorg Chem Res*, 11: 1781–1788.

Tang GH, Chen DM, Qiu BY, et al., 2011. Cytotoxic amide alkaloids from *Piper boehmeriaefolium. J Nat Prod*, 74: 45–49.

Tang J, Feng Y, Tsao S, et al., 2009. Berberine and Coptidis Rhizoma as novel antineoplastic agents: a review of traditional use and biomedical investigations. *J Ethnopharmacol*, 126: 5–17.

Tao J, Yang J, Chen C, et al., 2011. Evaluation of *Cynanchum otophyllum* glucan sulfate against human immunodeficiency virus and herpes simplex virus as a microbicide agent. *Indian J Pharmacol*, 43: 536.

Teoh ES, 2016. *Medicinal Orchids of Asia*. Cham, Switzerland: Springer Nature

Tian J, 2011. *Study on Chemical Constituents of Psammosilene tunicoides* W.C. Wu et C.Y. Wu. Ph.D Dissertation. Second Military Medical University, Shanghai, China.

Tian J, Shen Y, Li H, et al., 2012. Carboline alkaloids from *Psammosilene tunicoides* and their cytotoxic activities. *Planta Med*, 78: 625–629.

Tian J, Shen Y, Yang X, et al., 2010. Antifungal cyclic peptides from *Psammosilene tunicoides. J Nat Prod*, 73: 1987–1992.

Tian JM, Shen YH, Yang XW, et al., 2009. Tunicyclin A, the first plant tricyclic ring cycloheptapeptide from *Psammosilene tunicoides. Org Lett*, 11: 1131–1133.

Tseng YC, Xia NH, Gilbert MG, 1999. *Piperaceae*. Beijing: Science Press & St. Louis: Missouri Botanical Garden Press.

Wang J, Zhang Y, Gao X, 2009. Study on the antioxidant activities of *Arundina graminifolia* Dai medicines. *J Yunnan Nationalities Univers*, 2: 148–150.

Wang W, Fu D, Li L, et al., 2001. *Ranunculaceae*. Beijing: Science Press & St. Louis: Missouri Botanical Garden Press.

Wang W, Yuan L, Gu XZ, et al., 2012. New beta-carboline alkaloids contained in *Psammosilene tunicoides. China J Chin Mater Med*, 37: 3240–3242.

Wang X, Xu J, Qiu D, et al., 2006. Study on the anti-inflammatory effects and possible mechanism of total saponins of *Psammosilene tunicoides. Zhongguo Shiyan Fangjixue Zazhi*, 12: 56–58.

Wang YH, Goto M, Wang LT, et al., 2014a. Multidrug resistance-selective antiproliferative activity of *Piper* amide alkaloids and synthetic analogues. *Bioorg Med Chem Lett*, 24: 4818–4821.

Wang YH, Morris-Natschke SL, Yang J, et al., 2014b. Anticancer principles from medicinal *Piper* plants. *J Tradit Complement Med*, 4: 8–16.

Wen B, Tian JM, Huang ZR, et al., 2020. Triterpenoid saponins from the roots of *Psammosilene tunicoides. Fitoterapia*, 144**:** 104596.

Wen B, Yuan X, Zhang WD, et al., 2014. Chemical constituents from the aerial parts of *Psammosilene tunicoides. Phytochem Lett*, 9: 59–66.

Xia MY, Yang J, Zhang PH, et al., 2018. Amides, isoquinoline alkaloids and dipeptides from the aerial parts of *Piper mullesua. Nat Prod Bioprospect*, 8: 419–430.

Xu JJ, Xu WX, Li YP, 2018. Chemical constituents of essential oil from *Arundina graminifolia*. *Chem Nat Compd*, 54: 193–194.

Yan M, 1983. Clinical antiepileptic study of *Qingyangshen* Tablet. *Yunnan Zhongyi Xueyuan Xuebao*, 1: 16–19.

Yan M, 2017. Phytochemical, detoxification and anti-infective study on Arundina graminifolia (Dai medicine). Master thesis, South China University of Technology.

Yan M, Tang B, Liu M, 2018. Phenanthrenes from *Arundina graminifolia* and *in vitro* evaluation of their antibacterial and anti-haemolytic properties. *Nat Prod Res*, 32: 707–710.

Yang J, Huang XB, Wan QL, et al., 2017. Otophylloside B protects against Aβ toxicity in *Caenorhabditis elegans* models of Alzheimer's disease. *Nat Prod Bioprospect*, 7: 207–214.

Yang J, Wan QL, Mu QZ, et al., 2015. The lifespan-promoting effect of otophylloside B in *Caenorhabditis elegans*. *Nat Prod Bioprospect*, 5: 177–183.

Yang JX, Wang H, Lou J, et al., 2014. A new cytotoxic diphenylethylene from *Arundina graminifolia*. *Asian J Chem*, 26: 4517.

Yang ML, Chen J, Sun M, et al., 2016. Antifungal indole alkaloids from *Winchia calophylla*. *Planta Med*, 82: 712–716.

Yang Y, Tian K, Hao J, et al., 2004. Biodiversity and biodiversity conservation in Yunnan, China. *Biodivers Conserv*, 13: 813–826.

You Y, Sun X, Li F, et al., 2020. Research advances on chemical constituents and pharmacological effects of the synonym plant *Winchia calophylla* A. DC. *J Pharm Res*, 39: 160–167.

Yuan YX, Guo F, He HP, et al., 2018. Two new monoterpenoid indole alkaloids from *Alstonia rostrata*. *Nat Prod Res*, 32: 844–848.

Zhan ZJ, Bao SM, Zhang Y, et al., 2019. New immunomodulating polyhydroxypregnane glycosides from the roots of *Cynanchum otophyllum* CK Schneid. *Chem Biodivers*, 16: e1900062.

Zhang J, Cao W, Tian J, et al., 2012. Evaluation of novel saponins from *Psammosilene tunicoides* and their analogs as immunomodulators. *Int Immunopharmacol*, 14: 21–26.

Zhang K, Ni W, Chen CX, et al., 1998. Chemical constituents of *Piper mullesua*. *Plant Divers*, 20: 374–376.

Zhang P, Cong L, Ye J, 2016. Research progress of Dai medicine *Ya Jie Sha Ba*. *Chin J Ethnomed Ethnopharm*, 25: 7–8.

Zhao YB, Shen YM, He HP, et al., 2007. Antifungal agent and other constituents from *Cynanchum otophyllum*. *Nat Prod Res*, 21: 203–210.

Zhao ZM, Sun ZH, Chen MH, et al., 2013. Neuroprotective polyhydroxypregnane glycosides from Cynanchum *otophyllum*. *Steroids*, 78: 1015–1020.

Zheng S, An H, Shen T, et al., 2002. Studies on the new diterpene lactones from *Rodgersia sambucifolia* H. *Indian J Chem*, 41B: 228–231.

Zhong XH, Bao MF, Zeng CX, et al., 2017. Polycyclic monoterpenoid indole alkaloids from Alstonia *rostrata* and their reticulate derivation. *Phytochem Lett*, 20: 77–83.

Zhu C, Yin Z, 2017. Study on the resource of *Psammosilene tunicoides*, an endangered medicinal species. *Chin J Ethnomed Ethnopharm*, 26: 133–138.

Zhu WM., He HP, Fan LM, et al., 2005. Components of stem barks of *Winchia calophylla* A. DC. and their bronchodilator activities. *J Integr Plant Biol*, 47: 892–896.

Index

D

E

F